Environmental Security

M000239410

Economic development, population growth and poor resource management have combined to alter the planet's natural environment in dramatic and alarming ways. For over twenty years, considerable research and debate have focused on clarifying or disputing linkages between various forms of environmental change and various understandings of security. At one extreme lie sceptics who contend that the linkages are weak or even non-existent; they are simply attempts to harness the resources of the security arena to an environmental agenda. At the other extreme lie those who believe that these linkages may be the most important drivers of security in the twenty-first century; indeed, the very future of humankind may be at stake.

This book brings together contributions from a range of disciplines to present a critical and comprehensive overview of the research and debate linking environmental factors to security. It provides a framework for representing and understanding key areas of intellectual convergence and disagreement, clarifying achievements of the research as well as identifying its weaknesses and gaps. Part I explores the various ways environmental change and security have been linked, and provides principal critiques of this linkage. Part II explores the linkage through analysis of key issue areas such as climate change, energy, water, food, population and development. Finally, the book concludes with a discussion of the value of this subfield of security studies, and with some ideas about the questions it might profitably address in the future.

This volume is the first to provide a comprehensive overview of the field. With contributions from around the world, it combines established and emerging scholars to offer a platform for the next wave of research and policy activity. It is invaluable for both students and practitioners interested in international relations, environment studies and human geography.

Rita Floyd is a Birmingham Fellow in Conflict and Security at the Department of Political Science and International Studies at the University of Birmingham, UK.

Richard A. Matthew is a Professor in the Schools of Social Ecology and Social Science at the University of California at Irvine, and founding Director of the Center for Unconventional Security Affairs (www.cusa.uci.edu).

Environmental Security

Approaches and issues

Edited by
Rita Floyd and Richard A. Matthew

LONDON AND NEW YORK

First published 2013
by Routledge
2 Park Square, Milton Park, Abingdon, Oxon OX14 4RN

Simultaneously published in the USA and Canada
by Routledge
711 Third Avenue, New York, NY 10017

Routledge is an imprint of the Taylor & Francis Group, an informa business

British Library Cataloguing in Publication Data
A catalogue record for this book is available from the British Library

Library of Congress Cataloging in Publication Data
Floyd, Rita, 1977–
    Environmental security / Rita Floyd and Richard A. Matthew. — 1st ed.
    p. cm.
    Includes bibliographical references and index.
    1. Environmental policy—United States. 2. National security—United States.
    I. Matthew, Richard A., 1965– II. Title.
    GE180.F569 2012
    363.7'05610973—dc23
    2012023692

ISBN: 978-0-415-53899-2 (hbk)
ISBN: 978-0-415-53900-5 (pbk)
ISBN: 978-0-203-10863-5 (ebk)

Typeset in Times New Roman and Franklin Gothic
by Florence Production Ltd, Stoodleigh, Devon

We dedicate this book to our respective children Corin S. Floyd (born 2011), Ainslie C. S. Matthew (born 2001), Elissa A. Matthew (born 2003) and Liam A. Matthew (born 2003), in the hope that our natural environment will still be a place of beauty and a source of well-being for them to enjoy as adults.

CONTENTS

FIGURES

TABLES

ACKNOWLEDGEMENTS

This book is one of those notorious edited books that has been years in the making. Rita first wanted to read a book of this kind back in 2003 when she was a student trying to come to grips with environmental security, but none was available. At that time the idea of editing was daunting and it became more a concrete reality only after the International Studies Association's annual meeting in Chicago (2007), after which Richard joined the project as co-editor. Having taught environmental security for many years, he was equally frustrated with the absence of such a book. The initial enthusiasm for the book became slowly side-lined by other commitments (notably Richard's travels and Rita's maternity leave) and the not insignificant challenge of co-editing across the Atlantic. In the end we owe completion to Routledge's Andrew Mould's interest in, and enthusiasm for, this project, who first heard of it in the summer of 2011.

We are grateful to our contributors, some of whom have waited patiently for the completion of this book for many years. We count ourselves lucky that many have continued to believe in the project and also in us. We are also grateful to our two fabulous research assistants Karen An and Rosemarie (Rosie) Said (both from the University of California at Irvine), who have done a terrific job of formatting and checking the final version of each chapter. We would like to thank Faye Leerink and Stewart Pether from Routledge for their continuous support. And we are grateful for the helpful comments of three anonymous reviewers on a detailed proposal for this book.

More specifically Rita would like to thank the ESRC for financial support during 2008–2009 and the British Academy for financial support and the time to work on this book during 2009–2012. Stuart Croft for listening to grumbles about the project's progress, as well as for his continuous support. She would also like to thank her husband Jonathan Floyd for his advice, help and love.

Richard would like to thank the Samueli Foundation for supporting research carried out at the Center for Unconventional Security Affairs, including the research for this volume, and for supporting the Center's operations, making it possible to engage Rosemarie and Karen. He also would like to thank Rita, who first conceived of the volume, and whose tireless dedication to its completion sustained the project.

The book's cover was chosen deliberately. Instead of selecting a – for this kind of book – more standard picture of an already damaged environmental landscape, we wanted the cover to depict nature's beauty, and in that way remind everyone why we should care about environmental issues.

CONTRIBUTORS

Saleem H. Ali is Professor of Sustainability Science and Director of the Centre for Social Responsibility in Mining at the University of Queensland, Australia, and also a tenured professor of environmental studies at the University of Vermont (USA). His most recent book is *Treasures of the Earth: Need, Greed and a Sustainable Future* (Yale Univ Press, 2010). Dr. Ali was selected as a "Young Global Leader" by the World Economic Forum in 2011 and an "Emerging Explorer" by the National Geographic Society in 2010. He holds a doctorate in environmental planning from MIT, a Masters in Environmental Studies from Yale and a BSc in Chemistry (summa cum laude) from Tufts University. He can be followed on Twitter @saleem_ali.

Alexander Carius is director of adelphi and a senior policy analyst and public policy advisor on climate, environment and development. For almost two decades he has lead international expert teams to analyse security implications of global environmental and climate change. He regularly advises government agencies and international organisations on strategies to address emerging risks and threats and develop cooperative mechanisms towards resource governance and peacebuilding.

Jennifer Sciubba is an Assistant Professor in the International Studies department at Rhodes College in Memphis, Tennessee. Dr. Sciubba has studied at the Max Planck Institute for Demographic Research in Rostock, Germany, and is a former demographics consultant to the Office of the Secretary of Defense (Policy) in Arlington, VA. Her latest book is *The Future Faces of War: Population and National Security* (Praeger/ABC-CLIO 2011). She received her PhD and MA from the University of Maryland and her BA from Agnes Scott College.

Geoffrey D. Dabelko is a Professor in Ohio University's Voinovich School of Leadership and Public Affairs (http://www.ohio.edu/voinovichschool/) and director of the Environmental Studies Program. From 1997-2012 he was director of the Woodrow Wilson Center's Environmental Change and Security Program, a nonpartisan research-policy forum on environment, population, health, development, and security issues. He currently serves as a senior advisor to the Wilson Center's Global Sustainability and Resilience Program. His current research focuses on climate change, natural resources, and security as well as environmental pathways to confidence-building and peacebuilding, with a special emphasis on water resources. Geoff is co-editor with Ken Conca of *Environmental Peacemaking and Green Planet Blues: Four Decades of*

Global Environmental Politics (4th edition). He is an IPCC lead author for the 5th assessment and member of the UN Environment Programme's Expert Advisory Group on Environment, Conflict, and Peacebuilding. He holds a Ph.D. in government and politics from the University of Maryland.

Simon Dalby is CIGI Chair in the Political Economy of Climate Change at the Balsillie School of International Affairs, Waterloo, Ontario. He was formerly Professor of Geography, Environmental Studies and Political Economy at Carleton University in Ottawa. He is coeditor of *Rethinking Geopolitics* (Routledge 1998), *The Geopolitics Reader* (Routledge 1998, 2006), the journal *Geopolitics*, and author of *Creating the Second Cold War* (Pinter and Guilford, 1990), *Environmental Security* (University of Minnesota Press, 2002) and *Security and Environmental Change* (Polity, 2009).

Tom Deligiannis is completing his Ph.D. in the Department of Political Science at the University of Toronto. He is currently a Lecturer in the Department of Political Science at the University of Western Ontario in London, Ontario, an Adjunct Faculty member in the Department of Environment, Peace, and Security at the UN-mandated University for Peace (UPEACE) in Costa Rica, and an Associate Fellow of the Institute for Environmental Security in The Hague.

Nicole Detraz is an Assistant Professor of Political Science at the University of Memphis. Her research centers on exploring the multiple connections between security, the environment, and gender. She explores the implications of various actors linking broad issues like climate change to security discourses. Her work has recently appeared in *Global Environmental Politics*, *Security Studies*, and *International Studies Perspectives*. She is the author of *International Security and Gender* (Polity 2012).

Indra de Soysa is professor of political science at the Norwegian University of Science and Technology in Trondheim, Norway. His recent articles appear in *Journal of Peace Research, International Studies Quarterly, International Organization, Conflict Management and Peace Science*, and *International Interactions*. Currently, he serves as Warden of St. Thomas College Mt. Lavinia in Sri Lanka.

Rita Floyd is a tenure track Birmingham Fellow in Conflict and Security at the Department of Political Science and International Studies at the University of Birmingham (UK) and a Fellow of the Institute for Environmental Security, The Hague. She is the author of several peer-reviewed articles and of *Security and the Environment: Securitisation Theory and US Environmental Security Policy,* Cambridge University Press, 2010. Dr Floyd can be contacted at r.floyd@bham.ac.uk.

Carolyn Lamere works at the Environmental Change and Security Program at the Woodrow Wilson International Center for Scholars. She graduated from Rhodes College in Memphis, Tennessee in 2012 where she studied International Relations. She has focused her research on population and global health, including through onsite work in Nicaragua with GlobeMed, a student-run nonprofit.

Achim Maas works as coordinator for the cluster "Sustainable Interaction with the Atmosphere" at the Institute for Advanced Sustainability Studies (IASS) in Potsdam,

Germany. He primarily focuses on climate foreign policy, climate engineering and risk governance. Until May 2012 he worked at adelphi in Berlin as a senior project manager on interlinkages between natural resources, peace and conflict. He studied in Frankfurt am Main and Bradford and holds an MA in International Politics and Security Studies.

Patrick MacQuarrie is the Water Policy and Sustainability Advisor for the global IUCN Water Programme. His areas of expertise are resilience and adaptive capacity, transboundary water policy, conflict management, hydropolitics, hydrology and agricultural engineering. Patrick recently completed his PhD where he managed the Transboundary Freshwater Dispute Database (TFDD) and worked on several transboundary water projects with international partners including the Mekong River Commission, the Nile Basin Initiative, USAID, IRG, and the World Bank. Prior to his PhD, Patrick worked in the private sector as a fluidic research and development engineer. He holds a BS in Agricultural Engineering and MS in Engineering Science from Washington State University, an M.Phil in International Peace Studies at Trinity College, Dublin, and a PhD in Resource Geography at Oregon State University.

Richard A. Matthew (BA McGill; PhD Princeton) is a Professor in the Schools of Social Ecology and Social Science at the University of California at Irvine, and founding Director of the Center for Unconventional Security Affairs (www.cusa.uci.edu). He is also a Senior Fellow at the International Institute for Sustainable Development in Geneva; a Senior Fellow at the Munk School for International Affairs at the University of Toronto; a senior member of the UNEP Expert Group on Environment, Conflict and Peacebuilding; and a member of the World Conservation Union's Commission on Environmental, Economic and Social Policy. He has carried out fieldwork in conflict zones throughout South Asia and Sub-Saharan Africa, and has consulted widely with government agencies and the private sector. He has served on several UN missions, including two that he led to Sierra Leone, and he was the lead author of the UN technical report, *Sierra Leone: Environment, Conflict and Peacebuilding Assessment*. He has over 140 publications.

Bryan McDonald (BA/MA Virginia Tech; PhD UC, Irvine) teaches on American environmental history and environmental studies at Pennsylvania State University. He is the author of *Food Security* (Polity, 2010) and co-editor of *Global Environmental Change and Human Security* (MIT Press, 2009) and *Landmines and Human Security: International Politics and War's Hidden Legacy* (SUNY Press, 2004, paperback 2006). His research articles and reviews have appeared in *Environment, The Environmental Change and Security Project Report, Global Environmental Politics, Natural Resources Journal* and *Politics and the Life Sciences.* He is a Fellow at Penn State's International Center for the Study of Terrorism and a member of the Advisory Board of the Center for Unconventional Security Affairs at the University of California, Irvine.

Dennis Pirages is Professor of Political Science at the University of Nevada, Las Vegas. He received his doctorate from Stanford University, and spent three decades at the University of Maryland. He is a lifetime Fellow of the American Association for the Advancement of Science. Dr. Pirages has authored or edited fourteen books and numerous articles and book chapters. He was one of the first scholars to explore the sustainability concept and currently is working on alternative ways of thinking about security.

Adam Simpson is Lecturer in the International Relations program at the University of South Australia. He previously taught at the University of Adelaide where he remains an Associate in the Indo-Pacific Governance Research Centre. His research adopts a critical approach to environment, development and security in the South. He has published in journals such as *Pacific Review*, *Third World Quarterly* and *Environmental Politics* and is the author of *Energy, Governance and Security in Thailand and Myanmar (Burma): A Critical Approach to Environmental Politics in the South* (Ashgate, 2013).

Bishnu Raj Upreti is a senior researcher in conflict management. He has written and/or co-edited 26 books on conflict, peace, state-building, and security and frequently published in peer reviewed international journals. Besides research, he is also teaching at Kathmandu University. He has worked as Research Fellow at King's College London and Surrey University, UK. He is engaged with policy-makers, politicians and the national and international media on Nepal's armed conflict and peace process. Dr Upreti is member of advisory boards of different national and international organisations. He is currently the South Asia Regional Coordinator of NCCR North-South, a global research network active in addressing the challenges to sustainable development. Corresponding email address to Dr Upreti is bupreti@nccr.wlink.com.np.

Michael Watts is currently Class of 1963 Professor of Geography and Development Studies, at the University of California, Berkeley, where he has taught for over thirty years. A Guggenheim Fellow in 2003, he served as the Director of the Institute of International Studies from 1994-2004. He is the author of 14 books and over 200 articles, his latest book with photographer Ed Kashi is *The Curse of the Black Gold* (powerHouse, New York, 2008). Watts has consulted for a number of development agencies including OXFAM and UNDP and has provided expert testimony for governmental and other agencies, including most recently the Bowoto et al., v. Chevron case in the federal district court in San Francisco. Watts has served as a consultant to the Ford and Rockefeller Foundations and a number of NGOs and foundations. Watts is currently the Chair of the board of Trustees of the Social Science Research Council and serves on a number of boards of non-profit organizations including the Pacific Institute.

Mary C. Watzin recently became Dean of the College of Natural Resources at North Carolina State University, after spending 21 years in the Rubenstein School of Environment and Natural Resources as Professor, and for the last three years, as Dean. She holds a PhD in Marine Sciences from the University of North Carolina at Chapel Hill, and a BSc in Marine Science from the University of South Carolina. She has received numerous awards and other recognitions for her teaching, research, and service, including the Kroepsch-Maurice award for teaching excellence, the Teddy Roosevelt Conservation award, and the Ibakari- Kasumigaura Prize recognizing her work with colleagues in Macedonia and Albania on transboundary water management. Watzin specializes in lake and watershed ecology and management, with a wide-ranging program of research focused on understanding how human activities influence ecosystem health and how adaptive management can be used to improve environmental outcomes.

Anja Wittich is currently in charge of the IUCN Caucasus Cooperation Center located in Tbilisi, Georgia. She is responsible for realizing the IUCN Caucasus Strategy, for developing the overall IUCN programme in the region, negotiating with donors and partners as well as for supporting the growing member network of IUCN in the South Caucasus. Her position with IUCN has been supported by the German Centrum für Internationale Migration und Entwicklung (CIM) since 2009. Prior to that she worked for the Berlin-based environmental organization adelphi and coordinated a programme on environmental law and policy in the South Caucasus and focused on environmental cooperation and security issues in transition countries.

Aaron T. Wolf is a professor of geography in the College of Earth, Ocean, and Atmospheric Sciences at Oregon State University, USA. His research and teaching focus is on the interaction between water science and water policy, particularly as related to conflict prevention and resolution. A trained mediator/ facilitator, he directs the Program in Water Conflict Management and Transformation, through which he has offered workshops, facilitations, and mediation in basins throughout the world.

ABBREVIATIONS

AIDS	Acquired immunodeficiency syndrome
BALWOIS Project	Water Observation and Information System for Balkan Countries
CBC	Caucasus Biodiversity Council
CBDN	Caucasus Business Development Network
CBM	Confidence building measures
CI	Caucasus Initiative (of the German Federal Government)
CSR	Corporate Social Responsibility
DFID	United Kingdom's Department for International Development
DOD	US Department of Defense
DRC	The Democratic Republic of the Congo
DUSDES	Office of the Deputy Undersecretary of Defense for Environmental Security
ENP	European Neighbourhood Policy
ENVSEC	Environment and Security Initiative
EPA	Environmental Protection Agency
ESS	Environmental Security Studies
EU	European Union
FDI	Foreign direct investment
GECHS	Global Environmental Change and Human Security
GEF	Global Environment Facility
GLOFs	glacial lake outburst floods
HIV	Human immunodeficiency virus
IFI	International Financial Institution
IO	International organizations
IPCC	Intergovernmental Panel on Climate Change
IRFC	International Federation of the Red Cross and Red Crescent Societies
IWRM	Integrated Water Resource Management
LOCP	Lake Ohrid Conservation Project
LOMB	Lake Ohrid Management Board
MENA	Middle East and North Africa region
MoE	Ministries of the environment

MOU	Memorandum of Understanding
NATO	North Atlantic Treaty Organization
nef	new economics foundation
NGO	Non-governmental organization
NK	Nagorno-Karabakh
NRE	Non-recognized entities
OSCE	Organization for Security and Co-operation in Europe
PRIO	Peace Research Institute in Oslo
REC	Regional Environment Centre for the Caucasus
SARS	Severe acute respiratory syndrome
SCRM	South Caucasus River Monitoring
SDC	Swiss Development Cooperation
SO	South Ossetia
SSA	Sub Saharan Africa
STS	Science and Technology Studies
TJS	Trans-boundary Joint Secretariat
UN	United Nations
UNDP	United Nations Development Programme
UNSC	United Nations Security Council
USAID	United States Agency for International Development
USDAT	Undersecretary of Defense for Acquisition and Technology
WSSD	World Summit on Sustainable Development

ENVIRONMENTAL SECURITY STUDIES

An introduction

Rita Floyd and Richard A. Matthew[1]

Introduction

Over the past two decades a substantial body of literature has emerged in North America, Europe and throughout much of the developing world exploring the existing and potential linkages between security and the environment. While it explores some ideas that have roots in classical thought, this literature is largely a response to two almost simultaneous events: the end of the Cold War (1989–92), which compelled a rethinking of the concept of security, and the 1992 Rio Earth Summit, which mobilized scientific evidence of global environmental change into a global policy agenda widely regarded as urgent and vital.

The ensuing flood of academic enquiry has examined large-scale environmental processes such as climate change and deforestation. It has explored the socio-political effects of changes in natural resource assets at various scales of analysis. It has examined the activities of the military from an environmental perspective and sought to understand the role the environment plays in conflict and in conflict resolution. It has explored the interaction of environmental change with demographic and health factors from a security perspective. It has considered the environment in terms of different levels of security analysis including global, regional, national, subnational and human. It has studied vulnerability to environmental change from the perspective of equity and poverty. And it has sought to understand the impacts of strategies for conservation and adaptation to environmental stress in different security contexts.

This work has been important and influential. It has been the subject of considerable policy activity. It has been popularized by leading journalists and authors. It has been integrated into International Relations and security studies curricula at both undergraduate and graduate levels. Interest among the emerging generation of scholars appears widespread and strong, and many of the scholars who defined the field in the early post-Cold War years continue to be active in it.

Perhaps reflective of the complexity of our high-speed and densely interconnected world, environmental security has not evolved over the past two decades as a homogeneous field of analysis, but rather as a polysemous category encompassing a wide range of analytical and normative meanings and positions. Indeed, the field of environmental security is so fragmented that the term is used to refer to entirely different, and even oppositional concepts. For example, environmental security has been used, on the one hand, to formulate a critique of the military's misconduct during warfare and in preparation for war,[2] and, on the other, to describe the military's approach to greening defence.[3]

The environmental security literature includes considerable analytical diversity, such as in the ongoing debate over whether natural resource scarcity or natural resource abundance is more closely related to conflict. These arguments, in turn, have been challenged by scholars who contend that resource tensions and pressures tend to generate innovation and cooperation, rather than violent conflict. The term has been used as a platform for moving away from traditional understandings of national security, and hence linked to agendas such as human security (e.g., IHDP's Global Environmental Change and Human Security Project). It has generated fierce critiques by scholars such as Daniel Deudney, Marc Levy and Betsy Hartmann on analytical, methodological and ideological grounds respectively.[4] It has proven a remarkably creative paradigm, generating new interdisciplinary insights by, for example, reaching into the natural disasters literature to study vulnerability and adaptation, into epidemiology to look at the rise of zoonotic infectious disease, and into earth systems science to use climate science as a principal source of data. In view of the diversity of work it has been called on to perform, it is perhaps inevitable that the meaning of environmental security has remained diverse and often contested.

The existing environmental security literature reflects the fragmentation of the field. Books on the subject are usually written from within one intellectual approach, and rarely impart a broad sense of the field. This volume has been prepared to remedy that situation and to provide an inclusive overview of what we term "environmental security studies" (ESS). We regard diversity as a strength of the field and a necessary basis for intellectual and policy innovation, and we see no value in adopting a reductionist stance vis-à-vis a literature that has interdisciplinary, multidisciplinary and transdisciplinary attributes. The complex stresses of the twenty-first century – from global and local economic crises to climate change to unprecedented rates of urbanization – do not lend themselves to unified and parsimonious forms of analysis. Today, competing approaches provide different but important insights into a world of dense relationships, crumbling institutions, dramatic transformations and high levels of uncertainty.

By collecting the most influential approaches to environmental security in a single volume, we aim to offer a rich overview of the field, and a valuable resource for those interested in environmental security. In addition to presenting various intellectual approaches to environmental security, we include chapters on the most pertinent issues unfolding at the intersection of security and the natural environment – namely, water, conservation, food, sustainable development, demography, energy and climate change. As this is a book about the theoretical approaches that inform the study of environmental security and about key issues being studied, we have not included separate chapters by the critics of environmental security but asked authors to address their concerns as appropriate in each chapter.

In order to provide context for subsequent chapters, our introductory chapter offers a brief overview of environmental security's intellectual history. This history is organized into three parts: 1) early formulations of environmental security in the context of the environmental movement of the 1960s and 1970s; 2) the deepening of the concept in response to the end of the Cold War and the Rio Earth Summit, which we regard as the key defining moments of the field; and 3) the recent evolution of research and debate in response to the findings of the 2007 reports by the Intergovernmental Panel on Climate Change (IPCC).

We conclude our introductory chapter with a chapter-by-chapter review of the volume.

A history of environmental security

Early formulations

The earliest versions of thinking that might be framed as environmental security date back to antiquity. Thucydides' *The Peloponnesian War* and Plato's *Republic*, for example, compare the security of societies living within their limits, like Sparta, to those like Athens that rely heavily on imports.[5] While interdependence creates an exciting dynamism in Athens, it also creates vulnerabilities, and both authors concur that self-sufficient societies are more secure. Some variant of this thinking extends across the ages, receiving an especially forceful rendering in the famous work of the eighteenth century demographer Thomas Malthus, who contended that if human populations grew faster than their agricultural output, which he regarded as a likely scenario, then a gap between supply and demand would emerge that would result in famines, epidemics and wars. This theme resurfaced following World War II, when Fairfield Osborn, asked:

> When will it be openly recognized that one of the principal causes of the aggressive attitudes of individual nations and of much of the present discord among groups of nations is traceable to diminishing productive lands and to increasing population pressures?[6]

However, in large measure, the contemporary formulation of the idea that there is a connection between the health of our natural environment and the security of individual persons, societies and even the biosphere emerges in the context of the environmental movement of the 1960s and 1970s. During that period, environmentalism was mobilized around texts such as Rachel Carson's seminal book, *Silent Spring* (1962), which problematized the use and abuse of pesticides in the countryside by showing the devastating effects of such products on songbirds. This was followed by the influential works of authors such as Lynn White Jr. (*The Historical Roots of Our Ecological Crisis*, 1967), Garrett Hardin (*The Tragedy of the Commons*, 1968), Paul Ehrlich (*The Population Bomb*, 1968), and Donella Meadows *et al.* (*The Limits to Growth*, 1972). Their arguments combined into a compelling neo-Malthusian worldview, and they elicited a dramatic and influential policy response in the US that included establishing the Environmental Protection Agency and passing legislation to ensure clean air and clean water. At roughly the same time, 1972, high-level representatives from over 100 countries travelled to Stockholm to attend the first UN Conference on the Human Environment, and the US agreed to fund the creation of the UN Environment Programme in Nairobi, Kenya.

Throughout this period, security events provided fodder for this nascent field, such as the 1967 war between Israel and Jordan, which was widely regarded as being a resource conflict as it was ostensibly fought over water. Similarly, the oil shocks of 1973 and 1979 became elements of national security thinking as they highlighted the Western world's level of dependence on foreign oil. Among the responses was an increase in oil drilling closer to home in an effort to ensure national security. Also important was the public outcry against the deliberate manipulation of the natural environment in the name of national security. During the Vietnam War herbicides such as the infamous "Agent Orange" had devastating consequences on the environment and consequently on people who relied on it. These events, too, evoked considerable policy activity. For example, the United States Department of Defense (DOD) Directive 5100.50, "Protection and Enhancement of Environmental Quality", was written in 1973, and four years later many states signed two important international agreements: the Additional Protocol I to the 1949 Geneva Convention on the Protection of Victims of International Armed Conflicts (1977) and the Convention on the Prohibition of Military or Any Other Hostile Use of Environmental Modification Techniques (1977). Under President Reagan, the US Congress established the Defense Environmental Restoration Account to fund the decontamination of military sites.

Taken together these and similar events led some analysts to link environmental issues with the language of security. For example, in 1977 then President of the Worldwatch Institute Lester Brown was among the first to argue that environmental issues had become matters of national security.[7] Brown's insight received a more elaborate theorization in the work of Princeton University's Richard Ullman, who in 1983 identified a list of environmental problems, including earthquakes, conflicts over resources and territory, population growth and resource scarcity, which he believed made necessary a redefinition of national security.[8] Ullman's call was reiterated and expanded by a number of scholars, including the prominent environmental scientist Norman Myers,[9] who famously argued that without environmental security all other forms of security become worthless – in short, Myers argued that environmental security is our "ultimate security".[10]

The field matures

The end of the Cold War (1989–92) and the almost simultaneous global attention given to the environment at the Rio Earth Summit in 1992 provided the platform for a rapid expansion of the field and remarkable deepening of its concepts and key terms. The response moved along two interactive trajectories. On the one hand, the governments of the US and former USSR, and many of their neighbours, began to assess the toxic legacy of the Cold War and experiment with new forms of cooperation, new levels of commitment to base clean-up, and new attempts to integrate civilian expertise into the defence community.

On the other hand, the idea that environmental stress could be a source of national insecurity, introduced to many practitioners through the 1987 Brundtland Commission report, *Our Common Future*, began to attract attention. While the attention came from many quarters of the world, its scale and impact were greatest in the United States, where the vast security architecture built up during the Cold War suddenly lost much of its relevance and had to be repurposed. Thus even while the end of the Cold War was still being debated by scholars, a concept of environmental security was incorporated by the

Bush administration into the National Security Strategy of 1991. In subsequent years, under the Clinton administrations, the issue was successfully securitized as part of a larger attempt to broaden dramatically the content of this key national security document.[11] Indeed, the consecutive Clinton Administrations sought to both define a new approach to national security strategy and implement a deep restructuring of the Department of Defense (DOD).

On the latter front, Eileen Claussen, formerly head of atmospheric programmes at the Environmental Protection Agency (EPA), was appointed as special assistant to the president for global environmental affairs at the National Security Council, and new offices were created in the DOD, including that of the Undersecretary of Defense for Acquisition and Technology (USDAT), which housed the also newly created Office of the Deputy Undersecretary of Defense for Environmental Security (DUSDES), headed by Sherri Wasserman Goodman. Goodman's mandate included environmental planning, base clean-up, compliance with regulations, pollution prevention, education and training, technology development, military-to-military programmes, and resource conservation measures.

At the doctrinal level, Secretary of Defense Perry introduced the notion of "preventative defense" in 1996, a concept that included environmental security as a basis for military-to-military cooperation in areas such as decommissioning nuclear submarines, disposing of nuclear materials, and developing environmental standards for training and operations. Even more innovative, perhaps, was Vice-President Gore's effort to make the intelligence community's data gathering and data analysis skills accessible to the public.

The environmental security lens was adopted by the State Department as well, as when Secretary of State Warren Christopher announced in 1996 that:

> the environment has a profound impact on our national interests in two ways: First, environmental forces transcend borders and oceans to threaten directly the health, prosperity and jobs of American citizens. Second, addressing natural resource issues is frequently critical to achieving political and economic stability, and to pursuing our strategic goals around the world.[12]

Christopher argued that "environmental initiatives can be important, low-cost, high-impact tools in promoting our national security interests."[13]

Of course, the US government's intentions for including environmental security as part of its national security strategy can be questioned and ultimately the moral and practical value of the Clinton administrations' environmental security policies have been regarded as problematic by many analysts. For example, was the real beneficiary of such policies the American people, or rather the national security establishment, which benefited by gaining a much needed raison d'être and hence by continued federal funding?[14]

Although the early practice of environmental security might be regarded as very different from the favourable theoretical accounts described by scholars such as Myers and Ullman, the United States' operationalization of environmental security did lead to much debate and activity among the world's militaries on "greening defense", and it inspired military-to-military cooperation on sharing classified data to assess the trends and impacts of environmental degradation. Scholars working in Defense Academies continue to write about the environmental role of the military. Much of their work focuses on how the military can reduce its ecological footprint and why it should do so, what its role should be in the context of violent conflict and humanitarian crises where environmental factors are important to understanding and response, and what technologies and equipment military forces operating in extreme environments might require for success.[15]

In addition to environmental security becoming part of actual policymaking agendas, the end of the Cold War also had enormous consequences for the academic study of environmental security. The failure of neo-realism and other traditional approaches to predict the end of the Cold War – something neo-realism had explicitly claimed as part of its remit – resulted in the partial discrediting of traditional approaches to studying international relations and security and in the widening of security studies to include economic, political, societal and environmental security.[16]

Some, but not all wideners also pushed for fundamental changes to the concept of security. Here a prominent argument was that states – the traditional recipients of military security – can be a great source of insecurity for people living within them, and that therefore human beings are a more appropriate referent object of security. Some environmentalists seized upon this deepening of security and, in the spirit of Myers' "ultimate security" argument, proposed that the most appropriate referent object of security is the biosphere. Possible other referent objects of security include civilizations, identities, orders (i.e. institutions of and/or particular forms of international society), and regions.[17]

Resource scarcity and violent conflict thesis

Among the most prominent wideners of security were the early theorists of environmental conflict, including the so-called Bern-Zurich group headed by Günther Baechler, and the Toronto group headed by Thomas Homer-Dixon. The latter, for example, theorized that environmental scarcity has deleterious social consequences, which in turn could lead to violent conflict.[18] According to Homer-Dixon, resource scarcity arises in three ways: from a real decrease in the supply of a resource (for example, clear-cutting a forest); from an increase in demand due mainly to population growth or changes in consumption patterns; or from structural factors (for example, through the privatization of a resource such as water). Through a series of case studies the Toronto group undertook in the 1990s, Homer-Dixon developed an influential model that showed how, given certain social conditions, resource scarcity could contribute to violent civil conflict. Homer-Dixon concluded that in the future resource scarcity might weigh more heavily in the processes that determined violent civil conflict.[19] Research by Colin Kahl (2006) developed these ideas by exploring resource scarcity in relation to state failure (the collapse of functional capacity and social cohesion) and state exploitation (when a collapsing state acts to preserve itself by giving greater access to natural resources to groups it believes can prop it up).[20]

A somewhat related theme explores how environmental stress could affect the elements of national power. National power is the aggregation of several variables, such as geography, resource endowment, military capacity, intelligence gathering, population, social cohesiveness, type of regime and size of economy. Environmental change can affect all of these basic elements of national power.

For example, militaries may be less effective at projecting and exercising power if they have to operate in flooded terrain or during a heat wave. Warming that affects land cover could reduce a country's renewable resource base. Intelligence is difficult to gather and analyze in a domain marked by uncertainty about social effects.[21]

The impact of the Toronto group's work was such that a critical engagement with Homer-Dixon's environmental scarcity and violent conflict thesis might be regarded as the starting point for several competing approaches to environmental security. Before listing these, it is important to note that work in the vein of the Toronto group continues.[22] In chapter 2 of this volume Tom Deligiannis engages with the critics of environmental scarcity theses and argues that there is much room for re-engagement and consensus.

Be that as it may, the competing approaches the Toronto group's thesis has stimulated include: 1) environmental abundance and conflict theses; 2) political ecology; 3) human security; 4) ecological security; 5) feminist environmental security; and 6) environment, conflict resolution and peacebuilding theories. In the following we will elaborate each one of these briefly. It should be clear that each approach contains within it complementary but also competing variants, and that what we can provide here is the briefest and most general overview of this rich literature. Dividing the literature into these seven distinct theoretical approaches is an analytical device that aims to bring some order into the growing field of environmental security studies. A detailed examination of each approach can be found in chapters 2–8, each one of which engages with one approach only.

Environmental abundance and conflict theses: the resource curse

Resource curse theorists focus on the potential of environmental factors to lead to violent conflict.[23] Their focus is not, however, on the scarcity of renewables, but instead they argue that the biggest driver of environmental violence is the local abundance of non-renewable resources such as precious metals, diamonds and oil. As Indra de Soysa – who is one of the best-known proponents of this approach – argues in chapter 3, when environmental abundance coincides with bad governance, itself often a result of resource abundance, the latter is anything but a blessing. Instead it frequently leads to violent conflict, and consequently to widespread poverty and human insecurity. There are different views on why an abundance of certain natural resources may be – ironically – a curse. Having a large stock of a valuable commodity may, for example, reduce economic innovation and tend to create a very lopsided economy that is vulnerable to changes in the global price of the commodity, to foreign intervention and to local corruption. Also, there may be less investment into public services, like education, associated with economic growth, and fewer opportunities to develop political skills such as mediation and consensus-building.

Political ecology

Environmental conflict is one of four central themes informing the distinct subdiscipline of political ecology.[24] Although contemporary political ecology is a subfield of geography, it is interdisciplinary in its origin. Among its many influences are peasant studies, environmental history, cultural ecology, dependency theory, postcolonial theory and poststructuralism. According to the founding editors of the *Journal of Political Ecology*, James B. Greenberg and Thomas K. Park, the two most important theoretical influences are "political economy, with its insistence on the need to link the distribution of power with productive activity and ecological analysis, with its broader vision of bio-environmental relationships".[25] While not all political ecologists would agree with that assessment,[26] all agree that ecology is inherently "political". Consequently environmental issues, including environmental conflict, cannot be explained without accounting for political factors. Political ecologists have criticized "apolitical" environmental conflict theses (both the environmental scarcity and the environmental abundance and conflict theses) for their blindness to structural factors of the global political economy (for example, consumer demand in the Global North, along with distribution and control practices) which produce local scarcity and increase competition for globally scarce but locally abundant resources elsewhere (usually in the Global South). In addition to aiming to "unravel the political forces at work in environmental access, management and transformation", political ecology is informed by a normative agenda of transformation towards "better, less coercive, less exploitative, and more sustainable ways of doing things".[27] In chapter 4 of this volume Michael Watts offers an innovative political ecology of environmental security.

Human security

Proponents of a human security approach to environmental security tend to argue that the literature that links the environment to conflict "is theoretically rather than empirically driven, and is both a product and legitimation of the North's security agenda".[28] In the conflict literature, the Global South is routinely portrayed as the violent, threatening "other", exemplified in Robert Kaplan's polemic "The Coming Anarchy" (1994), while the natural environment is conflated with resources and the important question, why do people resort to violence, is systematically ignored.[29]

In the 1994 UNDP report that popularized the concept, human security "was said to have two main aspects. It means, first, safety from such chronic threats as hunger, disease and repression. And second, it means protection from sudden and harmful disruptions in the patterns of daily life".[30] The UNDP report explains four dimensions of human security as fundamental: it is universal, its components are interdependent, it is easier to protect through prevention than intervention and it is people-centred.[31] In response to this report, a group of scholars established the Global Environmental Change and Human Security programme in 1997 and defined human security:

> as something that is achieved when and where individuals and communities have the options necessary to end, mitigate or adapt to threats to their human, environmental and social rights; have the capacity and freedom to exercise these options; and actively participate in pursuing these options.[32]

Many of the key findings of this programme are presented in the volume entitled *Global Environmental Change and Human Security*.[33] In chapter 6 of this volume Simon Dalby discusses some of the latest attempts by scholars to link human security and environmental issues, including their policy recommendations.

Ecological security

Proponents of ecological security are concerned with the health and well-being of the biosphere, which is why this approach is sometimes also referred to as biosphere security. Proponents generally commence from deep green ecology's principle of biocentric equality, whereby all species are considered equal. Ecological security is believed to be a delicate balance rendered unstable by human action. Unless this balance is redressed human beings will suffer dire consequences, one being the spread of pandemic diseases; another is the emergence of new pathogens. This connection between ecological security and human well-being, together with the inescapable anthropocentricity of human writing on these issues, mean that ecological security can be understood as a variant of human security. In chapter 7 of this volume Dennis Pirages argues that ecological security is the natural successor of environmental security as human security. Yet this view is not shared by everyone in favour of ecological security. In green political thought, many deep green ecologists wish to safeguard the biosphere regardless of the consequences of such actions for human beings.

Feminist environmental security

Unlike most of the other approaches to environmental security discussed here, feminist environmental security is in its infancy. While only a few pieces have been published specifically on feminism and/or gender and environmental security, some writings on feminism and environment more generally (notably ecofeminism) can be counted as part of this approach.[34] Proponents of feminist environmental security are concerned with the gender-blindness of all other approaches to environmental security. A major concern is that women and men are affected differently by environmental damage, not, however, due to "natural" given features, but rather due to socially constructed roles they inhabit in a given society. In many parts of the world women have lower standing than men, culturally and socio-economically if not legally, and are more likely to be responsible for gathering water and fuel wood. This situation renders women especially vulnerable to global environmental change, including environmental conflict.

Given that a feminist approach to environmental security is at such an early stage of development, it is not yet clear if feminist environmental security will become a viable stand-alone approach or if it will simply reform the human security approach to environmental security. Although human security tends to be gender-blind, there is plenty of overlap between the two approaches. Like human security scholars, feminist environmental security scholars subscribe to a wide view of security (whereby security is about more than the absence of violent conflict, cf. chapter 1); they are concerned with the security of human beings as opposed to states, they reject the possibility and desirability of positivist value-free analysis and they are informed by an emancipatory goal of what true environmental security would look like. In chapter 8 Nicole Detraz examines the value of the perspective gender brings to environmental security studies.

Environment, Conflict Resolution, and Peacebuilding

A final group of scholars has focused on environmental peacemaking, peacekeeping, and peacebuilding.[35] This approach is intellectually rooted in peace research and conflict management; as such it focuses on the role of the natural environment in mitigating and transforming violent conflict.[36] Environmental cooperation takes different forms, but the general idea that the "environment can contribute to peace"[37] is taken from empirical evidence gathered in conflict zones or otherwise volatile areas. For example, natural resources have been identified as critical to many aspects of the peacebuilding process.[38] Natural resources support economic recovery and sustainable livelihoods, and the challenge they pose can contribute to multiple forms of dialogue, cooperation and confidence building. Post-conflict societies typically have resource-based economies, and therefore need to acquire the technical capacity to manage natural resources sustainably and cope with the additional stress of climate change impacts. Given that this approach remains little known among security studies scholars more generally,[39] we have dedicated one chapter to it, and another in the issues section to the link between conservation and environmental cooperation (see chapters 5 and 10).

The engagement with climate science

A new wave of development and expansion of the field of environmental security has been triggered by the publication of the IPCC's fourth assessment report (2007) on the causes and consequences of global climatic change. This report sounded an influential, science-grounded alarm for an issue that has been on the agenda of international society since at least the United Nations Framework Convention on Climate Change (UNFCCC) was established in 1992, and that received enormous attention in 1997, the year of the Kyoto Treaty. More recently, media reporting and high casualty environmental disasters including Hurricane Katrina in 2005 and the Asian Tsunami of 2004 have served to highlight threats originating from climate change – although neither has been classified as the direct result of global warming. In the 2007 report, climate scientists synthesized research showed significant changes in the distribution of water across the planet's surface, increases in the intensity of storms and other severe weather events, longer heat waves and droughts, gradual sea level rise and, related to this, aggressive flooding.

Climate change adds momentum to all of the existing approaches to environmental security as questions that have long been addressed by that literature (e.g. Who or what should be the appropriate recipient of environmental security? Are environmental issues likely to lead to conflict or cooperation?) are now being asked in the context of climate change. Indeed, climate change has become a paramount security concern for many states and the rhetorical link between climate change and security (national, human, regional) has been made by individuals from within the academy, by policymakers, and by those, like former US Vice-President Al Gore and the previous British chief scientific advisor Sir David King, formerly in government service.[40] The United States government, which had the most extensive environmental security policies of any country in the 1990s but which abandoned environmental security under the George W. Bush administrations,[41] is returning to this issue largely in response to concerns about the security implications of climate change. Scholars are considering the impact of climate change for national, international and human security. [42] The United Nations has become heavily involved in raising awareness of the connection between human insecurity and climate change.[43] Many traditional environmental conflict research centres are reconsidering established theories in the climate change context,[44] and the possible causal link between climate change and conflict is being heavily debated.[45]

Although environmental security has managed to establish itself as a subtheme in security studies, there is a risk that those with an interest in climate security, yet unfamiliar with the established environmental security literature, may overlook the latter, or at the very least, not regard it as a natural entry point to the climate security conundrum. This is worrying because more than twenty years of environmental security research has delivered some tangible results that would make an ideal starting point for climate security research. It is, however, a real possibility in part because of the allure of the new and in part because the environmental security literature is so diverse, multilayered and confusing that easy access is denied to new scholars.

In large measure concerns about climate change have reinvigorated the different theoretical approaches to environmental security sketched out above, and climate security is not so much a concept as it is a debate.[46] For example, some theorists of environmental scarcity identify climate change as a driver of violent conflict, while methodological critics such as Nils Petter Gleditsch argue that there is simply not enough evidence for this claim.[47] Within the UN, which is an influential proponent of human security, arguments have been formulated that our concern should not primarily be with national security and war, but instead with climate change's ability to erode the livelihoods of the world's poorest people.[48] The US military, which has re-emerged on the environmental security scene in this context, has raised concerns about military readiness, state failure and violent conflict in a climate changed world through devices such as a panel of retired military leaders.[49]

Popular awareness of global warming, of course, is not the only reason why a definitive guide to the literature is timely. Another is that the existing volumes on environmental security are either too dated, such as *Contested Grounds: Security and Conflict in the New Environmental Politics* which was published in 1999, or too narrow in their focus, as are, for example, *Environmental Conflict* (2001) and *Global Environmental Change and Human Security* (2010), to provide an adequate overview of environmental security studies.

Chapter-by-chapter review

Following this introductory chapter the volume is divided into two parts. Part one is dedicated to the different approaches to environmental security. Chapter 1 offers a framework for understanding and ordering the field of environmental security studies. This is followed by separate chapters on the seven approaches to environmental security as identified earlier in this chapter: environmental scarcity and conflict (chapter 2); environmental abundance and conflict (chapter 3); political ecology (chapter 4); environmental peacebuilding (chapter 5); human security (chapter 6); ecological security (chapter 7); and feminist environmental security (chapter 8).

Part two engages with some of the most pertinent issues to have arisen in the context of environmental security: water (chapter 9); conservation (chapter 10); demography (chapter 11); sustainable development (chapter 12); food (chapter 13); energy (chapter 14); and climate change (chapter 15). All of the authors in part two of the book utilize insights from one or more of the different approaches to environmental security as laid out in part one. In more detail, the chapters on water and conservation, though in different ways, utilize and contribute to the environmental peacebuilding approach. The chapters on sustainable development, food and energy are largely concerned with and contributions to the human security approach. The chapter on demography focuses on the security implications for states. Finally the chapter on climate change considers the threat from global warming to both national and human security, while it also considers the linkages between climate change and peacebuilding.

Overview

In chapter 1 Rita Floyd presents a framework that identifies and organizes the elements of unity and divergence in the polysemous field of environmental security studies. This framework is based on two epistemological lines that tend to divide scholars of environmental security: 1) the role of the analyst and his or her view of whether theory is mainly analytical or Critical, or some combination of analytical and normative elements; and 2) the analyst's view of the key terms, concepts and questions that define security. Floyd argues that this framework helps to understand four central debates that animate and divide ESS. These are: 1) the debate over whether scarcity, abundance or political factors lead to violent environmental conflict; 2) the debate over whether security, and the study of security should be about violent conflict only or also about reductions in human choice, welfare and well-being; 3) the cornucopian criticism of the resource scarcity and conflict thesis; and 4) the debate over whether under conditions of environmental stress cooperation or conflict is more likely.

In chapter 2, Tom Deligiannis examines the findings of the environmental scarcity theses as developed over the past fifteen years by the Toronto and also the Bern-Zurich Groups, and re-evaluates them in the light of their many critics. In an effort to reconcile the environmental scarcity theses with valid concerns raised by their critics – notably political ecologists and environmental abundance and conflict theorists – Deligiannis shows in what areas consensus can be reached. He argues that the environmental scarcity theses have been subjected to erroneous and largely polemical criticisms that need refocusing. The synthesized research project that emerges from this analysis highlights the continued relevance of the environmental scarcity thesis though it is perhaps more accurately labelled "qualitative environment-conflict research", a term Deligiannis employs throughout his chapter.

Deligiannis' view of the value of a synthesized research project is not shared by Indra de Soysa, who in chapter 3 argues that armed conflict results not from environmental scarcity, but from policy failure associated with the relative abundance of resource wealth. On this basis, de Soysa contends that the problem of violence can be solved by conscious policies in the realm of global governance, for it is a combination of developmental and governance traps and the lack of "institutional capital" that lead to persistent poverty, human insecurity and continued dependence on natural resources. In particular, he highlights how factors associated with globalization can help to mitigate social and economic breakdown by allowing the build-up of better institutional capital through international diffusion and conscious global governance. As an example of a region where the risk of violent conflict has been steadily decreasing, precisely because of these positive developments, he examines the case of Sub Saharan Africa (SSA), which – interestingly – is one of the areas where the effects of climate change are already being felt in dramatic and extensive ways.

In chapter 4 Michael Watts aims to provide a conceptual account of how political ecology approaches the sort of problems and questions posed by the rise of the broad field of environmental security. Beginning with a theoretical account of political ecology itself, he commences with the question of how politics is conceptualized within critical political ecology. Watts goes on to provide a brief account of how this approach differs from the Toronto and Bern-Zurich groups (complementing Deligiannis in chapter 2) and attempts to situate the rise of security as an environmental issue on a broader landscape of both post-Cold War politics and intellectual trends within the social scientific study of the environment. A key part of the story, he argues, is the rise of the security state – something which pre-dates September 11 – and the extent to which the rise of "biosecurity" provides a powerful discursive frame for thinking about a variety of issues, including environmental conflict, the "securitization" of climate and the ways in which diseases and threats of pandemics are linked to terror.

Watts elaborates by offering a new way of thinking about a political ecology which builds upon the lectures of Michel Foucault on security, territory and biopower. He argues that the rise of security as a way of thinking about modern rule – both as geopolitics and biopolitics – offers a way of rethinking environment as a security question.

In chapter 5, Achim Maas and Alexander Carius with the assistance of Anja Wittich provide a detailed overview of the environmental cooperation literature, including a section on method that details how an analyst interested in this approach needs to proceed. They argue that the environmental cooperation literature and practice is not one coherent school of thought, but rather a discourse informed by three major trends: 1) environmental peacebuilding can identify and address the causes of conflicts where the environment (due to either scarcity or abundance) plays a significant role; 2) dialogue over environmental issues between conflicting parties can lead to trust building and ultimately peace; 3) sustainable development itself is often a precondition for durable peace, a view that is shared by others in this book including de Soysa (chapter 3) and Bishnu Upreti (chapter 12).

In order to exemplify environmental peacebuilding in practice Maas *et al.* examine the case of the South Caucasus, where myriad organizations work on the nexus between peace, security and environment. This case study holds a number of valuable lessons for the three trends dominating environmental peacebuilding discourse, concerning the preconditions for cooperation (i.e. willingness to cooperate), the conditions of cooperation (i.e. that parties to a conflict must enter dialogue as equals), and the effect of sustainable development. While the success of environmental peacebuilding is highly context dependent, as a normative ideal it has universal applicability.

In chapter 6 Simon Dalby offers a review of the trajectory of environmental security as human security. He contends that although this approach has been around since the mid-1990s, the relevance of environmental security as human security has suffered because for a long time it failed to propose a coherent policy solution. He argues that (human) security needs to be understood and reconceptualized in the age of what earth system scientists[50] have come to call the Anthropocene, the condition whereby humanity is no longer separate from the natural environment but human behaviour directly influences and changes earth systems and functionings.[51] This, Dalby argues, means that we need to rethink the relationship between security and humanity. We need, Dalby argues, "a new 'political geoecology' for the Anthropocene, where security involves quite literally thinking about how we are making the planetary future, in whose interests, and how it might be appropriately shared so as to render most people free from the most obvious dangers to their security".

In chapter 7 Dennis Pirages argues that human security and human well-being can only be achieved in the context of ecological security, which he defines "as maintenance of dynamic equilibria in continually evolving relationships among human societies and between them and key components of the ecosystems in which they are embedded". Four critical relationships are identified: 1) between human societies and nature's resources and services; 2) between human societies and pathogenic microorganisms; 3) between human societies and populations of other animal species; and 4) among human societies. Pirages argues that these equilibria have been upset to such an extent that they cause the biggest threat to human security. Although he uses the term "ecological security", and he maintains that human beings are but one of many species, he recognizes that he cannot escape his own anthropocentric worldview and his primary concern is with the well-being of human beings. His main point is that unless ecological interdependence features as part of environmental security policies, human insecurity will persist.

In chapter 8 Nicole Detraz makes the case for the systematic inclusion of gender into ESS. She contends that one characteristic that all existing approaches to environmental security have in common is that they are gender blind. She argues that integrating "gender in ESS requires rethinking terms and concepts, problematizing power relations, and asking new questions". She goes on to examine the gender blindness of the environment-conflict approaches, human security and ecological security (which she understands in less anthropocentric terms than does Pirages in the previous chapter) respectively. Perhaps unsurprisingly, the environment-conflict theses (she discusses the scarcity and abundance theses only) are, with their narrow view of security and their close connection to the state security, furthest removed from feminist concerns. Feminist security studies – like human security – is about exposing the threats and vulnerabilities of individuals. However, unlike human security, gender theorists stress the extent to which men and women are affected differently by environmental change and conflict, experience environmental migration differently, and play different roles in the production of environmental damage.

Although Detraz emphasizes that much of ecofeminism overlaps with ecological security, insofar as nature is believed to have intrinsic value, she is adamant that neither women nor men inhabit simplistic roles within the environment–security nexus, whereby women are always victims and men are always "destroyers" of the natural world. Although women suffer disproportionally more from environmental damage and conflict than men, they do so because of the low social standing they inhabit in many parts of the world, a standing that leaves them dependent on natural resources – for example, in their role as firewood collectors. In addition to adding layers of new explanation to the causes and consequences of environmental damage, and highlighting new actors, Detraz's mission is also a normative one. She concludes that unless the gendered nature of environmental insecurity is recognized, genuine environmental security for all cannot be achieved.

Part two of this volume focuses on some of the most important issues that have arisen in the context of studying environmental security. In chapter 9 Patrick MacQuarrie and Aaron T. Wolf address the issue of water scarcity, which is often depicted as a likely cause of future violent conflict. Although MacQuarrie and Wolf find no evidence to substantiate predictions of imminent interstate water wars, they believe that water is likely to cause or exacerbate tensions among states. Moreover, at the substate level conflict over water is already a common occurrence. A decrease in water availability is expected to increase incidences of intrastate violence. Despite these negative findings, however, the authors are keen to point to the potential for cooperation over water; indeed, cooperation in this area is one of the empirical strongholds of the environmental peacebuilding and cooperation approach. The issue of water security thus simultaneously speaks to the environmental security as human security literature (as human well-being is strongly dependent on the availability of water), and also delivers important research findings for the conflict and cooperation literatures.

In chapter 10, Saleem Ali and Mary Watzin focus on the role of conservation in environmental peacebuilding. A review of conservation efforts at the Eastern European Lake Ohrid and Lake Presa watersheds highlights the crucial role scientists, external mediators and politics play in peacebuilding efforts. Although the authors acknowledge their findings may not translate to other regions, their case study suggests that scientists and conservation bodies can play an important role in the way towards a sustainable peace. In this particular case, over time the exchange of data between scientists has led to trust building between former (Cold War) enemies for the better of the people and the natural environment itself.

In chapter 11 Jennifer Sciubba, Carolyn Lamere and Geoffrey D. Dabelko examine the role that population/demographic change plays in national security. Population and demographic trends – for example, changing size of population, ethnic composition of states, ratio of immigration to emigration, gender and age structure – play out in various different ways and can be linked to national security in numerous ways. In an effort to bring order into a vast literature on population and national security, Sciubba *et al.* utilize Barry Buzan's idea of sectors of security as analytical lenses. The authors show that national security considerations connected to demography arise in all five sectors of security, which is to say in the military, political, environmental, societal and economic sectors. For example, in the military sector of security the unavailability of able-bodied young male soldiers in an aging population has the potential to seriously compromise military security. Economic security, in turn, may be compromised in states with both large numbers of retired elderly people and comparatively lower numbers of people in the active workforce while societal security issues may arise from immigration.

Although Sciubba *et al.* are concerned in their chapter with national security only, they recognize that population trends have the potential to severely compromise human security and biosphere security. Wherever possible they point out overlap between other conceptions of security and population.

In chapter 12, Bishnu Upreti turns to the issue of development and environmental security. He differentiates between environmental security at the macro and the micro levels, whereby the first refers to change at the planetary level, most notably global climate change, while the latter is about local environmental problems. Poor people, he argues, are disproportionally affected by both. They suffer from macro environmental insecurity because they are unable to adapt, while micro environmental insecurity is part of their daily lives. At the same time, however, underdevelopment and poverty are causes of micro environmental insecurity because poor people rely on natural resources for survival, without being able to reinvest into environmental management. In the past, top-down development practices such as structural adjustment have worsened this problem, as these practices have paid no attention to the connection between poverty and development. Development practice needs to be sustainable; indeed Upreti argues, sustainable development and environmental security as human security are complementary concepts that reinforce one another. Sustainable development, however, cannot be achieved in a reductionist, top-down fashion. What is needed is a holistic constructivist approach that focuses simultaneously on transdisciplinarity, social learning and communicative rationality. Overall Upreti strikes an optimistic tone: societies have the capacity to learn, adapt and change, with both environmental security and sustainable development as potential outcomes of this.

In chapter 13, Bryan McDonald considers the issue of food security, which exists when "all people at all times have access to sufficient, safe and nutritious food necessary to lead active and healthy lives". McDonald argues that there are three ways in which food relates to security. First, malnutrition is the world's largest single killer. Second, the issue of food safety and the fact that contaminated foods can sicken and even kill people both affect human security. Third, and perhaps most relevant for the purposes of this book, there is a critical relationship between food supply and environmental change. On the one hand, food production is a massive driver of global environmental change. Agriculture alone accounts for about 13 per cent of global greenhouse gas emissions, beef and soy production are major causes of deforestation, and overfishing and the use of fertilizers affect waterways and the oceans. In short, food production and the search for food security compromise ecological/biosphere security.

On the other hand, global environmental change – especially climate change – droughts, floods, storm surges and the like – pose a massive threat to food production, thus endangering human security. McDonald argues that the only way forward is the creation of a "sustainable global food system" that aims to meet the nutritional needs of all in an ecologically sensitive fashion, while providing food safety. While a sustainable food system seems a long way off, there are some signs for optimism, especially in the fact that people are increasingly sensitive to food safety and food origin and conscious of food miles.

In chapter 14, Adam Simpson examines the issue of energy security. He shows that when conceptualized as a national security issue, as it has been in industrial societies at least since the oil crises of the 1970s, energy security is about securing the "availability, affordability and reliability" of fossil fuels. Although recently some attention has been paid to sustainability, energy security as national security continues to compromise other aspects of environmental security, including food security and climate security as human security. On the basis of these observations Simpson does two things. First he proposes a definition whereby energy security is achieved "when there is sufficient energy available to satisfy the reasonable needs of the political community (the referent object) in an affordable, reliable and sustainable manner as long as pursuing it does not cause environmental insecurity to that or any other political community". Second, he advances a critical energy security perspective[52] that turns attention from the traditional concern with the states of the Global North to the poor and marginalized in the Global South.

In chapter 15, Richard Matthew examines the security implications of climate change. He argues that the early responses to the 2007 IPCC report tended to be highly speculative and alarmist, and he provides an explanation for why they failed to catalyse significant investments into climate change adaptation and mitigation. He argues that climate science data, however, do suggest that adverse climate change effects are taking place more quickly and on a larger scale than was imagined only a few years ago. Using the hydrological system of the Hindu Kush Himalaya region as his reference point, he suggests ways in which human and state security may be threatened and offers a series of recommendations for reducing vulnerability. He then expands upon these recommendations as examples of forms of cooperation that should be encouraged at different levels of human activity and in different institutional and cultural settings. He adds that a variant of this argument provides a rationale for integrating climate change adaptation and mitigation into post-conflict peacebuilding and disaster recovery. He also cautions against excessive reliance on top-down approaches to managing climate change, and points to the high levels of innovation evident at the very local sites where impacts are most advanced. Finally, he considers the arguments of skeptics of a climate change–security linkage, and concludes that while they raise some valid concerns, their historical data may not provide much insight into the future.

Chapter 16 serves as an afterword to the entire volume with Rita Floyd speculating on the future of environmental security studies. Her analysis is delivered in three parts. First, informed by the observation that climate change is increasingly becoming a central topic of environmental security studies, she engages with critiques of climate security and examines what these mean for ESS as a whole. Second, she aims to uncover what lessons, if any, the discipline of security studies can learn from the subfield of environmental security studies. Third and finally, she points to gaps in the literature and thus identifies topics and issues for future research in ESS.

Notes

1   The authors would like to thank Stuart Croft, Simon Dalby, Tom Deligiannis, Jonathan Floyd, Jonna Nyman and Adam Simpson for their comments on earlier drafts of this chapter, as well as three anonymous reviewers of a draft proposal of this chapter for their helpful comments.
2   See for example A. H. Westing, 'Environmental Warfare: Manipulating the Environment for Hostile Purposes', *Environmental Change & Security Project Report* 3, 1997, 145–49.
3   K. H. Butts, 'Why the military is good for the environment', In J. Käkönen (ed.), *Green Security or Militarized Environment*, Aldershot: Dartmouth, 1994, 83–109.

4   D. Deudney, 'The case against linking environmental degradation and national security', *Millennium* 19, 1990, 461–76; M. A. Levy, 'Is the Environment a National Security Issue?', *International Security* 20 (2), 1995, 35–62; B. Hartman, 'Population, Environment and Security: A New Trinity', *Environment and Urbanization* 10 (2), October 1998, 113–27.

5   For discussion see R. A. Matthew, *Politics Divided: Nation versus State in International Relations*, Lanham, MD: Lexington, 2002, chapter 2.

6   F. Osborn, *Our Plundered Planet*, New York: Grosset & Dunlap, 1948, pp. 200–01.

7   L. Brown, 'Redefining national security', Worldwatch Institute paper 14, Washington DC, 1977.

8   R. H. Ullman, 'Redefining Security', *International Security* 8, 1983, 133.

9   N. Myers, 'The environmental dimension to security issues', *The Environmentalist* 6, 1986, 251–57; 'Population, Environment and Conflict', *Environmental Conservation* 14, 1987, 15–22; 'Environment and Security', *Foreign Policy* 74, 1998, 23–41.

10   N. Myers, *Ultimate Security: The Environmental Basis of Political Stability*, New York: W.W. Norton & Company, 1993.

11   R. Floyd, *Security and the Environment: Securitisation theory and US Environmental Security Policy*, Cambridge: Cambridge University Press, 2010, 61ff.

12   W. Christopher, *In the Stream of History: Shaping Foreign Policy for a New Era*, Stanford: Stanford University Press, 1998.

13   Ibid.

14   For the argument that the real beneficiary of US environmental security was the national security establishment see Floyd, *Security and the Environment*, 116–21.

15   See, for example, C. Pumphrey (ed.), *Global Climate Change: National Security Implications*, Carlisle, PA: US Army War College, Strategic Studies Institute, 2008; CNA Corporation, *National Security and the Threat of Climate Change*, Alexandria, VA: CNA Corporation, 2007. For a critique see M. Pemberton, *The Budgets Compared: Military vs Climate Security*, Washington DC: Institute for Policy Studies, 2008. Post-Cold War attempts to "green" the military are important subjects of analysis, but because they do not constitute an approach or distinctive meaning in the sense noted below they are not featured as a stand-alone chapter in this volume.

16   B. Buzan, *People, States and Fear: An agenda for International Security Studies in the Post-Cold war era*, 2nd edition, Hemel Hempstead: Harvester, 1991.

17   The most systematic work on broadening and deepening has been carried out by the Copenhagen school. See, for example, B. Buzan, O. Wæver and J. de Wilde, *Security: A New Framework for Analysis*, Boulder, CO: Lynne Rienner, 1998.

18   T. Homer-Dixon, 'On the threshold: Environmental changes as causes of acute conflict', *International Security* 16, 1991, 76–116; 'Environmental scarcities and violent conflict: Evidence from cases' *International Security* 19, 1994, 5–40; 'Debate between Thomas Homer-Dixon and Marc A. Levy', *Environmental Change and Security Project Report*, Washington DC: The Woodrow Wilson Center, 1996, 49–60; *Environment, Scarcity, and Violence*, Princeton, NJ: Princeton University Press, 1999.

19   Homer-Dixon, *Environment, Scarcity, and Violence*, 177–182.

20   C. H. Kahl, *States, Scarcity, and Civil Strife in the Developing World*, Princeton, NJ, Oxford: Princeton University Press, 2006.

21   R. A. Matthew, 'Is Climate Change a National Security Issue?', *Issues in Science and Technology* 27 (3), 2011, 60.

22   See most notably, Kahl, *States, Scarcity, and Civil Strife in the Developing World*; T. Deligiannis, 'The Evolution of Environment-Conflict Research: Towards a Livelihood Framework', *Global Environmental Politics* 12 (1), 2012, 78–100.

23   See, for example, I. de Soysa, 'The Resource Curse: Are Civil Wars Driven by Rapacity or Paucity?', in M. Berdal and D. Malone (eds), *Greed & Grievance: Economic Agendas*

in Civil War, Boulder,CO: Lynne Rienner, 2000, 113–35; M. T. Klare, *Resource Wars: The New Landscape of Global Conflict*, New York: Henry Holt and Company, 2001; P. Collier and A. Hoeffler, 'Greed and Grievance in Civil War', Working Paper, Oxford: Centre for the Study of African Economies, 2002; P. Le Billon (ed.), *The Geopolitics of Resource Wars: Resource Dependence, Governance and Violence*, London: Frank Cass, 2004; P. Le Billon, *Fuelling War: Natural Resources and Armed Conflict*, Adelphi paper no. 373, Oxford: Routledge, for the International Institute of Strategic Studies, 2005; and more recently, P. Le Billon, *Wars of Plunder: Conflicts, Profits and the Politics of Resources*, C. Hurst & Co., 2012.

24   According to Paul Robbins, the other three are degradation and marginalization; conservation and control; and environmental and social movement. P. Robbins, *Political Ecology: A Critical Introduction*, Oxford: Blackwell Publishing, 2004.

25   J. B. Greenberg and T. K. Park, 'Political Ecology', *Journal of Political Ecology* 1, 1994, 1.

26   See Robbins, *Political Ecology: A Critical Introduction*, chapter 1, who provides an overview of different definitions of political ecology.

27   Robbins, *Political Ecology: A Critical Introduction*, p. 12.

28   J. Barnett, 'Destabilizing the environment-conflict thesis', *Review of International Studies* 26, 2000, 271.

29   Ibid., pp. 271–88.

30   UNDP (United Nations Development Program), *Human development report*, New York, Oxford: Oxford University Press, 1994 (23).

31   UNDP, *Human development report* 1994 (22).

32   GECHS, *Global Environmental Change and Human Security: GECHS Science Plan*, Bonn: IHDP, 1999.

33   R. A. Matthew, J. Barnett, B. McDonald and K. L. O'Brien (eds), *Global Environmental Change and Human Security*, Cambridge, MA: MIT Press, 2009.

34   N. Detraz, 'Environmental Security and Gender: Necessary Shifts in an Evolving Debate', *Security Studies* 18, 2009, 345–69; Ú. Oswald Spring, *Gender and Disasters. Human, Gender and Environmental Security: A HUGE Challenge*, Bonn: UNU-EHS, 2008; H. Goldsworthy, 'Women, Global Environmental Change, and Human Security', in R. A. Matthew, J. Barnett, B. McDonald and K. L. O'Brien (eds), *Global Environmental Change and Human Security*, pp. 215–35. On eco feminism see, for example, C. Merchant, *Earthcare: Women and the Environment*, New York: Routledge, 1996; K. J. Warren (ed.), *Ecofeminism: Women, Culture, Nature*, Bloomington: Indiana University Press, 1997; K. J. Warren, *Ecofeminist Philosophy: A Western Perspective on What It Is and Why It Matters*, Boulder, CO: Rowman & Littlefield, 2000.

35   Notable works include S. H. Ali, *Peace Parks: Conservation and Conflict Resolution*, Cambridge, MA: MIT Press, 2007; K. Conca and G. D. Dabelko (eds), *Environmental Peacemaking*, Washington: Woodrow Wilson Center Press, 2002; R. A. Matthew, M. Halle and J. Switzer (eds), *Conserving the Peace: Resources, Livelihoods and Security*, IUCN/IISD E&S Task Force Report, 2002; S. Dinar (ed.), *Beyond Resource Wars: Scarcity, Environmental Degradation, and International Cooperation*, Cambridge, MA: MIT, 2011. See also the website of the Institute for Diplomacy and Environmental Security, at the James M. Jeffords Center, University of Vermont. Online: www.uvm.edu/ieds/node/2 (accessed August 2012).

36   Cf. Maas *et al.*, this volume, chapter 5.

37   Maas *et al.*, this volume, chapter 5.

38   UNEP, *From Conflict to Peacebuilding: The Role of Natural Resources*, United Nations Environment Programme, Geneva: UNEP, 2009.

39   The growing literature on desecuritization in securitization studies, for example, makes virtually no reference to this approach.

40   D. King (13 February 2009) 'Sir David King warns of Climate Change Conflict', University of Oxford. Online: www.ox.ac.uk/media/news_stories/2009/090213.html (accessed August 2012); For a review of twenty-four different national security strategies and the issue of climate change see M. Brzoska, 'Climate Change as a Driver of Security Culture', in J. Scheffran, M. Brzoska, H. G. Brauch, P. M. Link and J. Schilling (eds), *Climate Change, Human Security and Violent Conflict: Challenges for Societal Stability*, Hexagon Series on Human and Environmental Security and Peace, Volume 8, Berlin, Heidelberg, New York: Springer Verlag, 2012.

41   Floyd, *Security and the Environment,* chapter 5.

42   See, for example, J. W. Busby, 'Who Cares about the Weather? Climate Change and U.S. National Security', *Security Studies* 17, 2008, 468–504; C. Paskal, *Global Warring: How environmental, economic and political crises will redraw the world map*, London: Routledge, 2010; D. Moran, *Climate Change and National Security: A country level analysis*, Washington DC: Georgetown University Press, 2011; C. Parenti, *Tropic of Chaos: Climate Change and the New Geography of Violence*, Nation Books, 2011; S. Angenendt, S. Dröge and J. Richert (eds), *Klimawandel und Sicherheit: Herausforderungen, Reaktionen und Handlungsmöglichkeiten*, Baden-Baden: Nomos Verlagsgesellschaft, 2011.

43   Human Development Report 2007/2008, *Fighting Climate Change: Human solidarity in a divided world*, New York: United Nations Development Programme Publication, 2007.

44   For example, The Research Council of Norway's NORKLIMA programme has decided to allocate 7.5 million NOK to support a three-year CSCW project on Security Implications of Climate Change within PRIO.

45   N.P. Gleditsch, 'Whither the weather? Climate change and Conflict', *Journal of Peace Research* 49 (1), 2012, 3–9. See also the debate between O. M. Theisen, H. Holtermann and H. Buhaug, 'Climate Wars? Assessing the Claim That Drought Breeds Conflict', *International Security* 36 (3), Winter 2011/12, 79–106; M. Burke, J. Dykema, D. Lobell, E. Miguel and S. Satyanath, 'Climate and Civil War: Is the Relationship robust?', National Bureau of Economic Research Working Paper No. 16440. Online: www.nber.org/papers/w16440.pdf (accessed August 2012). See also S. M. Hsiang, K. C. Meng and M. A. Cane, 'Civil Conflicts associated with global climate change', *Nature* 476, 2011, 438–41.

46   For further elaboration see R. Floyd, 'The Environmental Security Debate and its Significance for Climate Change', *The International Spectator* 43, 2008, 51–65.

47   N. P. Gleditsch and O. M. Theisen, 'Resources, the environment and conflict', in M. D. Cavelty and V. Mauer (eds) *The Routledge Handbook of Security Studies*, Abingdon: Routledge, 2010, pp. 221–31.

48   United Nations, *Human Development Report: Fighting climate change: Human solidarity in a divided world*, New York: UNDP, 2007–08. Online: http://hdr.undp.org/en/reports/global/hdr2007-2008/ (accessed August 2012).

49   CNA Corporation, 'National Security and Threat of Climate Change,' Alexandria, VA: CNAC, 2007.

50   W. Steffen, A. Sanderson, P. D. Tyson, J. Jäger, P. A. Matson, B. Moore III, F. Oldfield, K. Richardson, H. J. Schellnhuber, B. L. Turner and R. J. Wasson, *Global Change and the Earth Systems: A Planet under Pressure*, Berlin, Heidelberg, New York: Springer.

51   See also S. Dalby, *Security and Environmental Change*, Cambridge: Polity, pp. 105–59.

52   Simpson does not distinguish between capital and lower case critical theory in the way we do, on the grounds that Marxists and those working in the Marxist tradition do not have a monopoly on emancipation, but also because he wants to keep his approach as inclusive and broad as possible.

1

ANALYST, THEORY AND SECURITY

A new framework for understanding environmental security studies

Rita Floyd[1]

The history of environmental security studies presented in the introduction reveals that environmental security means different things to different people. My task in this chapter is to bring order to this messy and opaque field. To this end I offer a new framework for understanding environmental security studies that serves to interrelate the chapters in the book on the basis of key areas of division.

Different schools of thought in ESS are often divided in accordance with the entity deemed threatened (the referent object of security), and thus it is possible to divide the field into human-, state- and biosphere- centric approaches.[2] Ultimately, however, this categorization is more helpful when discussing the actual practice of environmental security, as the United Nations, individual states and green activist groups tend to promote distinct referent objects of environmental security. It fails, however, to reveal the nuances between the different and competing theoretical approaches in ESS.

A better way of ordering environmental security studies is to separate the field along two dividing lines. The first of these pertains to the role of the analyst and his or her view of the nature of theory. The second dividing line pertains to his or her view of security. Let me begin with the first.

Dividing line I: the role of the analyst and the nature of theory

In security studies, the appropriate role of the security analyst is contested. Some analysts are *primarily* concerned with how security is practised while others are *primarily* concerned with the condition of security, and how this can be achieved.[3]

Traditional security analysts, for example, do not give much thought to what it means to be secure and who or what should be secured. They simply assume the primordial right of states to security and are concerned with studying the practice of national and military security. According to Stephen Walt, for example, "[security studies] explores the conditions that make the use of force more likely, the ways that the use of force affects individuals, states and societies, and the policies that states adopt in order to prepare for, prevent, or engage in war."[4] By comparison, consider how different the theoretical literature on human security is. Much of that literature is consumed by the debate over the nature of human security, with the literature generally split into whether (human) security is and/or should be about freedom from violent conflict only, or whether it should be about both freedom from fear and freedom from want. In short, human security scholars are concerned with the condition of human security and how it can be achieved.[5]

A similar divide is apparent in the subfield of ESS. On the one hand there are those whose *primary* interest lies with specifying an ideal condition of environmental security and how this can be achieved. The now considerable literature on environmental security as human security falls into that category. Thus, it is in this literature where one finds definitions of the actual meaning of environmental security. One of the most insightful contributors to this approach, for example, tells us that environmental security is "the process of peacefully reducing human vulnerability to human-induced environmental degradation by addressing the root causes of environmental degradation and human insecurity".[6]

How different scholars consider the appropriate role of the analyst is inseparable from their view of theory. The *majority* of scholars *primarily* interested in the condition of environmental security are informed by a Critical epistemology.[7] They believe that theory is never value-neutral and that "all theory is *for* someone *for* some purpose".[8] As such they do not differentiate between theory and practice and theoretical analysis is considered a form of practice.[9] The task or role of the environmental security analyst, then, is to imagine, teach and ultimately bring about a better future in which those identified as most vulnerable (environmentally insecure) are rendered more secure.

On the other hand, there are those whose *primary* interest is with studying how environmental security plays out in practice. Here the environmental security literature is *dominated* by positivist causal studies, concerned with the role of the environment in generating violent conflict. For these analysts theory simply serves to explain the world. And even when this kind of analytical and explanatory work gives way to policy recommendations, these remain non-normative. Thus these scholars do not advance a comprehensive case for why one set of policies should be adopted over another. Instead, their recommendations are fact-based, which is to say they are focused on what is happening and why it is happening rather than on what should happen. Considering, however, that the motives for research are rarely ever value-free these kinds of policy recommendation do not remain without problems and they must be treated accordingly. Moreover, this kind of research will normally be improvable methodologically, while a weak or questionable methodology has implications for the validity of policy recommendations.

Given all I have said so far it is possible to argue that the two dominant types of theory in contemporary environmental security studies are "Critical" and "analytical". Applying descriptive labels such as these is always a difficult endeavour in the social sciences. It is important to note therefore that Critical Theory is not purely and solely normative, but also may contain within it an empirically grounded explanatory analytical element. Indeed, Max Horkheimer, one of its founding fathers, maintained that Critical Theory is viable only if it can explain the circumstances of oppression and enslavement.[10]

Throughout this section I have stressed the importance of *primary interest* in either the condition or the practice of environmental security. This is important because the Critical Theorist's work on the condition of environmental security will often begin by demonstrating that the current *practice* of environmental security leads to widespread human (or biosphere) insecurity. On that basis, s/he then will advance a proposal for what environmental security should ideally entail.[11] Similarly, works on the *practice* of environmental security are – just as realism and other traditional approaches to security studies – themselves already informed by a view of what security is (see below for an elaboration on this theme and whether this renders all approaches normative). While the difference between works on the condition of environmental security vis-à-vis its practice then fails to conclusively distinguish between Critical Theory and analytical theory, the *primary interest* or the preponderance of the argument is a good indication of the theory used.[12]

However, not all analysts within ESS will limit themselves to just one of these two types of theory. Not only are other types of theory conceivable and used (for example, some scholars within ESS are influenced by poststructuralism – see below), but also, perhaps most importantly, Critical Theory is not the only theory through which a normative claim can be made. Some analysts even combine analytical with normative theory, but not as emancipatory theory. This in itself is only possible when proponents do not subscribe to the Critical Theorist's view that theory is tantamount to practice; or, in other words, when scholars differentiate between the role of the analyst and that of the securitizing actor.

One example of this type of work is provided by some environment, conflict resolution, and peacebuilding scholars. By and large these scholars follow the positivist mainstream with their focus on hypothesis testing and falsification, and seemingly strive to explain the world only (see below for more on method). However, given that the focus of this literature is on "conflict transformation"[13] this literature is also explicitly normative. The normative aspect of this work is summarized in one influential book, which "hope[s] to nudge governments, intergovernmental organizations, social movements, and other actors who have not been aggressive about environmental cooperation to be somewhat more so, by pointing to credible possibilities of peacemaking spin-offs, if they are willing to act to realize them".[14]

Another example is provided by scholars working with (versions of) securitization theory as developed by Ole Wæver and the Copenhagen School. Securitization theory holds that security is a performative speech act, because issues become matters of security policy only when they become recognized as existentially threatening by a suitably powerful securitizing actor and are declared security threats. Provided that the issue is accepted as threatening also by a relevant, sanctioning audience the issue is then moved out of the normal/democratic policymaking realm into a state of emergency politics where it can be addressed by the use of extraordinary measures.[15] Securitization theory is a postpositivist analytical theory that aims at explaining[16] who can securitize, on what issues, under what conditions, and to what effect.[17] Although the Copenhagen school somewhat misleadingly claims that "security means survival",[18] it is important to note that this says nothing about the circumstances under which a referent object is actually safe from harm (secure); or in other words, about the condition of (environmental) security. Instead, to "qualify" for securitization an issue needs to be perceived to be existentially threatening in severity by the securitizing actor, but also by a relevant audience. In line with this the Copenhagen school clearly specifies that the role of the security analyst is *functionally distinct* from that of the securitizing actor and that only the latter can be the architect of successful securitization.[19] This said, however, it is also the case that because the securitization analyst believes in the securitizing force of language (security is after all considered a performative speech act) his or her written work "is at least partly responsible for the co-constitution of social reality, as by means of the securitizaton analyst's own text this reality is (re)produced. In other words, in writing or speaking security the [securitization] analyst him/herself executes a speech act."[20] In an effort to overcome this "normative dilemma of speaking and writing security"[21] the securitization analyst is then practically forced to make normative recommendations. Wæver does so by promoting desecuritization as *ceteris paribus* preferable to securitization. The former refers to the process whereby an issue is moved out of the realm of security, emergency politics and exception (that defines securitization) and back into ordinary politics.[22] Interestingly, while arguing that the environment should be desecuritized,[23] he manages to make a normative argument without saying a word about what environmental security should ideally be. Or in other words, nowhere does he write about the actual condition of (environmental) security.[24]

Wæver's normative preference for desecuritization is informed by the idea that desecuritization and securitization tend to have the same outcome, whereby desecuritization is believed to lead to politicization, dialogue and an opening up of debate, while securitization is believed to lead to depoliticization, dedemocratization and potentially the security dilemma. Not only has it been shown that desecuritization can lead to depoliticization[25] but also as a normative strategy desecuritization is feasible only when one ignores – as he does – the existence of objective existential or real threats.[26] Yet, in such cases – for example, when climate change induced global sea-level rise endangers the physical existence of small island states – securitization may well be the more appropriate solution. On the basis of these and similar ideas, I have suggested elsewhere that a better strategy for escaping securitization theory's "normative dilemma" is to differentiate between (morally) right and wrong securitizations. "Just securitization theory" identifies a set of criteria that determine the morality of securitization across all sectors of security, including for environmental security. The criteria are: 1) there must be an objective existential threat; 2) the referent object of security must be morally legitimate; and 3) the security response must be appropriate to the threat in question.[27]

From theory to method

Before turning to the second dividing line, it is useful to include a section on method. Simply put, an analyst's view of theory, as well as his or her epistemological position, informs the choice of methods he or she deems appropriate. With the exception of political ecologists, environmental conflict scholars are informed by a positivist epistemology that aims to emulate the natural science. Their research methods typically include quantitative work. Recent work on the interrelationship between demographic factors and violent conflict utilizes and cross-references often large sets of statistical data such as those on human-induced soil degradation or on freshwater availability with data on armed conflicts.[28] The predictive value of this type of analysis is unclear for at least two reasons. First, environmental change endlessly creates a new context for social behaviour and therefore the future is always unlike the past. Second, the density of connections in the human world suggest to many analysts that outcomes will tend to be more comparable to the non-linear models of quantum physics than the linear models of Newtonian physics. These reasons also explain why study results often are not easy to replicate with new data.[29]

While quantitative studies are increasingly popular,[30] the bulk of positivist environmental conflict research, however, is made up of qualitative case study research. This usually takes one of two forms. First, empirical observation leads to generalization and consequently theory development (induction). Second, on the basis of specific theoretical observations the analyst begins by formulating predictive hypotheses (deduction). Depending on whether these are found to be true or false in subsequent empirical analysis, they serve to prove or disprove (parts of) the initial theory. An example of the latter is "Causal Pathways to Conflict" by Wenche Hauge and Tanja Ellingson, in which the authors devise and test predictive hypotheses on the basis of Homer-Dixon's environmental scarcity and conflict model.[31]

A perennial disagreement between different environmental conflict theorists and their methodological critics concerns the independent variable – importantly not only what it should be (notable examples include environmental scarcity and environmental discrimination), but also whether it is actually possible to identify an independent variable at all. Tom Deligiannis' chapter in this volume (chapter 2) offers a detailed discussion of this issue.

Critical Theorists (including political ecologists and many human security scholars), poststructuralists and other critical scholars (for example, feminists and securitization scholars) reject the idea that the social world can and should be studied in the same ways as the natural world. The rejection of Enlightenment positivism is driven among other things by the idea that many occurrences in the social world are not directly observable in the way positivism demands; that is, the social world is more complex than research methods favoured in the natural sciences permit. In addition, postpositivism rejects positivism's insistence that the researcher be value-free and detached on the grounds that, at a minimum, the values of the observer influence the choice of case study and research topic.

Unlike other approaches to environmental conflict, political ecology is dominated by geographers, many of whom regard political science as unable to understand "the sheer complexity of the relationships between environment and violence in many places".[32] Their field research and detailed case study analysis utilizes insights from many different disciplines across the social and natural sciences, all of which have influenced the emergence of political ecology more generally.[33] Again, unlike other environmental conflict theorists, political ecologists do not advance "a single theory of violence".[34] War and physical violence are only one form of violence; focusing on this alone ignores that much violence is symbolic or cultural. Examples of symbolic violence include social humiliation and reduced access to resources due to a lower social standing.

Case study research is informed by the aim to "examine how causal powers, located in the two spaces of production and power relations, create forms of social mobiliza-tion and conflict in specific circumstances".[35] This is perhaps most easily explained by considering the questions that motivate a political ecologist's research. As Peluso *et al.* put it:

It is important to ask [. . .] why violence occurs in some places and not in others, why some factors are more important than others, and why brutal acts define some conflicts and not others. The purpose of rich empirical case studies is to reveal how these causal forces articulate in specific circumstances.[36]

Case study analysis by human security scholars is equally rich. By drawing on a host of tools and insights from a wide variety of disciplines including environmental history, environmental science, anthropology and development economics, human security scholars aim to show which humans in what places are affected most severely by global environmental change. Analysis typically concludes with arguments encouraging normative change, often inspired by research findings from a large number of disciplines including development economics, global political theory (especially cosmopolitanism), moral philosophy and green theory. Thus a recent collection of essays on global environmental change and human security includes pieces on inequality as a determinant of vulnerability, the particular vulnerabilities of women, the links between environmental change and health, and the significance of environmental security for sustainable development. While not cumulative in the natural science sense, the overall effect is to demonstrate the multiple ways in which environmental change results from social structures, values and processes, and in turn affects people differentially based on their locations within these.[37]

Many proponents of human security and political ecologists, who, as I argue, can be situated in the camp of Critical Theory, use the preferred research method of post-structuralism – discourse analysis. Discourse refers to "the language and representations through which we describe the world".[38] As part of discourse analysis, researchers can study anything including official policymaking documents, interviews, polling data and unofficial documents. They also may study the premises and consequences of competing theoretical approaches. Discourse analysis takes many forms and it is used for different ends.[39] Poststructuralists, for whom the world becomes knowable through discursive representations only, are skeptical of metanarratives as they reject the very possibility of objectivity, universal truth and reason. The aim of such discourse analysis is to show that all representations are necessarily power-laden, subjective and biased. Although there is concern for those that are side-lined (commonly referred to as "othered") by these ostensibly totalizing metanarratives, poststructuralism's inbuilt rejection of universal truth or emancipation leaves it unable to suggest what a better world order (or in this case, what ideal environmental security) would look like. This is perhaps the reason why poststructural ideas are often supplemented with other theories in ESS.

Critical Theorists also believe that discourses are power-laden – not because they are skeptical of objectivity and reason, but because the world is unequal, due to the prevalence of factors such as patriarchy, capitalism, statism and consumerist democracy.[40] Quite unlike poststructuralists, they believe in universal emancipation, and environmental security as human security is one such ideal form of environmental security.

In environmental security studies the boundaries between poststructuralism and Critical Theory are sometimes blurred, with researchers taking cues from both traditions. Simon Dalby's work, for example, is heavily influenced by poststructuralist writings, yet by siding with the human security approach he ultimately advocates an emancipatory ideal.[41]

The same is true for political ecologists, whose work on the structurally produced conditions for environmental violence is heavily influenced by Marxism, yet – as, for example, the introduction to Nancy Lee Peluso and Michael Watts' *Violent Environments*[42] – also draws very heavily on Foucault, who warned against repoliticization as this was thought to lead to a repetition of exclusion and dominance.[43] Generally speaking, some proponents of Critical Theory and poststructuralists study discourses to show how existing theories (especially mainstream environmental conflict approaches) and practices of environmental security (especially the military's interpretation of environmental security, but also national security) render people (and often the world's most vulnerable people) more insecure. The aim is, in short, to show the moral and practical poverty of competing narrow[44] approaches to environmental security and – for Critical Theorists at least – to propose alternative versions of environmental security.

The Copenhagen school's securitization theory also relies on discourse analysis. Analysts using this theory study publicly available documents only, in order to trace the social and political construction of issues as security threats. Although securitization theory has nothing to say in particular on environmental security (e.g. what form it should take, or the possibility of environmental conflict vis-à-vis environmental cooperation), it is the only theory in existence that can grasp that environmental security takes the many forms it takes today. Thus, it allows one to differentiate between (potential) securitizing actors (e.g., states, the UN, critical environmental theorists), referent objects of security (e.g., states, individuals, the biosphere) and the nature of the threat (e.g., long term, human environmental misconduct, environmental conflict, climate change). As such securitization theory is or could be a key tool for scholars of environmental security.[45]

Dividing line II: the view of security

The second line dividing in the field of environmental security studies pertains to the view researchers hold of security itself. Two different ideal-typical positions are possible. First, security can be defined in narrow terms as the absence of violent conflict only. This position is shared by many theorists of environmental conflict, who are interested in the environment *primarily* because it has the potential to generate violent conflict.

The second position is that security can be about more than the absence of violent conflict – it also can be about the presence of certain basic capabilities. This wide view of security is supported by human security scholars, feminists, but also by proponents of ecological security.

I realize that some scholars might object to this categorization, and that position one could be viewed as a subset of position two.[46] The consequence of this is that not everyone within ESS will be happy, or indeed able, to identify neatly with just one of the two categories I have laid out. Notably, political ecologists are interested in environmental conflict and violence, but because of the way they define violence their view of security is much wider than their interest in environmental conflict suggests and in effect closer to that of human security scholars. I am also aware that many researchers concerned with human/ecological security are also interested in the causes of violent conflict[47] (see, for example, Pirages' chapter in this volume) and that some of those studying violent conflict are concerned with (some of) the causes of human insecurity, as opposed to the insecurity of states (see, for example, de Soysa's chapter in this volume). Nevertheless, I am interested here in developing a framework that enables me to bring order into a vast and diverse literature, and this can only be done with the help of some cautious and well-intended simplifications.

A potential additional problem is that each of my two positions on the appropriate area of research for the subject of environmental security holds within it a view on the condition of security. This raises an obvious question: is the earlier explained dividing line pertaining to the role of the analyst redundant? Or in other words, are not all approaches to environmental security necessarily normative? In International Relations more generally, this point has been advanced by Critical Theorists who have accused their positivist mainstream colleagues of being normative without owning up to it, because all theory rests on pre-made choices of what is relevant and what therefore is included and excluded.[48] While the latter is indisputable it merely means that the motivations for adopting one view of security over another are rarely value-neutral, but it does not amount to normativity as I understand it, which lies in the nature and intention of the argument. Normative theory (which includes but is not exhausted by Critical Theory) engages in arguments regarding what should and should not be the case in the world and what we should and should not do within it, while analytical theory engages in arguments regarding what does exist and why it exists, and what people do and why they do it.[49] I believe this categorization is useful even though at a higher level of abstraction the normative and empirical can blur into and in fact become the same reality-generating discourse.

In addition to the two well-known views on security, it is also necessary to include the view of those scholars who question the appropriateness of security framing altogether. Some environmental cooperation/peacebuilding scholars, for example, stress well-known empirical observations that environmental problems tend to lead to cooperation and "desecuritization"[50] between conflicting parties and may even lead to a sustainable peace. These scholars usually focus on the potential for conflict resolution and do not endlessly theorize about the diminished likelihood of violent environmental conflict.

Four debates defining contemporary environmental security studies

As well as bringing order into the field of environmental security studies, the here proposed framework also goes some way towards understanding the fault lines that structure some of the main debates defining contemporary environmental security studies. Some of these debates can be reduced to the debating party's view of security as well as their view of the role of the analyst and his or her view of theory. As I show below, in some cases, once the arguments produced on either side of any given debate are understood along my two dividing lines, it appears that regardless of existing and perhaps even future empirical evidence produced to support the arguments on either side, the debates will remain in large measure irresolvable.

There are at least four debates that dominate the existing environmental security literature. The first is the debate internal to the environmental conflict literature on whether environmental conflict is scarcity, abundance or structurally motivated. Proponents of scarcity and abundance theses are united by a narrow view of security and by their belief that the role of the analyst is to explain the world only. They differ only insofar as they locate the sources of violent environmental conflict in either the local abundance of natural resources or in the scarcity of natural resources. Yet, as Colin Kahl has pointed out, the resource abundance thesis ultimately collapses into the scarcity thesis, as locally abundant resources are only valuable because they are scarce on the global scale.[51] If this is so, then perhaps more acute fault lines lie between scarcity and abundance theses on the one hand, and structurally (politico-economical) induced conflict on the other. The latter view is advanced by political ecologists.

Although political ecologists seemingly advance a narrow view of security as the absence of violent conflict, theirs is in fact a fairly wide view of security in that violence is redefined to include non-physical violence against individual persons. As such their view of security is much broader than that of other environmental conflict theorists. What is more, their (partially) Marxist heritage obliges them to envisage alternative and better world orders; in short, theirs is a Critical Theory, not an analytical one. These epistemological differences make it unlikely that this debate between scholars of environmental conflict will ever be resolved.[52]

The second debate is about whether security concerns the absence of violent conflict only or whether true security is "freedom from want" in addition to the absence of violent conflict. This debate is at the heart of disputes over the nature of security studies. Though outright criticism of a widened concept of security may be diminishing, it continues to have its articulate and influential proponents such as Walt. One of his several well-known arguments is that security studies should be about the phenomena of war and things directly related to war such as alliance formation, and that redefining the boundaries of the discipline "would destroy its intellectual coherence and make it more difficult to devise solutions to any of these important problems."[53] Similarly, and with explicit reference to environmental security, the American political scientist Daniel Deudney has argued that the environment is not a national security issue because environmental degradation is not a likely cause of interstate conflict.[54] He contends that the notion of environmental security rests on a number of analytical errors, most of which could be redressed by a more careful analysis of history. States are far more likely to innovate and trade than fight, for example, and today the disincentives to the use of force make innovation and trade almost certain. Of course, both Walt and Deudney acknowledge that environmental threats can have lethal repercussions, but these are better cast – and addressed – as natural disasters threatening human well-being rather than as matters of national security.[55] An underlying concern here is that if everything is framed as a security issue, then "security" loses its analytical and policy significance. This view corresponds to that of Marc Levy, who has argued that because of the failure of environmental security analysts to define both "environment" and "security" sufficiently, the concept possesses no analytical credibility and is simply inappropriate.[56]

This second debate is part of the quintessential and much wider debate between non-traditional security studies and their traditional counterparts. Proponents of the first prefer not only a suitably widened and broadened academic field of security, but they are concerned as well with the meaning of security, as opposed to taking it as a given.

They also focus on the co-constitutive role of the security analyst vis-à-vis security policies.[57] Deep epistemological, ontological and methodological divisions between traditional and non-traditional security studies render this debate practically irresolvable.

Third, is the debate between environmental scarcity theorists and their cornucopian critics. As argued in the introduction to this volume, climate change has brought with it renewed interest in the possibility of overt violent conflict – boosting the research profile of environmental conflict theses. In particular, the environmental scarcity and conflict thesis is bolstered because many believe that climate change will lead to a scarcity of resources (for example, of freshwater or food due to weather related crop failure), with people resorting to violence to fight over the remaining scarce resources. Another way in which climate change is presented as a contributor to violent conflict is more indirect. For example, environmental scarcity could force people to leave their homelands, generating conflict if they migrate to areas where they are not welcome. Not everyone is convinced by this logic and this renewed interest in environmentally induced conflict has brought with it a new wave of critics of environmental scarcity/conflict theses. Critiques of "neo-Malthusian" approaches to environmental security are well known.[58] With a view to climate change related conflict, those influenced by cornucopian ideas now argue not only that "we are far from being able to account reliably for the negative and positive effects on human affairs" but also that "climatic change will require adaptation, which while costly, may also lead to innovation."[59] The link between environmental migration and violent conflict is largely dismissed on the grounds that historically "most migration does not lead to conflict".[60] Ragnhild Nordås and Nils Petter Gleditsch, for example, point out that although migration numbers predicted as a result of climate change seem "frightening" they are very small compared to the numbers of economic migrants.[61]

With regards to the possibility of violent conflict, cornucopian inspired critics argue that "there is not yet much evidence for climate change as an important driver of conflict [. . . at most] environmental change *may* under certain circumstances increase the risk of violent conflict, [and] the existing evidence indicates that this is not generally the case."[62] This claim is substantiated by a host of very thorough, peer-reviewed and mainly statistical investigations into the relationship between climate change and armed conflict.[63]

Although the arguments of scholars with cornucopian leanings are important and balance the sensational arguments of impending climate wars that dominate the popular press, the last word on climate change is not yet spoken. Thus we must not forget that climate change and its consequences could become much larger problems in the future,[64] and we must remind ourselves that prediction in International Relations theory has not served us well in the past. In fairness, however, it does seem that this cautious view is actually shared by many of the critics. As it stands, and in spite of these scholars' skepticism, they are unable to completely dismiss the link. The evidence they themselves have found shows that in some cases environmental factors do lead to violent conflict. What remains unclear is to what extent it is possible to extrapolate from individual studies and develop wider theories of conflict and climate change. It is likely, however, as Gleditsch argues, that "interaction with exogenous conflict-promoting factors" as they admit Homer-Dixon and others have always stressed, do matter more than large-N studies have hitherto acknowledged.[65] Above all else, what seems to matter is the nature and stability of political institutions.[66] The interesting thing about this claim for environmental conflict research is that there seems to be some room for convergence between the research agendas of environmental scarcity and conflict scholars (because they stress this interaction), resource abundance and conflict scholars (because they stress good governance), and their cornucopian critics.

Fourth and finally is the debate over whether environmental stress will lead to cooperation or conflict. Empirical evidence supplied by scholars of environmental peacebuilding seems to bolster the cornucopian argument that fears about environmentally induced violent conflict often are misplaced or overstated, as environmental stress has been shown frequently to lead to cooperation between conflicting parties instead. As Patrick MacQuarrie and Aaron Wolf show in their chapter on water security (chapter 9) the popular claim that water wars will be the definitive wars of this century may prove to be an exaggeration. Many, even rival, states cooperate peacefully on water issues. However, as Saleem Ali and Mary Watzin point out in chapter 10, experiences from individual cases may not necessarily translate to all regions. The same can be said about the enduring and contemporary value of the findings of historical analysis of a world with far fewer people and far fewer political actors than is now the case.

A major systematic review with the research question "What is the evidence that scarcity and shocks in freshwater resources cause conflict instead of promoting collaboration?", carried out by the new economics foundation (nef), funded by the United Kingdom's Department for International Development (DFID) shares this careful view. Largely because of the heterogeneous nature of different studies, this study was unable to answer the research question in a definitive way.[67] The authors recommend that future studies on conflict and cooperation need to be more homogeneous in nature, and use the same definitions of conflict, collaboration and scarcity, and also the same theoretical frameworks of additional explanatory variables, as otherwise the directional relationship cannot be conclusively established and the debate will not be resolved.

Conclusion

The fact that only the latter two of the four debates within ESS have some hope of being resolved does not spell doom for the field as a whole. Ultimately the different approaches are trying to achieve different things. As long as environmental security in whichever form remains a concern in the policymaking world, the issue will not go away and security analysts will continue to need analytical theory in order to explain the world. Moreover, given that not all of us will find it sufficient to act as normatively silent interpreters of (official) policy, we will also want to continue thinking about what environmental security should ideally mean and how it can be achieved. In short, although some of the debates that define the field of environmental security studies are pretty much irresolvable, all sides to the debates matter and should be heard. The overriding objective of this volume is precisely to give voice to a range of approaches and issues in environmental security studies.

Notes

1   I would like to thank Richard Matthew for his advice and help with this chapter. I would also like to thank Stuart Croft, Simon Dalby, Tom Deligiannis, Jonathan Floyd, Jonna Nyman, Adam Simpson and Achim Maas for their comments on, all or parts of, earlier drafts of this chapter, as well as three anonymous reviewers of a draft proposal of this chapter for their helpful comments.

2   See, for example, J. Barnett, 'Environmental Security', in A. Collins (eds), *Contemporary Security Studies*, Oxford: Oxford University Press, 2007, pp. 182–203.

3   J. Herington, 'Security and the anatomy of value', unpublished paper delivered at the *7th Pan-European International Relations Conference*, Stockholm, Sweden, 9–11 September 2010. Online: http://stockholm.sgir.eu/uploads/Herington%20-%20Draft.pdf (accessed August 2012).

4   S. Walt, 'The Renaissance of Security Studies', *International Studies Quarterly* 35, 1991, 212.

5   For an elaboration on this theme, see R. Floyd, 'Conceptualising human security as a securitising move', *Human Security Journal/Revue de la Sécurité Humaine* 5, 2007, 38–49.

6   J. Barnett, *The Meaning of Environmental Security*, London: Zed Books, 2001 p. 129.

7   I use "Critical Theory" (upper case) and "critical theory" (lower case) in line with convention. This convention is explained well by Columba Peoples and Nick Vaughan-Williams: "Critical Theory (upper case) is conventionally used to denote a Marxian tradition of theorising that includes elements of Marx's philosophy – most notably his invocation to not only 'interpret the world' but to 'change it', [while the] use of the lower case 'critical theory' is generally used in the social sciences to identify a more diverse range of ideas and approaches that includes Marxian-inspired thought but is far from limited to it and even challenges it in some respects. Whereas the former has a particular (emancipatory) purpose, the latter is more heterogeneous in its concerns and goals" (*Critical Security Studies: An Introduction*, Abingdon: Routledge, 2010, p. 18).

8   R. W. Cox, 'Social Forces, States and World Orders: Beyond International Relations Theory', in R. Keohane (ed.), *Neorealism and Its Critics*, New York: Columbia University Press, 1986, p. 207 (emphases in original).

9   R. W. Jones, '"Message in a Bottle?" Theory and Practice in Critical security studies', *Contemporary Security Policy* 16 (3), 1995, 299–319.

10   J. Bohman and E. N. Zalta (eds), 'Critical Theory', *The Stanford Encyclopedia of Philosophy (Spring 2012 Edition)*. Online: http://plato.stanford.edu/archives/spr2012/entries/critical-theory/ (accessed August 2012).

11   See, for example, Barnett, *The Meaning of Environmental Security*, or more recently M. McDonald, *Security, The Environment and Emancipation: Contestation over Environmental Change*, Abingdon: Routledge, 2012.

12   The exception are political ecologists, who study with environmental conflict environmental practice, but who are informed by a Critical epistemology.

13   See Achim Maas *et al.*, this volume, chapter 5.

14   K. Conca, G. D. Dabelko (eds), *Environmental Peacemaking*, Washington, DC: Woodrow Wilson Center Press, 2002, p. 15.

15   B. Buzan, O. Wæver and J. de Wilde, *Security: A New Framework for Analysis*, Boulder, CO: Lynne Rienner, 1998.

16   I use the term "explaining" fully aware of the connotations this has in IR. Thus I realize that securitization theory clearly does not comply with the logic of explaining as outlined in M. Hollis and S. Smith, *Explaining and Understanding International Relations*, Oxford: Clarendon Press, 1990. This said, however, International Relations research is moving away from these kinds of rigid distinctions as evinced by, for example, Milja

Kurki's impressive reworking of causation in International Relations (M. Kurki, *Causation in International Relations*: *Reclaiming Causal Analysis*, Cambridge: Cambridge University Press, 2008). For more on causation and the explanatory value of securitization theory see S. Guzzini, 'Securitization as a causal mechanism', *Security Dialogue* 42 (4–5), 329–41 and R. Floyd, *Security and the Environment: Securitisation theory and US Environmental Security Policy*, Cambridge: Cambridge University Press, 2010, pp. 188–89.

17   O. Wæver, *Concepts of Security*, Copenhagen: Institute of Political Science, University of Copenhagen 14, 1997, 48; Buzan *et al.*, *Security: A New Framework for analysis*, p. 27.

18   Buzan *et al.*, *Security: A New Framework for Analysis*, p. 27.

19   Buzan *et al.*, *Security: A New Framework for Analysis*, pp. 33–34.

20   R. Taureck, 'Securitisation Theory and Securitisation Studies', *Journal of International Relations and Development* 9 (1), 2006, 53–61 at p. 57.

21   J. Huysmans, 'Defining social constructivism in security studies. The normative dilemma of writing security', *Alternatives* 27 supplement, 2002, 41–62.

22   O. Wæver, 'Securitization and Desecuritization', in R. D. Lipschutz (ed.), *On Security*, New York: Columbia University Press, 1995, pp. 46–86.

23   Wæver, 'Securitization and Desecuritization', p. 63.

24   This is a deliberate choice because for Wæver, security and insecurity are not considered exhaustive polar opposites, instead "insecurity is the situation when there is a threat and no defence against it; security is a situation with a threat *and* a defence against it" (O. Wæver, 'Securitization: Taking stock of a research programme in Security Studies', unpublished manuscript, 2003, p. 13).

25   Floyd, *Security and the Environment*, pp. 166–70; on a similar theme see also S. Cui and J. Li, '(De)securitizing frontier security in China: Beyond the positive and negative debate', *Cooperation and Conflict* 46 (2), 144–65.

26   R. Floyd, 'Can securitization theory be used in normative analysis? Towards a just securitization theory', *Security Dialogue* 42 (4–5), 2011, 427–39 at 436; see also T. Balzacq, 'The three faces of securitization: Political agency, audience and context', *European Journal of International Relations* 11 (2), 171–201, at 181.

27   For more detail, see Floyd, 'Can securitization theory be used in normative analysis?'.

28   See the various contributions to 'New Directions in Demographic Security', *Environmental Change and Security Program Report*, Issue 13, 2008–09, pp. 2–47.

29   I am grateful for these points to Richard Matthew

30   See, for example, O. M. Theisen, 'Other pathways to conflict? Environmental scarcities and domestic conflict', *Journal of Peace Research* 45 (6), 2008; H. Urdal, 'People vs. Malthus: Population pressure, environmental degradation, and armed conflict revisited', *Journal of Peace Research* 42 (4), 2005, 417–34.

31   W. Hauge and T. Ellingson, 'Causal Pathways to Conflict', in P. F. Diehl and N. P. Gleditsch (eds), *Environmental Conflict*, Boulder, CO: Westview Press, 2001, pp. 36–57.

32   S. Dalby, 'Environmental Security and Climate Change', in R. A. Denemark (ed.) *International Studies Online/International Studies Encyclopaedia*, Volume III, Oxford: Blackwell, 2010, pp. 1580–97. Online: www.isacompendium.com/ (accessed August 2012). See also, B. Korf, 'Cargo Cult Science, Armchair Empiricism and the Idea of Violent Conflict', *Third World Quarterly* 27 (3), 2006, 459–76.

33   For an insightful overview that highlights in particular the controversial absence of biophysical ecology and environmental science in contemporary (poststructuralist) political ecology, see P. A. Walker, 'Political Ecology: Where is the ecology?', *Progress in Human Geography* 29 (1), 2005, 73–82.

34   N. L. Peluso and M. Watts, 'Violent Environments', in N. L. Peluso and M. Watts (eds), *Violent Environments*, Cornell: Cornell University Press, 2001, p. 29.

35   Ibid.

36   Ibid.

37   Matthew *et al.*, *Global Environmental Change and Human Security*.

38   T. Dunne, M. Kurki and S. Smith, *International Relations Theory: Discipline and Diversity*, Oxford: Oxford University Press, 2007, p. 333.

39   See, for example, L. Hansen, *Security as Practice: Discourse Analysis and the Bosnian War*, Abingdon: Routledge, 2006; N. Fairclough, *Analysing Discourse: Textual analysis for social research*, London: Routledge, 2003. For an overview, see D. Howarth, *Discourse*, Buckingham: Open University Press, 2000.

40   K. Booth, *Theory of World Security*, Cambridge: Cambridge University Press, 2007, pp. 21ff.

41   S. Dalby, *Environmental Security,* Minneapolis: University of Minnesota Press, 2002.

42   N. L. Peluso and M. Watts (eds), *Violent Environments*, Cornell: Cornell University Press, 2001.

43   M. Foucault, 'The History of Sexuality', in C. Gorden (ed.), *Power/Knowledge: Selected Interviews and Writings 1972–1977*, New York: Pantheon Books, 1980 [1975], p. 190.

44   Narrow as defined in the next section.

45   Examples of works on environmental security that utilize aspects of securitization theory include: O. Corry, 'Securitisation and "Riskification": Second-order Security and the Politics of Climate Change', *Millennium: Journal of International Studies* 40 (2), 2012, 235–58; M. McDonald, *Security, the Environment and Emancipation: Contestation over Environmental Change*, London and New York: Routledge, 2012; S. Stetter, E. Herschinger, T. Teichler and M. Albert, 'Conflicts about water: Securitizations in a global context', *Cooperation and Conflict* 46, 2011, 441–59; N. Detraz, 'Threats or Vulnerabilities? Assessing the Link between Climate Change and Security', *Global Environmental Politics* 11 (3), August 2011, 104–20; N. Detratz and M. Betsill, 'Climate Change and Environmental Security: For Whom the Discourse Shifts', *International Studies Perspectives* 10, 2009, 303–20; M. J. Trombetta, 'Environmental security and climate change: analysing the discourse', *Cambridge Review of International Affairs* 21 (4), 2008, 585–602; J. de Wilde, 'Environmental Security Deconstructed', in H. G. Brauch, J. Grin, C. Mesjasz, P. Dunay, N. C. Behera, B. Chourou, U. O. Spring, P. H. Liotta and P. Kameri-Mbote (eds), *Globalisation and Environmental Challenges: Reconceptualising Security in the 21st Century*, Berlin-Heidelberg, New York: Springer-Verlag, 2008, pp. 595–602; Floyd, *Security and the Environment*; and various authors in H. G. Brauch, U. O. Spring, J. Grin, C. Mesjasz, P. Kameri-Mbote, N. C. Behera, B. Chourou and H. Krummenacher (eds), *Facing Global Environmental Change: Environmental, Human, Energy, Food, Health and Water Security Concepts*, Berlin-Heidelberg, New York: Springer-Verlag, 2009.

46   I would like to thank Tom Deligiannis for bringing this point to my attention. Indeed his own work is a case in point, see T. Deligiannis, 'The Evolution of Environment-Conflict Research: Towards a Livelihood Framework', *Global Environmental Politics* 12 (1), 2012, 78–100.

47   J. Barnett and N. W. Adger, 'Environmental Change, Human Security, and Violent Conflict', in R. A. Matthew, J. Barnett, B. McDonald and K. L. O'Brian (eds), *Global Environmental Change and Human Security*, Cambridge, MA: MIT Press, 2009, pp. 119–36.

48   Cf. S. Smith, 'The contested Concept of Security', in K. Booth (ed.), *Critical Security Studies and World Politics*, Boulder, CO: Lynne Rienner, 2005, pp. 27–62.

49   J. Floyd, *The impossibility thesis: a methodological explanation of interminability in contemporary political philosophy*, University of Oxford: Dept. of Politics and International Relations, Social Sciences Division, 2009.

50 Desecuritization here in a generic sense and not necessarily in the way envisaged by the concept's originator Ole Wæver. For an excellent overview see: L. Hansen, 'Reconstructing desecuritisation: the normative-political in the Copenhagen school and directions for how to apply it', *Review of International Studies* 38 (3), 2012, 525–46.

51 Kahl, *States, Scarcity, and Civil Strife in the Developing World*, p. 18.

52 Though see Tom Deligiannis' contribution to this volume (chapter 2).

53 Walt, 'The Renaissance of Security Studies', 213.

54 D. Deudney, 'The case against linking environmental degradation and national security', *Millennium* 19, 1990, 461–76 at 461.

55 Ibid., 463.

56 M. A. Levy, 'Is the Environment a National Security Issue?', *International Security* 20, 1995, 37ff.

57 O. Wæver, 'Aberystwyth, Paris, Copenhagen New Schools in Security Theory and the Origins between Core and Periphery', unpublished manuscript, 2004, p. 13; see also R. Floyd and S. Croft, 'European non-traditional security theory: From theory to practice', *Geopolitics, History, and International Relations* 3 (2), 2011, 152–79.

58 See, for example, J. L. Simon, 'Pre-debate Statement: Julian Simon', in N. Myers and J. L. Simon (eds), *Scarcity or Abundance? A Debate on the Environment*, New York: Norton, 1994; J. L. Simon, 'Closing Statement by Julian Simon', in Myers and Simon, *Scarcity or Abundance?*; B. Lomborg, *The Skeptical Environmentalist*, Cambridge: Cambridge University Press, 2001.

59 N. P. Gleditsch and M. Theisen, 'Resources, the Environment and Conflict', in M. Dunn Cavelty and V. Mauer (eds), *The Routledge Handbook of Security Studies*, London: Routledge, 2010, p. 227.

60 I. Salehyan, 'From Climate Change to Conflict? No Consensus Yet', *Journal of Peace Research* 45 (3), 2008, 315–26 at 319.

61 R. Nordås and N. P. Gleditsch, 'Climate change and conflict', *Political Geography* 26, 2007, 627–38 at 632.

62 N. P. Gleditsch, 'Whither the weather? Climate change and conflict', *Journal of Peace Research*, 2012, 49 (1), 3–9 at 7

63 See, for example, C. Raleigh and H. Urdal, 'Climate Change, Demography, Environmental Degradation, and Armed Conflict', *Environmental Change and Security Program Report* 13, 2008–09, 27–33; C. S. Hendrix and S. M. Glaser, 'Trends and triggers: climate, climate change and civil conflict in sub-Saharan Africa', *Political Geography* 26 (6), 2007, 695–715. See also R. Nordås and N. P. Gleditsch (eds), Special Issue on Climate Change and Conflict, *Political Geography* 26, 2007.

64 Though see, for example, C. Webersik, *Climate Change and Security: A Gathering Storm*, Santa Barbara, CA: Praeger Publishers, 2010; N. Mabey, *Delivering Climate Security: International Security Responses to a Climate Changed World*, Whitehall Paper No. 69, London: Royal United Services Institute, 2007, who argue that climate change and its consequences are already major issues of concern.

65 Gleditsch, 'Whither the weather? Climate change and conflict' p. 6.

66 Salehyan, 'From Climate Change to Conflict? No Consensus Yet'.

67 V. Johnson, I. Fitzpatrick, R. Floyd and A. Simms (2011), 'What is the evidence that scarcity and shocks in freshwater resources cause conflict instead of promoting collaboration?', CEE review 10-010, Collaboration for Environmental Evidence. Online: www.dfid.gov.uk/r4d/PDF/Outputs/SystematicReviews/SR10010.pdf (accessed August 2012).

2

THE EVOLUTION OF QUALITATIVE ENVIRONMENT-CONFLICT RESEARCH

Moving towards consensus

Tom Deligiannis

The question of whether human-induced environmental change should be considered a security threat has been an important part of the post-Cold War debate about redefining security.[1] Those arguing that security should be redefined to include environmental factors argue that conventional definitions place undue emphasis on the zero-sum character of relative power gain at the expense of potential threats that can have a positive- or negative-sum impact on the welfare of states and of the people in them.[2] Opponents, on the other hand, argue that such a broad definition of security is conceptually weak – to the extent that it is almost vacuous – and motivated by politics rather than analysis.[3] Several researchers have chosen to side-step this debate and narrow the analytical focus to the possible relationship between human-induced environmental and demographic change and violent conflict. During the 1990s, qualitative research projects in Canada, led by Thomas Homer-Dixon at the University of Toronto (Toronto Group), and in Switzerland, led by Günther Baechler (Bern-Zurich Group), provided a wealth of case studies and hypotheses for researchers to consider.[4] This research has been strongly criticized by some scholars.[5] Others have proposed alternative hypotheses that they feel better explain the linkages advanced by the Toronto Group and the Bern-Zurich Group.[6] In a few cases, scholars have refined and continued research in the tradition of these 1990s qualitative projects.[7]

Twenty years after the publication in 1991 of Homer-Dixon's seminal article, "On the Threshold"[8] and the beginning in earnest of research on environment and conflict, little, if any, consensus exists about qualitative environment-conflict research. Disputes remain unresolved about whether linkages exist, how they operate, which factors and processes should be emphasized, and the direction of future research. Basic ontological, epistemological, and methodological disagreements and, in some cases, notably harsh polemics have paralyzed discussion.[9] Qualitative research seeking to build on the 1990s work is largely moribund, with little agreement on fresh questions that will move inquiry forward. The focus in environment-conflict research has shifted away from qualitative studies to quantitative examinations of linkages, econometric studies of high value resource conflicts, and demographic security studies.[10]

While study of the original questions addressed by Homer-Dixon and Baechler's projects – that is, of the particular connections between environmental change or environmental scarcity and conflict – has progressed little, the legacy of this research is substantial. Insights have filtered into the highest levels of national and international peace and security policymaking.[11] As well, following the release of the 2007 IPCC report on climate change,[12] a flurry of studies have emerged about the security implication of climate change which draw upon many of the hypothesized findings of earlier environment-conflict research.[13] However, given the discord over qualitative environment-conflict research, there is a danger that once sustained and detailed examination of the climate change-security research begins this work may fall prey to a new round of polemical critiques that mirrors past disputes. Qualitative environmental-conflict researchers need to explicitly acknowledge the limitations of earlier work and craft a research agenda to build on areas of consensus in previous work.

In the next sections, I examine the findings of the two biggest qualitative environmental-conflict research projects over the last fifteen years. I identify areas in which consensus can be reached, pinpoint polemical discussions that should be refocused, and isolate fundamental disagreements where a more sophisticated ontology may be necessary. The conclusion of this review is that the stalled evolution of qualitative environment-conflict research can be traced to the polemical approaches of critics and advocates of research in the field who have failed to adequately synthesize areas of agreement, and instead focused on points of clarification and rebuttal. This chapter seeks to assess the disagreements among scholars in order to synthesize points of agreement on the conceptual and empirical record of qualitative environment-conflict research.

I begin with a brief review of the projects of the Toronto Group and the Bern-Zurich Group to highlight areas of agreement and divergence in their models, in light of criticisms about the definition of the independent variable. Disputes about the nature of the independent variable in qualitative environment-conflict research provide insights into controversies over the role of inequality, population factors, and consumption influences. The independent variable used by Homer-Dixon and Baechler also influences the way in which critics have interpreted the results of their research. This has resulted in polemical and overly simplified interpretations by some, as is evidenced in debates about Neo-Malthusianism and 'greed vs. grievance', discussed below.

The Toronto Group and the Bern-Zurich Group

Concerns about the security implications of human-induced environmental change have a long and contentious history.[14] In the 1990s, a number of scholars examining this relationship chose to focus on those areas where both the local environmental relationships were crucial for people's survival, and the opportunities and capabilities to forestall negative implications were weakest – in the world's poorest, developing states.[15] People who are heavily reliant on natural resources for their survival– particularly renewable resources like land, water, and forests – and who are limited in their ability to sustainably manage these resources are particularly at risk of the impacts of human-induced environmental transformation. Today, almost half of the 7 billion people on the planet rely upon local natural resources for a large part of their well-being.[16] Those living in developing countries are particularly tied to their local natural resources and thus vulnerable to human-induced pressure on these resources. Investigating the material impact of changes in these key resources is thus highly relevant. Both the Toronto Group and the Bern-Zurich Group recognized this reality as they set out to conduct a series of qualitative case studies on environmental change-conflict linkages in the 1990s. Each hypothesized that human pressure on natural resource endowments could affect the material well-being of developing societies and increase the risk of conflict.

Recognizing the methodological problems involved in testing hypotheses related to human-ecological systems, namely the futility of trying to control for confounding variables, the Toronto Group rejected the quasi-experimental methodology that comparativists in political science typically use to produce generalizations.[17] Instead, the Group adopted a case-study approach wherein cases were selected explicitly on the basis of observed change in both the independent variable "environmental scarcity" and the dependent variable violent conflict.[18] Using a process tracing methodology,[19] the Group addressed the question *Can environmental change cause conflict, and, if so, how?* This question focused on the hypothesized causal role of a specific independent variable, environmental scarcity. Homer-Dixon defines environmental scarcity as a tripartite variable – a composite of three factors: degradation or depletion of the resource (supply-induced scarcity), increased demand for the resource due to population growth or increased per capita consumption (demand-induced scarcity), and changes in access to the resource due to skewed distribution among social groups (structural scarcity).[20] The question driving this project was therefore narrower and more tightly defined than a more general question – of a type commonly asked by researchers – like: *What causes civil conflict?*[21]

The Toronto Group's research suggests that environmental scarcities indirectly help to generate various forms of civil conflict, like insurgencies, group conflict, coup d'etats, and so on.[22] Their research did not support a link between human-induced environmental and demographic scarcities and inter-state conflict. Homer-Dixon hypothesized that environmental scarcities influence the incidence of violent civil conflict through a series of intermediate social effects, like constrained economic productivity, intra- or inter-state migration, the creation and aggravation of group tensions and divisions, and the weakening of institutions and the state's capacity to respond to public needs and effectively deliver public goods. As well, scarcities often interact in particularly important ways to cause resource capture and ecological marginalization.

> *Resource capture* occurs when the degradation and depletion of a renewable resource (a decrease in supply) interacts with population growth (an increase in demand) to encourage powerful groups within a society to shift resource access (that is, to change the resource's distribution) in their favor. These groups tighten their grip on the increasingly scarce resource and use this control to boost their wealth and power. Resource scarcity intensifies scarcity for poorer and weaker groups in society.[23]

Ecological marginalization is often interlinked with resource capture and often a consequence of resource capture.

> Ecological marginalization occurs when unequal resource access (skewed distribution) combines with population growth (an increase in demand) to cause long-term migration of people to ecologically fragile regions such as steep

upland slopes, areas at risk of desertification, tropical rain forests, and low-quality public lands within urban areas. High population densities in these regions, combined with a lack of knowledge and capital to protect the local ecosystem, cause severe resource degradation (a decrease in supply).[24]

In all cases, Homer-Dixon and his colleagues emphasized that scarcities never act alone to cause conflict, but instead interact with a wide range of contextual factors, operating across multiple levels and multiple scales.[25] (See Figure 2.1.)

Günther Baechler's Zurich-based Project on Environment and Conflict (Bern-Zurich Group)[26] examined a much broader selection of case studies, but came to similar conclusions as the Toronto Group in the end. While sharing a similar concern with the Toronto Group about the impact of environmental change on the material well-being of people in developing countries,[27] Baechler's focus on the *transformation* of human–environment relationships as a starting point of analysis, results in a much broader independent variable than Homer-Dixon's focus on environmental scarcities. Though environmental transformation encompasses both negative and positive consequences, the Bern-Zurich Group's focus is the negative consequences of human-induced environmental transformation. It can frequently lead to "environmental discrimination," which "occurs when distinct actors – based on their international position and/or their social, ethnic, linguistic, religious, or regional identity – experience inequality through systematically restricted access to natural capital (productive renewable resources) relative to other actors."[28] Baechler takes a similar multi-causal approach to explaining how human

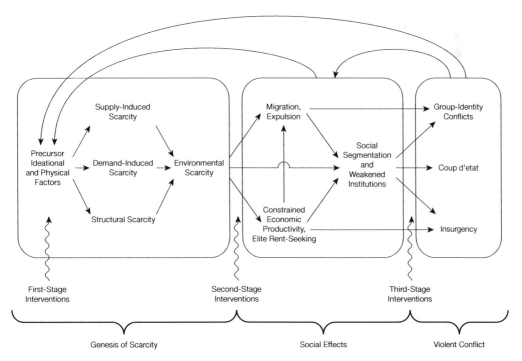

Figure 2.1 Toronto Group's core models of causal links between environmental scarcity and violence.

pressure on the natural environment can help to cause conflict. Environmental transformation combines with various factors to result in different types of sub-state conflict, such as ethnopolitical conflicts, center–periphery conflicts, migration conflicts, or international environmental conflicts.[29]

Defining the independent variable – critics and the role of inequality

The definition of the independent variable in environment-conflict research has long been a source of dispute among researchers. Homer-Dixon focuses on *environmental scarcity* as the independent variable. This tripartite variable has been criticized for including distributional and demographic dimensions. For those affected, environmental scarcity essentially describes a net decrease in the per capita availability of renewable resources within a system (where the system is usually taken as the whole territory of a given country or a sub-region of that country). By contrast, many ecologists and environmentalists focus on environmental change, a term "that refers only to a human-induced decline in the quantity or quality of a resource – that is to worsening supply-induced scarcity."[30] Incorporating unequal resource distribution into the independent variable, Homer-Dixon argues, allows for a more complete examination of the causes of change in resource availability.[31]

Homer-Dixon's inclusion of inequitable distribution in the independent variable *environmental scarcity*, however, has been criticized by scholars. The criticisms hinge on conceptual differences, and the belief among some scholars in the primacy of certain causal explanations. Some, like political ecologists James Fairhead, Nancy Lee Peluso, and Michael Watts, argue that processes like supply and demand reductions in renewable resources cannot be combined into an umbrella term "environmental scarcity" with the seemingly different political-economic processes that lead to inequitable distribution of resources.[32] Processes of culture, power, and political economy that shape inequality are causally prior, Peluso and Watts argue, and are more important than supply and demand changes as the causes of reduced resource availability. In fact, the former often lead to the latter, in their view. Homer-Dixon and Baechler err by starting their analysis at the genesis of scarcity, they argue, instead of examining the processes that created scarcity in the first place, what Homer-Dixon calls the "factors producing scarcity." By doing so, the Toronto Group is "privileging" resource scarcity in their causal framework, and creating "analytical obfuscation."[33] Similarly, after reviewing the Toronto Group's cases, Gleditsch and Urdal argue that "the greater problem [in many of the conflict cases studied by the Toronto group] . . . lies in unequal distribution rather than availability of natural resources,"[34] suggesting that the sources of inequality are the key causal variables.[35] Gleditsch would instead exclude distributional issues from the independent variable and restrict analysis of "environmental conflicts" to cases of supply and demand scarcity.[36] When the links between environmental change and conflict are examined using this definition of scarcity, according to Gleditsch, the relationship is questionable. He points to cross-national research on the relationships between supply and demand scarcities and violent conflict which has found weak influence compared to political and economic factors.[37] Although the environmental data used in this work is relatively crude, the fact that minimal evidence can be found to link supply and demand scarcities to conflict suggests in the eyes of critics like Gleditsch and Jack Goldstone that environmental causes of conflict are weak compared to other causes.[38]

Baechler's conceptualization of the independent variable largely agrees with the political ecology critique.[39] *Environmental transformation*, according to Baechler, is both broader and often causally prior in his conceptualization to supply, demand, and distributional scarcities examined by Homer-Dixon. It encompasses a variety of social and cultural transformations to the environment which affect resource availability, including both supply and demand changes in renewable resources, which are subsumed under the term *environmental degradation* – a consequence of human environmental transformation and disturbance of the environment.[40] Importantly, the mal-distribution of resources is often both a consequence of pre-existing structural inequalities and a consequence of human transformation of the environment, according to Baechler's model. Scarcities are essentially described as a social effect of human environmental transformation.[41] Structural patterns of socio-ecological inequality and discrimination can lead to negative reinforcing patterns of environmental scarcities and further marginalization for many in developing countries.[42] Although global structural inequities in markets and between developed and developing countries largely condition patterns of discrimination and inequality around resources according to Baechler, the impacts of poverty, high population growth rates, and environmental discrimination can have such strong transformative impacts that they should be conceptualized as an exogenous variable.[43] Baechler, like Homer-Dixon, thus accepts the causal importance of inequality as a factor causing scarcity, but would also point scholars to the important underlying structural inequalities in global capitalist economic relationships.

Disputes about including "inequitable distribution" in the independent variable underscore a deeper divide, particularly with political ecologists like Peluso and Watts. To these critics, beginning the analysis of environmental change-conflict linkages by examining the impacts of three types of scarcities misdiagnoses the nature of the independent variable. Rather than looking at the discrete and proximate mechanisms that are creating a decrease in available resources in any situation (the three sources of scarcity) as Homer-Dixon would, they would instead focus on the factors behind these mechanisms that are driving the processes of scarcity in the first place. In terms of Homer-Dixon's causal model, the dispute is essentially whether to locate the independent variable with the supply, demand, or distributional scarcities or within what he calls "precursor ideational factors" (or the causes of structural inequality, according to Baechler). [T]he emphasis on so-called scarce resources occludes the real sources of such problems/ conflicts and in so doing makes them more difficult to solve," Peluso and Watts write. "The best example of this point," they continue, "is perhaps the way Homer-Dixon describes his view of how appropriations of land/resources by elites create scarcity. The focus of his analysis is subsequently on the *scarcities produced* – not on the mechanisms of appropriation and exclusion from access at the heart of that process."[44]

Political ecologists like Peluso and Watts are correct in criticizing these projects for "theoretical underspecificity" in analyzing the deep factors producing scarcity. To be fair, however, the Toronto Group's case studies did examine, in varying degrees of thoroughness, the historical patterns of "appropriation and exclusion" which form the basis of distributional scarcities.[45] Homer-Dixon also clearly stresses the importance of "ideational factors" for making up the "broad and complex social and psychological context" for the relationship between societies and environmental change.[46] The Toronto Group similarly notes in various cases how such inequality is expressed in institutions, laws, and social relations in ways that produce scarcities for certain groups.[47] However, with the exception of resource capture or ecological marginalization, little attempt is made to theorize across cases, patterns, or processes among precursor ideational factors, or how they might drive scarcities later in the causal process.[48] A rich body of political ecological literature exists which suggests the origins and analytical significance of scarcities go beyond invocations of class, colonial legacies, or laws, norms, institutions, or rights, and that patterns exist across contexts and societies.[49] Political ecological explanations locate the cause of scarcity in the "social relations of production and the social fields of power" and theorize how "various systems of access to and control over resources emerge and are reproduced," resulting in scarcities.[50] The brief discussion of the theoretical dimensions of structural scarcity in Homer-Dixon's 1999 book,[51] by contrast, gives the impression that inequality is a given across societies, rather than a process, pattern, and outcome, constantly formed and recreated through the interactions of actors, access, and regimes of accumulation in a global capitalist economic system. The Bern-Zurich Group more thoroughly examined the political ecological footprint in environment-conflict linkages, as outlined above, though still insufficiently for political ecologists like Peluso and Watts.[52]

In light of this critique, however, future research must take care not to let the pendulum swing too far in the opposite direction. There is a danger that political ecological analyses of environment-conflict linkages will endogenize the causes of scarcity to political economic factors, and lose sight of the impact of "natural factors." Scholars from the Toronto Group have argued that political ecological critiques of environment-conflict research, by suggesting the causal primacy of political-economic factors and their relationship and conditioning of human-natural systems, are downplaying or ignoring the importance of environmental factors or quasi-naturalistic factors like population growth in these systems.[53] As well, there are differences among political ecologists on this point, with some political ecologists endogenizing environmental scarcity or environmental factors to political-economic factors and processes,[54] while others see varying degrees of interactivity between natural and political economic factors that are difficult to separate.[55] Even among the latter, however, causal primacy is often accorded to political economic influences in their accounts, with environmental factors playing some vaguely necessary but often causally undefined role.[56] Certainly, when scholars such as Homer-Dixon attempt to bridge the divide between political ecology analysis and Neo-Malthusian analysis, they are derided as naive and their work is twisted to a simplistic Neo-Malthusian caricature.[57] The independent causal potential of some factors, such as population growth, is discounted or ignored, as is discussed below. While Peluso and Watts are probably correct in claiming the lack of sophistication of political ecological analysis in the Toronto Group's work, many in their own field recognize the necessity of integrated natural and political-economic perspectives.

Future research needs to critically build on past environment-conflict research and strike a balance between sophisticated political ecological analysis that also integrates insights from environmental-conflict research. There are at least two reasons why the political ecology critique fails to convince that researchers should abandon the focus on environmental scarcities as the independent variable. First, political ecologists have never been able to resolve the argument from scholars like Homer-Dixon that some sources of environmental scarcities are independent of political-economic factors, and that causal interactivity makes it extremely difficult to separate out political-economic factors as more important than other natural factors in many situations.[58] Both Homer-Dixon and Baechler argue that environmental scarcities are never sufficient to cause conflict, but that they interact with multiple causes and often multiple forms of scarcity in many cases.[59] The agreement from some political ecologists about the role of natural factors[60] in helping to cause conflicts suggests a certain degree of consensus with the views of Homer-Dixon and Baechler which needs to be built upon in future research. In fact, there is now widespread recognition that the long history of human interaction with natural systems requires a new integrated framework to study "coupled human and natural systems."[61] New research needs to move beyond debates about whether social or natural factors are more important and instead develop comprehensive explanations that also grapple with the difficult analytical problem of determining the relative importance of various interacting causes and processes. This may require a new approach to model the complex interactive systems.[62]

Second, despite criticisms by political ecologists that environment-conflict research mistakenly examines the immediate circumstances (or proximate drivers) producing scarcity and their social effects, rather than distant drivers behind such processes, there are sound reasons to continue exploring how scarcity-induced social effects help to cause conflict.[63] For those interested in the downstream violent conflict processes of human–environmental interactions, a research strategy that focuses on the causal effects is exceptionally important because these outcomes need to be thoroughly examined to understand how they help cause violent conflict.[64] In many political ecology analyses of human–environmental change interactions, by contrast, violent conflict is treated less as the object of analysis, and more of an unfortunate outcome – an indicator of the political-ecological consequences of processes of exclusion, control, and appropriation. The analytical bias of such accounts – both in terms of focus and policy intervention – is on the political economic drivers and processes believed to be at the start of the causal process, and less on possible violent conflict outcomes. As Peluso and Watts conclude, "to say that environmental scarcity can contribute to civil violence is to state the obvious."[65] But in claiming that such an outcome is obvious, they also imply that the relationship between environmental transformation and violent conflict is simple, well understood, and unimportant. However, this is not the case, and many political ecology accounts of the genesis of violent conflict lack sophisticated analyses sensitive to social science work on the causes of civil conflict and revolution. As well, given the tendency to employ expansive definitions of "violence" by many political ecologists, their accounts complicate efforts to understand how particular patterns of human–environmental interaction cause different conflict outcomes. As Kahl has noted in reference to the cases in the book *Violent Environments*, "the causal logic whereby political, economic, and discursive practices and structures constitute particular environments and patterns of violence is underspecified."[66] Future research needs to more closely examine the relationship between the social effects of environmental transformation and violent conflict, both to test and refine existing hypotheses, and to better integrate social science research on violent conflict with more sophisticated analyses of environmental transformation and scarcity. This approach may also yield important insights into where policy interventions can forestall or prevent violent conflict, and complement the policy interventions suggested by political ecology analysis of deep processes and regimes of accumulation, power, and access to essential human environments. A research strategy that locates policy interventions only in underlying social, political, and economic processes is unnecessarily restrictive.

There are thus sound reasons for keeping inequality as a fundamental part of the independent variable, as Homer-Dixon and Baechler do, but also broadening the analysis in order to understand the broader processes, patterns, regimes, and actors that condition and create inequality and help cause conflict. A comprehensive, tripartite independent variable acknowledges that the inequitable distribution of resources rarely acts alone to help cause conflict; its impact is frequently "a function of its interaction with resource supply and demand."[67] While there may be cases of strictly distributional conflicts or conflicts based only on demand-induced scarcity, the possibility of multiple sources of scarcity should lead analysts to investigate the resource's supply *relative* to, first, demand on the resource, and, second, the social distribution and control of the resource. "The relationships between supply and demand and between supply and distribution determine people's actual experience of scarcity, and it is these relationships that . . . influence the probability of violence."[68] Such a focus is reasonable for any research program interested in environmental change–conflict links.

Defining the independent variable and characterizing outcomes in environment-conflict research: debates over population and consumption, Neo-Malthusianism

Another source of dispute over the definition of the independent variable by the Toronto Group and the Bern-Zurich Group revolves around the inclusion of demographic factors such as population growth as a source of environmental scarcity. Some critics and even some supportive commentators now commonly apply the label "Neo-Malthusian" to the research programs, models, and empirical findings of the Toronto Group and Bern-Zurich Group.[69] Once thus labeled, critiques of Neo-Malthusianism are employed to discredit the empirical findings of the Toronto Group and the Bern-Zurich Group.[70] Is there some truth to labeling the Toronto Group and Bern-Zurich Group findings as Neo-Malthusian?

A careful examination of these critiques reveals that many of these arguments employ straw-man Neo-Malthusian arguments. Painting the findings of the Toronto Group and Bern Zurich Group as Neo-Malthusian hinders attempts to deepen our understanding of environment–conflict linkages. It has led to a failure to recognize that multiple pathways of human–environment interactions exist in the real world – both local or national scarcity-induced social effects *and/or* the forestalling of these effects through the intervention of institutions, the state, or ingenuity. This emphasis on discursive labeling has retarded progress in identifying useful interventions to forestall or alleviate the impacts of scarcities, and obscures our understanding of when and why these interventions sometimes fail.

Some critics claim that Homer-Dixon's and Baechler's research programs are Neo-Malthusian because they adhere to deterministic single-factor explanations of the role of environmental scarcity – population factors, in particular – as a cause of violence. Hartmann, for instance, claims that population growth is the "single largest causal factor of environmental scarcity" in the Toronto Group's work, blamed "disproportionately for environmental degradation, poverty, migration, and ultimately political instability."[71] She argues that the Group's link between population growth and resource demand betrays the Group's determinism, because "[i]t does not necessarily follow that if there are more people, they will consume more – per capita consumption could fall for a variety of reasons."[72]

Yet Homer-Dixon and his co-authors go out of their way to eschew deterministic single-factor explanations. Their key independent variable, environmental scarcity, incorporates three factors – supply, demand, and distributional scarcities. At every subsequent stage in their model, their research showed that intervening socio-economic variables act to create causal contingencies. Indeed, the Group concluded that socio-economic factors can intervene at any stage to mitigate the effect of scarcity on conflict or to move the pathway away from conflict altogether. The Group also identified numerous examples of the interaction of multiple causes as well as feedback loops that cycle back to affect earlier variables, including the causes of scarcity.

Baechler similarly argues that population and environmental factors always operate with important intervening variables to produce conflictual outcomes: "The environmental conflict program does not lead to mono-causal explanations of violent conflicts or war. Instead, environmental disruption is embedded in a syndrome of factors complicating any conflict analysis."[73] Population dynamics, according to Baechler, combine with other factors like "poverty, inadequate land-use and land-tenure systems, environmental transformation, and poor state performance" to stimulate local conflicts and migration – migration which can be cross border migration or rural–urban migration, possibly leading to "conflicts in the area of destination."[74] Hartmann is therefore incorrect to assert that these projects put greater – if not primary – weight on the population factor (or any single factor) in their theoretical frameworks.[75]

The Toronto Group and Bern-Zurich Group models can be criticized, however, for not sufficiently emphasizing that demand-induced scarcities are strongly influenced by changes in consumption patterns in local, national, or international markets. As Hartmann notes, increased resource consumption may have "little to do with demographic factors but instead with increased demand in external markets for a particular product."[76] These consumption changes may be far removed from the location of the resource, with economic changes or cultural changes thousands of kilometers away triggering market signals that increase the rate of use of a resource, even if the population levels remain stable or decline in the areas under study. Underemphasizing consumption could appear to some critical scholars as overemphasizing population factors.

Homer-Dixon and Baechler do recognize that the consumption of resources is a crucial part of demand-induced pressure on resources. As Homer-Dixon notes, "Demand-induced scarcity is a function of population size multiplied by per capita demand for a given resource; an increase in either population or per capita demand increases total resource demand."[77] Because demand-induced scarcity is a product of such an interaction, it is impossible to say that one component factor is more important than another. Consumption and population change thus always make up the determination of demand-induced scarcity.[78] Furthermore, Homer-Dixon notes that population growth and consumption are influenced by a range of "ideational factors" and "economic preferences," which account for how and what people use and consume.[79]

However, the influence and causes of consumption are not always adequately expressed in the Toronto Group's causal frameworks and case studies alongside population factors, and its influence is not adequately explained in scarcity interactions like resource capture or ecological marginalization. Nor are the influences of consumption changes outlined in Baechler's conception of environmental discrimination, which is surprising, given the prominent political ecological footprint in Baechler's work. For example, Homer-Dixon explains that resource capture often happens when population growth combines with a fall in the supply and demand of a resource. This shift, Homer-Dixon argues, "can produce dire environmental scarcity for poorer and weaker groups in the society."[80] While this pattern of interaction is certainly plausible, consumption changes can trigger demand-induced scarcity and elite resource capture *irrespective* of any demographic changes.

There is ample evidence that resource capture happens without population growth-led demand changes, but through consumption-led demand-induced scarcity. Valuable renewable resources – valuable because of their proximity to local markets and/or local scarcity – are at risk of resource capture irrespective of demand-side increases. In the decades following Spanish conquest of the Andes, for example, Peru's indigenous population collapsed as a result of exposure to new diseases against which the population had no immunity and from the crushing exploitation of the Spaniards.[81] By the mid-sixteenth century, the indigenous population had fallen to just over 1 million from a pre-conquest population of around 9 million.[82] Throughout Peru, Spanish colonial elites began to systematically appropriate the best agricultural lands as their own – fertile areas with ready access to emerging markets – setting in motion decades of legal and social conflicts between elites and local communities.[83] In this case, regime change enabled the new colonial elites the opportunity to capture valuable land, even in the context of demographic collapse. In other cases, demand influences come from international markets. In the mid-1800s and in the early part of the twentieth century, increasing international demand for wool led to rising prices and stimulated wool exports throughout Latin America. Repeated cases of local resource capture were part of Latin America's wool boom. In Peru, powerful elites and petty elites reacted to these international price signals by seeking to expand domestic wool production.[84] Research shows that in the southern Peruvian altiplano, elites expanded their holdings "by a mixture of volition and coercion" – purchasing land outright, manipulating laws and institutions to capture the pasture land, entrapping peasants through debt, or using sheer force to gain control of the grazing lands of indigenous small-holders and communities.[85] The social impacts of this resource capture were aggravated by the slowly expanding highland populations at the time.

But it was the international price signals and the consumption changes driving them that were at the start of the causal process leading to the resource capture of the wool producing lands – especially after the outbreak of World War I stoked the demand for wool uniforms. Similar patterns were evident with cattle production in Central America between the 1950s and 1970s, as a result of demand for the US and domestic markets in the region.[86]

A close reading of the research of Homer-Dixon and Baechler demonstrates that population growth is a more frequently cited source of demand-induced scarcity than consumption-driven demand changes in their models. While possibly a function of the cases they examined, emphasizing demographic trends without also focusing on consumption influences could appear to preference population as the key variable for demand-induced scarcity. Consumption-driven demand signals must be recognized as important sources of demand-induced market impacts on scarcity in many areas. Corrections are needed to their models to highlight the negative impacts of consumption, as Figures 2.2 and 2.3 do in correcting Homer-Dixon's resource capture and ecological marginalization models. Similarly, consumption influences must be recognized as important drivers of environmental discrimination patterns described by Baechler. Modern markets often spread into new production areas as a result of consumption – or demand-induced – signals, at the expense of the more traditional or small-scale agriculturalists in those areas. Figures 2.2 and 2.3 modify Homer-Dixon's diagrammatic representation of resource capture and ecological marginalization to take into account possible consumption influences.

Other critiques of the work of the Toronto Group and the Bern-Zurich Group have set out to discredit their findings by categorizing them as part of widely critiqued Neo-Malthusian arguments on the consequences of population growth for environmental change. Homer-Dixon's use of the term "scarcity" (a buzzword for Neo-Malthusians) for

Resource Capture

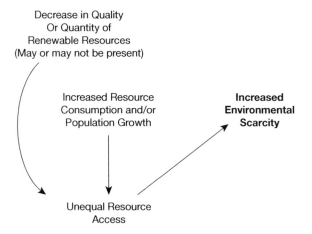

Figure 2.2 Homer-Dixon's diagrammatic representation of resource capture.

Source: adapted from Homer-Dixon, *Environment, Scarcity, and Violence*, p. 74.

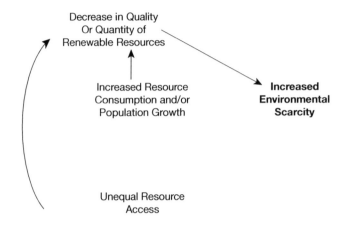

Adapted from Homer-Dixon, *Environment, Scarcity, and Violence*, p. 74.

Figure 2.3 Homer-Dixon's diagrammatic representation of ecological marginalization.

Source: adapted from Homer-Dixon, *Environment, Scarcity, and Violence*, p. 74.

his independent variable encourages the impression among some that his model is Neo-Malthusian.[87] Gleditsch and Urdal's critique of the Toronto Group's work, for example, emphasizes repeatedly that Homer-Dixon's thinking is Neo-Malthusian.[88] They support this characterization by noting similarities with Neo-Malthusian thinking and by selectively choosing elements in Homer-Dixon's theoretical model that correspond to Neo-Malthusian thought. For example, they argue that Homer-Dixon's pessimism about the relationship between population change and natural resource availability demonstrates his Neo-Malthusian provenance.[89] They strip his model to its core and describe it in strikingly similar terms to Paul Ehrlich's IPAT equation, one of the cornerstones of Neo-Malthusian thought.[90]

In spite of explicit attempts to differentiate the research of the Toronto Group and the Bern-Zurich Group from Neo-Malthusianism[91] and its "focus on the absolute physical limits to growth in a society,"[92] Gleditsch and Urdal classify these models as slight variants of this approach. In their models, both Homer-Dixon and Baechler emphasize the crucially important mediating role played by the state, to intervene to disrupt scarcity-conflict processes, or to alleviate the social consequences of human-induced environmental scarcity.[93] Their emphasis on the intervening role played by a society's social and technical capacity to overcome scarcities – its "ingenuity," to use Homer-Dixon's term – appears to clearly differentiate their positions from Neo-Malthusian positions, because the application of human ingenuity to overcome scarcity is a central position of the Cornucopian response to Neo-Malthusianism. However, to Gleditsch and Urdal, these arguments merely distinguish traditional Neo-Malthusian thought from Homer-Dixon's Neo-Malthusian thought.[94] They conclude that Homer-Dixon's pessimism about the ability of developing countries to come up with the necessary ingenuity to overcome the consequences of resource scarcities betrays a simplistic understanding of how societies throughout history have eventually overcome their negative impacts on the natural environment through economic development.[95]

Gleditsch and Urdal's critique raises important questions about whether the focus on population variables and the impact of reduced resource availability in the models of the Toronto Group and the Bern-Zurich Group, at a basic level, necessitate grouping their work with Neo-Malthusians. More importantly, is there some analytical relevance for those interested in environmental change-conflict research to deciding whether their models are Neo-Malthusian?

Branding the work of the Toronto Group and the Bern-Zurich Group as Neo-Malthusian is useful to critics because abundant evidence exists to discredit many general Neo-Malthusian claims.[96] This work can help to undermine credibility in the research of these two groups if the Neo-Malthusian label can be hung on their findings. Many analysts acknowledge that institutions, the state, or "the human ability to [apply] technology and . . . knowledge"[97] can interrupt and alleviate the processes and social effects of scarcity, thereby forestalling or heading off conflict further down the hypothesized causal chain. While this is true, it is not a complete explanation of possible outcomes. Scholars also admit that in the absence of these interventions, certain environmental scarcity-induced social effects can and do occur. As John Pender's careful analysis of research on population growth and agriculture concludes:

> population growth may stimulate a wide variety of responses at the household and collective level. Many of these responses are strongly conditioned by the nature of technology, infrastructure, institutions, and organizations. *In the absence of development of these factors, population growth is likely to lead to declining labor productivity and human welfare, as a result of diminishing returns.*[98]

We must recognize that there are, in fact, two possible idealized causal outcomes of the impacts of environmental change and population growth. (See Figures 2.4 and 2.5.) One pathway describes how environmental scarcities can lead to negative social effects like those described by Homer-Dixon and Baechler, with outcomes consistent with certain Neo-Malthusian claims. A large body of detailed empirical and case study research informs these linkages, such as the "*vicious circle model*" and its decendents.[99] The second possible pathway describes how the impacts of institutional, state, or social ingenuity interventions forestall or mitigate negative social consequences before they contribute to other conflict-generating processes like grievance formation or collective mobilization. This work is descended from Boserupian hypotheses about agricultural intensification patterns, and Cornucopian hypotheses about the application of ingenuity.[100] In both cases, the context of particular situations is crucially important, and highly variable; feedback loops operate, and causal interactivity makes the relationships complex. Importantly, recognizing one possible pathway does not preclude the other pathway from also operating, particularly because they could be operating at different scales.[101] Between these two poles (where most of the real world cases probably lie) are a range of outcomes depending upon the constellation of factors at play – the degree of state or institutional intervention, the degree of supply, demand, or distributional scarcities, and so on. Once we appreciate the variety of contextual situations and the different pathways of impacts of environmental scarcity or ingenuity, we begin to account for the wide variation in real world cases, which have been fodder for competing scholarly positions on human–environmental change impacts. These contrasting positions can and should be unified into one theoretical model, and both are possible outcomes depending upon the particular circumstances, as represented in the idealized figures 2.4 and 2.5.

Most cases probably exhibit some combination of scarcity-induced social effects, negative feedbacks, and interventions of varying effectiveness. States or other powerful actors are making their presence felt, though perhaps not in ways sufficient to ameliorate

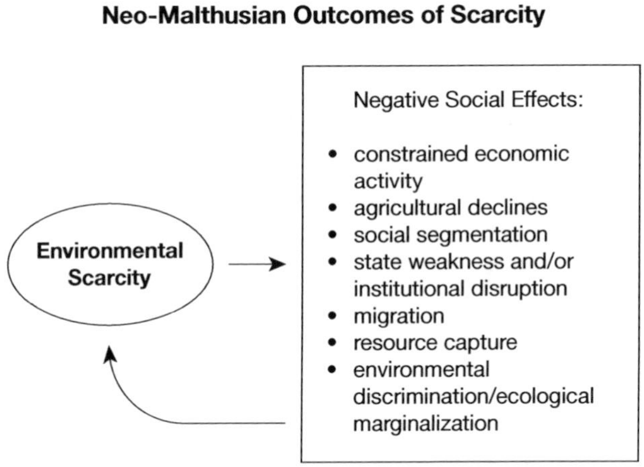

Neo-Malthusian Outcomes of Scarcity

Negative Social Effects:

- constrained economic activity
- agricultural declines
- social segmentation
- state weakness and/or institutional disruption
- migration
- resource capture
- environmental discrimination/ecological marginalization

Environmental Scarcity

Figure 2.4 Neo-Malthusian outcomes of scarcity.

Ingenuity Interrupts for a Different Causal Pathway

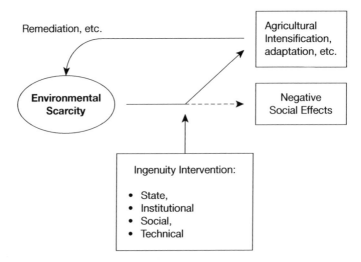

Remediation, etc.

Agricultural Intensification, adaptation, etc.

Environmental Scarcity

Negative Social Effects

Ingenuity Intervention:

- State,
- Institutional
- Social,
- Technical

Figure 2.5 Ingenuity interrupts for a different causal pathway.

negative consequences, or in ways which outright exacerbate negative impacts. Contextual and scale factors unique to each case add a dizzying layer of complexity to attempts to discover common causal patterns across time and space, and between cases. Of particular interest to researchers is identifying how and why scarcities arise in the first place, what particular interventions are effective, how and why the conditions for effective interventions are created, and the intervening variables and processes which are required to translate detrimental social effects from environmental scarcities into different kinds of violent and non-violent conflict. Answers to these questions will guide choices about effective interventions. An impressive body of research already exists in various disciplines to provide answers to some of these questions. But researchers must resist the temptation of constructing simplistic, polemical comparisons, and focus on identifying and verifying commonalities in each other's findings.

Characterizing outcome in environment-conflict research: greed vs. grievance

Greed versus grievance debates over the findings of the Toronto Group and the Bern Zurich Group similarly result in simplistic polemical analysis, which creates a false dichotomy around the state of abundance or scarcity of resources and their connections to civil conflict.[102] The Toronto Group and the Bern-Zurich Group are often said to employ a "grievance" hypothesis of conflict.[103] Real or perceived deprivation produces a psychological state of grievance that leads people to "want to engage in violent protest."[104] Some economists studying civil violence in poor countries, on the other hand, argue that conflict is motivated by "greed": people engage in violence when they rationally estimate that such behavior will allow them to seize or exploit a "lootable" source of wealth – that is, when the expected benefits of such behavior outweigh the expected costs.[105] The expectation of benefits from violence thus condition the opportunities available for actors.

According to these researchers, a number of variables in a given society affect the relative balance of benefits and costs, including low economic growth, low educational attainment, large proportions of unemployed male youth in the population, and heavy dependence on primary resource exports.[106] Lootable resources are usually extractive, non-renewable resources like minerals that have a high value-to-volume ratio and are easily seized and converted into currency in the absence of a strong state. Diffuse renewable resources like pulp timber (as opposed to valuable hardwoods), on the other hand, offer less opportunity for large-scale harvesting and sale in the absence of a functioning state.[107]

Empirical evidence supports the hypothesis that some poor countries suffer from a "resource curse": those that have high concentrations of lootable resources, measured as a function of their primary resource exports, are more likely to experience conflict.[108] Also, an abundance of valuable extractive resources helps create a domestic economic structure that shifts the balance of benefits and costs in favor of greed-motivated rebellion.[109] Richard Auty points out, however, that the resource curse hypothesis is not deterministic: even when all the economic and resource precursors are in place, in most cases greed-motivated violence does not occur.[110] Instead, many variables intervene, such as the geographical proximity of the lootable resource to the center of political power and the would-be rebels.[111]

Advocates of the resource curse perspective often argue that the abundance/greed hypothesis and the scarcity/grievance hypothesis of civil conflict are mutually exclusive. Either one or the other has to be right, but not both. Thus de Soysa compares Collier and Hoeffler's resource-curse model with the Toronto Group's environmental scarcity model.[112] He concludes that the empirical evidence supports the hypothesis that state dependence on natural resources is the main explanation of civil conflict and that greed is the main motive for violent behavior. To the extent that grievance motivates some actors to violence, these actors are generally the victims of the greedy behavior of others and are acting against them. De Soysa concludes that scarcity, if it plays any role at all, is wholly subordinated to resource abundance and the greed that this abundance evokes.

De Soysa's arguments – and those of like-minded researchers – are wrong in three important ways. First, these researchers set up a misleading contrast between abundance and scarcity: they say they are investigating violence that arises from resource abundance, while others are investigating violence that arises from scarcity. This is a false dichotomy. Both groups of researchers are, in fact, investigating violence that arises from scarcity. The lootable resources that de Soysa and others study only stimulate greed because they are valuable, and they are valuable only because they are *scarce* relative to demand. They may be locally abundant in one region or part of a country – a phenomenon sometimes labeled a "honey pot" – but they are globally scarce. If they were truly abundant, they would not be valuable, and people would not have a powerful incentive to loot them.

As discussed above, Baechler and Homer-Dixon make a similar point regarding the consequences of certain kinds of environmental scarcity. As cropland, forest, and fresh water resources become more scarce relative to demand in a poor country powerful elites often find it easier – and more profitable – to seize the remaining pools of these resources in order to extract enormous resource rents, the "resource capture" process. Such relationships can also operate across borders, as the scarcity of renewable resources such as productive cropland in one area of the globe can stimulate powerful investors to purchase or gain control of land in developing countries where land is more abundant.[113] In these circumstances environmental resources are locally abundant within a more general situation of scarcity – a situation that stimulates greed and can provoke violence between elite groups competing for rents or between elites, investors, and non-elites seeking to retain traditional access to these resources.

Second, resource curse researchers also often conflate the issues of resource dependence and resource scarcity and in doing so create a false divide between themselves and "scarcity" researchers. De Soysa writes that "proponents of both sides of the debate have assumed that resource dependence signifies objective abundance or scarcity."[114] This is clearly incorrect, at least when it comes to the Toronto Group. The Toronto Group's researchers have never argued that resource dependence signifies scarcity. They argue, instead, that resource dependence affects *vulnerability* to scarcity – an entirely different proposition.

For the Toronto Group, high resource dependence occurs when a large proportion of a given population depends on local renewable resources like cropland, forests, or fresh water to survive. Proponents of the resource curse hypothesis, in contrast, focus on the dependence of a national economy on revenues from the export of non-renewable resources. This kind of resource dependence, which is hypothesized to encourage greed-motivated violence, can exist in a given country at the same time that a large proportion of that country's population is heavily dependent on local renewable resources. There is no fundamental contradiction between the two research camps here.

Third and finally, de Soysa and others misrepresent the position of the Toronto Group and Bern-Zurich Group when they reduce their thesis to simplistic linear links between scarcity, grievance, and violence. In the process they create a straw man that is easily demolished. As discussed above, both the Toronto Group and Bern-Zurich Group models incorporate at multiple points grievance, greed, opportunity, and mobilization processes. Homer-Dixon argues, for example, that environmental scarcity can erode social capital and thereby deepen identity cleavages, raising the risk of grievance-motivated intergroup conflict. Simultaneously, the erosion of social capital can weaken a state's legitimacy and increases the opportunity of powerful groups, often motivated by greed, to challenge the state or capture valuable environmental resources.

Recently, scholars have begun to erase these false dichotomies. They are studying how the variables of greed, grievance, scarcity, and abundance can interact within the same conflict system. Korf, for instance, examines these four variables in the context of the relationship between the war economy of combatants (or potential combatants) and the survival economy of civilians. He suggests that greed-motivated conflict can create or reinforce local resource control patterns that lead to grievance-driven social, political, and economic processes that in turn perpetuate conflict – eventually even replacing greed as the main cause.[115] Many extractive industry conflicts – around mining or oil exploration, for example – display similar patterns. Scholars have usefully begun to distinguish between different types of lootable resources and the different types of conflict dynamics resulting from exploitation of these resources.[116] Lootable resources are much more easily transmitted into cash for rebel groups and to aid recruitment. But valuable minerals that require heavy industrial operations to remove them result in much longer term, slow acting conflict dynamics – either through corruption, temporally slow resource capture processes, or in conflicts over the distribution of rents or with nearby communities as a result of secondary environmental effects (from toxic pollution, for instance) on livelihoods. These combined greed and grievance dynamics in extractive disputes have not been sufficiently highlighted by many environmental conflict researchers, though other scholars of extractive industry disputes have outlined the deleterious impacts on local livelihoods.[117]

Conclusion

A detailed review of environment-conflict research since the early 1990s finds a great deal of consensus on hypothesized linkages between human-induced transformation of the natural environment, social effects of these changes, and linkages to various forms of violent conflict. While debates will continue about how to specify the independent variable and the role of demographic change, there is agreement among many that, at times, political-economic and social factors combine with other forms of human-induced scarcity, including demand-induced scarcities, to immiserate people in ways that can undermine social stability. The pathways from these social effects to the outbreak of violent conflict of various forms are complicated and multicausal, and a great deal of additional work needs to be done to understand both bottom-up and top-down expressions of violence. Qualitative studies may need to lower the level of analysis in future environment-conflict research, to better model the complexities of local-level dynamics.[118]

Certainly, there are also significant points of controversy around these findings, as has been outlined above. Many of the hypothesized linkages remain just that – hypotheses in need of further refinement and testing by a new generation of research with more detailed and comprehensive approaches. However, future research needs to abandon the polemical debates of the past around environment-conflict research projects, and strive to deepen the areas of consensus among many divergent approaches to studying human–environmental change linkages, including political ecological research, abundance/scarcity studies, demographic change studies, and so on. Constructive criticism is absolutely crucial to moving our understanding of these linkages forward, but not at the expense of simplifying or distorting competing approaches. This chapter has also emphasized that scholars need to be careful about specifying what their research does and does not seek to explain – that the dependent variable needs to be clearly defined in order to eliminate criticisms of explanatory over-reach. Similarly, critics need to carefully assess what conflict processes and types of violence environment-conflict research is seeking to explain, and resist the temptation to criticize studies for not explaining every form of violence associated with human environmental transformation.

Notes

1 M. Renner, 'National Security: The Economic and Environmental Dimensions', *Worldwatch Paper* 89, 1989; N. Myers, 'Environment and Security', *Foreign Policy* 74, 1989, 23–41; J. T. Mathews, 'Redefining Security', *Foreign Affairs* 68, 1989, 162–77. The debate is covered extensively in G. D. Dabelko, 'Tactical Victories and Strategic Losses: The Evolution of Environmental Security', unpublished PhD thesis, University of Maryland, College Park, 2003.
2 Mathews, 'Redefining Security'.
3 D. Deudney, 'The Case Against Linking Environmental Degradation and National Security', *Millennium* 19, 1990, 461–76; M. Levy, 'Is the Environment a National Security Issue?', *International Security* 20, Fall 1995, 35–62; D. Deudney, 'Environmental Security: A Critique', in D. H. Deudney and R. A. Matthew (eds), *Contested Grounds: Security and Conflict in the New Environmental Politics*, Albany: State University of New York Press, 1999, pp. 187–222.
4 The "Toronto Group" refers to the team of researchers from institutions on four continents that participated in three research projects headed by Thomas Homer-Dixon of the University of Toronto and carried out in collaboration with several American research groups. Projects focused on environmental change and acute conflict; environmental scarcities, state capacity, and civil violence; and on environment, population, and security. See T. Homer-Dixon, *Environment, Scarcity, and Violence*, Princeton, NJ: Princeton University Press, 1999; and T. Homer-Dixon and J. Blitt, *Eco-violence: Links Among Environment, Population, and Security*, Lanham, MD and Oxford: Rowman and Littlefield, 1998. The "Bern-Zurich Group" refers to the Zurich-based Project on Environment and Conflict, led by Günther Baechler and other scholars from the Swiss

Peace Foundation and the Centre for Security Studies and Conflict Research. For publications produced by the Bern-Zurich group, see G. Baechler, *Violence Through Environmental Discrimination: Causes, Rwanda Arena, and Conflict Model*, London: Kluwer Academic Publishing, 1999; G. Bächler and K. R. Spillman (eds), *Kriegsursache Umweltzerstörung. Band I. Ökologische Konflikte in der Dritten Welt und Wege ihrer friedlichen Bearbeitung*, Chur/Zurich: Verlag Ruegger, 1996; G. Bächler and K. R. Spillman (eds), *Kriegsursache Umweltzerstörung/Environmental Degradation as a Cause of War. Vol. II. Regional – und Länderstudien von Projektmirtarbeitern/Regional and Country Studies of Research Fellows*, Chur/Zurich: Verlag Ruegger, 1996; and G. Bächler and K. R. Spillman (eds), *Kriegsursache Umweltzerstörung/Environmental Degradation as a Cause of War, Vol. III, Länderstudien von externen Experten/Country Studies of Outside Experts*. Chur/Zurich: Verlag Ruegger, 1996.

5 M. A. Levy, 'Is the Environment a National Security Issue?', *International Security* 20 (2), 1995, 35–62; B. Hartmann, 'Will the Circle be Unbroken? A Critique of the Project on Environment, Population, and Security', in N. L. Peluso and M. Watts (eds), *Violent Environments*, Ithaca, NY: Cornell University Press, 2001, pp. 39–64.

6 J. Goldstone, 'Demography, Environment, and Security', in P. F. Diehl and N. P. Gleditsch (eds), *Environmental Conflict*, Boulder, CO: Westview Press, 2001, pp. 84–108; N. P. Gleditsch, 'Armed Conflict and the Environment', in Diehl and Gleditsch, *Environmental Conflict*, pp. 251–72; I. de Soysa, 'Paradise is a Bazaar? Greed, Creed, and Governance in Civil War, 1989–99', *Journal of Peace Research* 39, 2002, 395–416.

7 C. Kahl, *States, Scarcity, and Civil Strife in the Developing World*, Princeton, NJ: Princeton University Press, 2006; L. Ohlsson, *Environmental Scarcity and Conflict: A study of Malthusian concerns*, PhD Thesis, Göteborg, Sweden: Department of Peace and Development Research, Göteborg University, 1999; and L. Ohlsson, *Livelihood Conflicts: Linking Poverty and Environment As Causes of Conflict*, Stockholm: SIDA Environmental Policy Unit, 2000.

8 T. Homer-Dixon, 'On the Threshold: Environmental Changes as Causes of Acute Conflict', *International Security* 16, 1991, 76–116.

9 T. Homer-Dixon, 'Debating Violent Environments', *Environmental Change and Security Project Review* 9, 2003, 89–92; N. L. Peluso and M. Watts, 'Violent Environments: Responses', *Environmental Change and Security Project Review* 9, 2003, 93–96.

10 See special section, 'New Directions in Demographic Security', in *Environmental Change and Security Program Report*, Issue 13, 2008–09, pp. 2-47. Online: www.wilsoncenter.org/topics/pubs/ECSPReport13_DemographicSecurity.pdf (accessed August 2012).

11 United Nations Environment Program, *From Conflict to Peacebuilding: The Role of Natural Resources and the Environment*, Nairobi: UNEP, 2009; United Nations, *A More Secure World: Our Shared Responsibility*, Report of the Secretary-General's High-Level Panel on Threats, Challenges and Change, New York: United Nations General Assembly, 2004; G. D. Dabelko, 'Planning for climate change: the security community's precautionary principle', *Climatic Change* 96 (1–2), 2009, 13–21.

12 United Nations Intergovernmental Panel on Climate Change, *Climate Change 2007: Synthesis Report*, Contribution of Working Groups I, II and III to the Fourth Assessment Report of the Intergovernmental Panel on Climate Change, R. K. Pachauri and A. Reisinger (eds), Geneva: IPCC, 2007.

13 See, for example, German Advisory Council on Global Change, *Climate Change as a Security Risk* , London: Earthscan, 2008; D. Smith and J. Vivekananda, *A Climate of Conflict: The links between climate change, peace, and war*, London: International Alert, 2007; and Global Humanitarian Forum, *Human Impact Report: Climate Change – The Anatomy of a Silent Crisis*, Geneva: Global Humanitarian Forum, 2009.

14 S. Dalby, *Security and Environmental Change*, Cambridge: Polity Press, 2009: chapter 1.

15 Homer-Dixon, *Environment, Scarcity, and Violence*, pp. 4–5.

16 Millennium Ecosystem Assessment, *Eco-Systems and Human Well-Being: Synthesis*, Washington, DC: Island Press, 2005, p. 49. The report is available online: www.millenniumassessment.org/documents/document.356.aspx.pdf (accessed August 2012).

17 Homer-Dixon , 'On the Threshold'; T. Homer-Dixon, 'Environmental Scarcities and Violent Conflict: Evidence From Cases', *International Security* 19, 1994, 144–79; T. Homer-Dixon, 'Strategies for Studying Complex Ecological-Political Systems', *Journal of Environment and Development* 5, 1996, 132–48; Homer-Dixon, *Environment, Scarcity, and Violence*.

18 Homer-Dixon, 'Strategies for Studying Complex Ecological-Political Systems'. The results of several of these case studies are collected in Homer-Dixon and Blitt, *Eco-violence: Links Among Environment, Population, and Security*, and in Homer-Dixon, *Environment, Scarcity, and Violence*.

19 A. L. George and T. McKeown, 'Case Studies and Theories of Organizational Decision Making', in R. Coulam and R. Smith (eds), *Advances in Information Processing in Organizations*, London: JAI Press, 1985, pp. 21–58.

20 Homer-Dixon, *Environment, Scarcity, and Violence*, pp. 47–52.

21 The Toronto Group recognized that a process-tracing methodology produces different findings than a traditional quasi-experimental methodology; this approach has limited capacity to make causal power claims, but excels in discovering causal mechanisms. Many critics have failed to note that the Group explicitly acknowledged this limitation. Critics have incorrectly implied that the Toronto Group asserted general causal power claims in its research, and that its findings would have predictive power equivalent to those of large-N quantitative studies. See D. Schwartz, T. Deligiannis, and T. Homer-Dixon, 'The Environment and Violent Conflict', in P. F. Diehl and N. P. Gleditsch (eds), *Environmental Conflict*, Boulder, CO: Westview Press, 2001, pp. 273–94.

22 Homer-Dixon, *Environment, Scarcity, and Violence*, p. 177.

23 Schwartz *et al.*, 'The Environment and Violent Conflict', p. 295.

24 Ibid., p. 275; Homer-Dixon, *Environment, Scarcity, and Violence*, p. 177.

25 Homer-Dixon, *Environment, Scarcity, and Violence*, pp. 104–06, 169–76.

26 The Bern-Zurich Group's was initially known as the Environment and Conflicts Project (ENCOP). I refer to their work as the Bern-Zurich Group because there were additional projects after ENCOP.

27 Baechler, *Violence Through Environmental Discrimination*, p. xi.

28 Ibid., p. 87.

29 Ibid., p. 87. See Fig. 6.1, p. 180. The[0] two projects differ somewhat in their findings about interstate conflict, especially war over water resources. The Toronto Group found little evidence to support common post-Cold War predictions that the twenty-first century would be characterized by widespread war over freshwater. The Group concluded that their model was most relevant to intrastate-level behavior. Homer-Dixon's pronounced skepticism about the links between environmental scarcity and interstate war have been almost entirely ignored in later discussion of his work. Baechler, by contrast, included interstate water wars in a typology of environment-related conflicts that ranges from the domestic to the global levels of analysis. But he and Homer-Dixon agree that water-related conflicts are generally limited to disputes between upper and lower riparian states and will turn violent only when certain specific power asymmetries exist between states.

30 Homer-Dixon, *Environment, Scarcity, and Violence*, p. 48.

31 Ibid., p. 48.

32 J. Fairhead, 'International Dimensions of Conflict over Natural and Environmental Resources', in N. L. Peluso and M. Watts (eds), *Violent Environments*, p. 215; N. L. Peluso and M. Watts, 'Violent Environments', in their *Violent Environments*, pp. 18–19; N. P. Gleditsch and H. Urdal, 'Ecoviolence? Links Between Population Growth, Environmental Scarcity and Violent Conflict in Thomas Homer-Dixon's Work', *Journal of International Affairs* 56 (1), 2002, 288–89.

33 Peluso and Watts, 'Violent Environments', p. 18. The latter quotation is from Fairhead, 'International Dimensions of Conflict over Natural and Environmental Resources', p. 217.

34 Gleditsch and Urdal, 'Ecoviolence?', p. 289.

35 Gleditsch and Urdal's conclusion about the role of inequality does not come from research into the Toronto Group's cases, but instead appears to be drawn from their reading of the cases.

36 Gleditsch's views can be found in, 'Armed Conflict and the Environment', p. 258. Stephen Libiszewski's definition of environmental conflicts, which separates conflicts that erupt over the distribution of resources like land – what he calls *socio-economic scarcity* conflicts – from conflicts that result from processes of supply and demand scarcity, seems to have been influential in convincing some critics of the Toronto Group, like Gleditsch, tend to exclude distributional issues from the independent variable. Supply and demand scarcities must be operating for an incident to be called an environmental conflict, in Libiszewski's view:

> Conflicts over agricultural land, for example, which we defined as a renewable resource, have to be seen as *environmental* only if the land becomes an object of contention as a result of soil erosion, climate change, changes of river flows or any other environmental degradation. They are not environmental conflicts in the case of simply territorial conflicts like both World Wars and most colonial and decolonization wars. And they are neither necessarily environmental conflicts in the case of an anti-regime war with the goal of a more equal land distribution. This does not diminish the importance and the gravity of the conflict. And such a war can even be an environmental conflict, if unequal land distribution becomes for example a source of soil overuse. But it does not have to in every case.

> S. Libiszewski, 'What is an Environmental Conflict?', *Occasional Paper*, Bern: Swiss Peace Foundation and Zurich: Centre for Security Studies and Conflict Research, Swiss Federal Institute of Technology, 1992, pp. 6–7. However, Libiszewski does acknowledge that inequitable distribution frequently interacts with supply and demand scarcities, and that conflicts that appear to be based on distributional disputes may actually be rooted in supply and demand-side changes.

37 Gleditsch, 'Armed Conflict and the Environment', p. 256; W. Hauge and T. Ellingsen, 'Beyond Environmental Scarcity: Causal Pathways to Conflict', *Journal of Peace Research* 35 (3), 1998, 299–317.

38 Goldstone, 'Demography, Environment, and Security', pp. 88–91.

39 Baechler, *Violence Through Environmental Discrimination*, pp. 5–6.

40 "Degradation is used exclusively as an indicator of the degree of environmental transformation," Baechler writes. Baechler, *Violence Through Environmental Discrimination*, p. 5.

41 Ibid., pp. 179–80.

42 Ibid., p. 11.

43 Ibid., p. 11.

44 Peluso and Watts, 'Violent Environments: Responses', p. 94 Emphasis added.

45 See, for example, Homer-Dixon and Howard's Chiapas case study.

46 Homer-Dixon, *Environment, Scarcity, and Violence*, p. 51.

47 In addition to the discussion in the Chiapas case, examples from other case studies of the Toronto Group's include the discussion of Apartheid in South Africa, the structural scarcity around water allocation in Gaza, and discussion of political, economic, and military structures of authority and political-economic control in Pakistan. See Homer-Dixon and Blitt, *Ecoviolence.*

48 Baechler has a somewhat longer examination of deep-rooted structural patterns of socio-economic inequality in the global capitalist system and its relationship to environmental conflict, though these relationships are not a key focus in his model. See Baechler, *Violence Through Environmental Discrimination*, pp. 11–14.

49 S. Dalby, *Environmental Security*, Minneapolis: University of Minnesota Press, 2002, p. 88.

50 Peluso and Watts, *Violent Environments*, p. 29 and p. 19.

51 Homer-Dixon, *Environment, Scarcity, and Violence*, p. 52.

52 Peluso and Watts, *Violent Environments*, p. 18. Cursory research may explain why Peluso and Watts are equally dismissive of Baechler's approach, in spite of his obvious use of similar theoretical insights to frame his concept of environmental transformation. The critique of the Bern-Zurich Group's work by Peluso and Watts fails to examine his 1999 book, where he offers a much richer discussion of the political ecological roots of his conception of the independent variable, environmental transformation, as described above.

53 Homer-Dixon, 'Debating Violent Environments', p. 92; Schwartz *et al.*, 'The Environment and Violent Conflict', p. 277; Kahl, *States, Scarcity, and Civil Strife in the Developing World*, p. 24. See also D. H. Deudney, 'Bringing Nature Back In: Geopolitical Theory from the Greeks to the Global Era', in D. H. Deudney and R. A. Matthew (eds), *Contested Grounds: Security and Conflict in the New Environmental Politics*, Albany: State University of New York Press, 1999, pp. 25–60. I label population growth a "quasi-naturalistic variable" because demographic change has historically been a function of ecological conditions, epidemiological changes *and* societal transformations. Explanations for population growth do not solely rest on cultural or political-economic processes of accumulation, transformation, power formation, and control.

54 See, for example, Hartmann, 'Will the Circle Be Unbroken?'; Aaron Bobrow-Strain's critique of the Toronto Group's Chiapas study; and Dalby, *Environmental Security*, p. 88. A. Bobrow-Strain, 'Between a Ranch and a Hard Place: Violence, Scarcity, and Meaning in Chiapas, Mexico', in N. L. Peluso and M. Watts (eds), *Violent Environments*, Ithaca, NY: Cornell University Press, 2001, pp. 155–85. Bobrow-Strain argues (p. 157) that land conflicts in Chiapas are not a result of environmental scarcities, but "arise from the confluence of national economic reforms, changes in international commodity markets, and local histories of violence and insecurity that reduce *both* ranchers and peasants' capacity to use land intensively and effectively." However, there is a logical fallacy at the basis of Bobrow-Strain's denial of the importance of environmental factors in Chiapas which undermines his "social-social" political ecology position. He claims that the real scarcity in Chiapas was the lack of resources from the Mexican state to make better use of existing lands. But there would be no need for state resources if there was enough land in Chiapas of sufficient quality in the first place, or if the confluence of demand-induced and structural scarcities had not exacerbated the land-poor status of households.

55 Peluso and Watts, 'Violent Environments', pp. 25–6 and 27; Ibid., 'Violent Environments: Responses', p. 95. Other political ecologists similarly integrate "natural" factors into their political ecological analyses, and some of this work was influential on Homer-Dixon's work. See S. Stonich, *'I Am Destroying the Land!': The Political Ecology of Poverty*

and Environmental Destruction in Honduras, Boulder, CO: Westview Press, 1993, p. 165; W. H. Durham, 'Political Ecology and Environmental Destruction in Latin America', in M. Painter and W. H. Durham, *The Social Causes of Environmental Destruction in Latin America*, Ann Arbor: The University of Michigan Press, 1995, pp. 252–56; K. Jansen, *Political Ecology, Mountain Agriculture, and Knowledge in Honduras*, Amsterdam: Thela Publishers, 1998.

56 Peluso and Watts ('Violent Environments', p. 27) write:

> Nature itself is an important actor in the transformative or metabolic process [of the appropriation of nature]. The properties of natural resources and environmental processes shape, in complex ways, both the transformation process and the social relations of production. Nature enters in an active way into the production process just as soil degradation demands a human response and reaction to the production process. Forests and minerals and water are, in ways that have to be demonstrated, co-constitutive of the forms of use and disposition of resources.

57 Kahl notes Homer-Dixon's attempt to bridge both schools. Kahl, *States, Scarcity, and Civil Strife in the Developing World*, p. 22.

58 Schwartz *et al.*, 'The Environment and Violent Conflict', pp. 277–78; Homer-Dixon, *Environment, Scarcity, and Violence*, pp. 104–06.

59 Homer-Dixon, *Environment, Scarcity, and Violence*, pp. 178–79; Baechler, *Violence Through Environmental Discrimination*, p. 167.

60 Peluso and Watts, 'Violent Environments: Responses', p. 95.

61 J. Liu *et al.*, 'Coupled Human and Natural Systems', *Ambio* 36, December 2007, 639–48; see also L. H. Gunderson and C. S. Holling, *Panarchy: Understanding Transformations in Human and Natural Systems*, Washington DC: Island Press, 2002. Interestingly, among many of these scholars, the pendulum is swinging from the other direction, as natural scientists recognize the importance of a more thorough integration of social analysis in human-environmental management studies.

62 Many political ecologists reject this type of "systems" analysis, however, certainly complicating attempts to reach consensus on next steps.

63 Peluso and Watts do not use the terms "proximate" or "distant" in their critique, but the usage is consistent with their charge that Homer-Dixon is focused on the scarcities produced, and the immediate causes of these, rather than long-standing political economic impacts and processes.

64 Political ecologists also object to a focus on physical violent conflict, because their dependent variable defines "violence" much more expansively, with many including a variety of structural and discursive forms of violence. Peluso and Watts, 'Violent Environments', p. 23.

65 Peluso and Watts, 'Violent Environments', p. 23.

66 Kahl, *States, Scarcity, and Civil Strife in the Developing World*, p. 25.

67 Schwartz *et al.*, 'The Environment and Violent Conflict', p. 275.

68 Ibid.

69 Peluso and Watts, *Violent Environments*; Gleditsch and Urdal, 'Ecoviolence?' pp. 283–302; Kahl, *States, Scarcity and Civil Strife in the Developing World*.

70 Gleditsch and Urdal, 'Ecoviolence?', p. 287.

71 Hartmann, 'Will the Circle be Unbroken?', pp. 45–46. Similarly, Peluso and Watts, in describing Homer-Dixon's model, write that, "Population growth appears centrally as *the* driving force in all of these causal claims." Peluso and Watts, 'Violent Environments', p. 13. Emphasis in original.

72 Hartmann, 'Will the Circle be Unbroken?', p. 45.

73 Baechler, *Violence Through Environmental Discrimination*, p. 167.

74   Ibid., p. 96.

75   The related question of weighing causal variables in environmental conflict research has been a source of significant debate since the mid-1990s. See Gleditsch, 'Armed Conflict and the Environment', pp. 251–72; and Schwartz, *et al.*, 'The Environment and Violent Conflict,' pp. 273–94.

76   Hartmann, 'Will the Circle be Unbroken?', p. 45; Fairhead, 'International Dimensions of Conflict over Natural and Environmental Resources', p. 218.

77   Homer-Dixon, *Environment, Scarcity, and Violence*, p. 51. Baechler addresses the issue of consumption indirectly in various places, such as pp. 32–3, often in terms of the impacts of modernization and market penetration. He is far less explicit in noting the causal role of consumption than Homer-Dixon, however.

78   Homer-Dixon speaks of "per capita demand-induced changes," rather than using the term "consumption." His research largely pre-dates the growing body of research on the normative and empirical consequences of consumption. See, for example, K. Conca, T. Princen and M. F. Maniates, 'Confronting Consumption', *Global Environmental Politics* 1 (3), August 2001, 1–10.

79   Ibid.

80   Homer-Dixon, *Environment, Scarcity, and Violence*, p. 73.

81   P. Flindell Klarén, *Peru: Society and Nationhood in the Andes*, New York: Oxford University Press, 2000, pp. 48–49.

82   Noble David Cook, *Demographic Collapse: Indian Peru, 1520–1620*, New York: Cambridge University Press, 2004, p. 114.

83   S. J. Stern, 'The Social Significance of Judicial Institutions in an Exploitative Society: Huamanga, Peru, 1570–1640,' in G. A. Collier, R. I. Rosaldo and J. D. Wirth (eds), *The Inca and Aztec States, 1400–1800: Anthropology and History*, New York: Academic Press, 1982, p. 294. The exploitation of valuable non-renewable resources can similarly stimulate resource capture of valuable, but scarce renewable resources. In some parts of Peru, the exploitation of valuable minerals like silver and mercury stimulated the expansion of local and regional agricultural markets, further spurring the capture of good agricultural land nearby. See S. Stern, *Peru's Indian Peoples and the Challenge of Spanish Conquest: Huamanga to 1640*, Madison: The University of Wisconsin Press, 1993, p. 109.

84   N. Jacobsen, *Mirages of Transition: The Peruvian Altiplano, 1780–1930*, Berkeley: University of California Press, 1993, especially chapters 5 and 6.

85   Jacobsen, *Mirages of Transition*, pp. 227, 226–41.

86   Stonich, *'I Am Destroying the Land!'*, pp. 67–73; and M. Edelman, 'Rethinking the Hamburger Thesis: Deforestation and the Crisis of Central America's Beef Exports', in Michael Painter and William H. Durham (eds), *The Social Causes of Environmental Destruction in Latin America*, Ann Arbor: The University of Michigan Press, 1995, pp. 25–62. Stonich and Vandergeest's chapter on shrimp farming in Honduras offers another more recent example of this pattern: S. Stonich and P. Vandergeest, 'Violence, Environment, and Industrial Shrimp Farming', in N. L. Peluso and M. Watts (eds), *Violent Environments*, pp. 261–86.

87   Peluso and Watts, 'Violent Environments', p. 13.

88   Gleditsch and Urdal, 'Ecoviolence?'. Elsewhere, Gleditsch has been similarly explicit in branding Homer-Dixon a Neo-Malthusian:

> The Neo-Malthusian model of scarcity-driven conflict envisions population pressures and a high level of resource consumption combining to overexploit, degrade, and deplete resources, leading to competition and eventually to violent conflict. Thomas Homer-Dixon (1999), a prominent advocate of this school of thought, asserts that environmental scarcity is more likely to provoke internal conflict than interstate war.

N. P. Gleditsch, 'Beyond Scarcity vs. Abundance: A Policy Research Agenda for Natural Resources and Conflict', in United Nations Environment Programme, *Understanding Environment, Conflict, and Cooperation*, Nairobi: UNEP, 2004, p. 16.

89 Gleditsch and Urdal, 'Ecoviolence?', p. 287.

90 Ibid., pp. 292–93. On IPAT, see M. R. Chertow, 'The IPAT Equation and Its Variants', *The Journal of Industrial Ecology* 4 (4), 2001, pp. 13–29.

91 See, for example, Homer-Dixon, *Environment, Scarcity, and Violence*, chapter 3.

92 Gleditsch and Urdal, 'Ecoviolence?', p. 287.

93 See the role of the state section below.

94 They write: "While most Neo-Malthusians focus on the absolute physical limits to growth in a society, Homer-Dixon is more concerned about those societies that are 'locked into a race between a rising requirement for ingenuity and their capacity to supply it.'" Gleditsch and Urdal, 'Ecoviolence?', p. 287.

95 Ibid., p. 293.

96 A useful introduction to critiques of general Neo-Malthusian thought can be found in Homer-Dixon, *Environment, Scarcity, and Violence*, chapter 3.

97 Gleditsch and Urdal, 'Ecoviolence?', p. 288.

98 Emphasis added. J. Pender, 'Rural Population Growth, Agricultural Change, and Natural Resource Management in Developing Countries: A Review of Hypotheses and some Evidence from Honduras', in N. Birdsall, A. C. Kelley, and S. W. Sinding (eds), *Population Matters: Demographic Change, Economic Growth, and Poverty in the Developing World*, Oxford: Oxford University Press, 2003, p. 355.

99 A. de Sherbinin *et al.*, 'Population and Environment', *Annual Review of Environment and Resources* 32, 2007, 349–50. See also the 'Population, Environment, Development, Agriculture' model (PEDA). W. Lutz *et al.*, 'Population, Natural Resources, and Food Security: Lessons from Comparing Full and Reduced Form Models', *Population and Environment: Methods of Analysis. Special Supplement to the Population and Development Review* 28, 2002, 199–224.

100 See Pender, 'Rural Population Growth, Agricultural Change, and Natural Resource Management in Developing Countries', p. 325.

101 de Sherbinin *et al.*, 'Population and Environment', p. 350; S. R. Templeton and S. J. Scherr, 'Effects of Demographic and Related Microeconomic Change on Land Quality in Hills and Mountains of Developing Countries', *World Development* 27, 1999, 903–18. Templeton and Scherr argue that both processes are not only possible, but that processes of demographic change and human-induced land degradation may be followed by productivity investments that later ameliorate degradation. Some critics of environment-conflict research, such as de Soysa, argue that governance factors or state intervention are *always* intermediate between environmental scarcity and negative social effects, and as a result focusing on the causes of scarcity mistakenly focuses attention away from the real causal mechanism at work in such cases – inadequate governance. I. de Soysa, 'Ecoviolence: Shrinking Pie or Honey Pot?', *Global Environmental* 2, November 2002, p. 11. However, it is important to recognize that a vicious circle dynamic can happen regardless of whether institutions or social interventions are available, as is evident in isolated or impoverished regions where the state's presence is either non-existent or extremely limited. Some paleo-historical studies of declining island societies in the South Pacific demonstrate similar dynamics. In these cases, little if any "state" or "governance" capacity ever existed. Yet, environmental scarcities led to a vicious circle of poverty, civil decline, and violence. P. V. Kirch, 'Microcosmic Histories: Island Perspectives on "Global Change"', *American Anthropologist* 99, March 1997, 30–42. Thus, the causes and effects of environmental scarcities cannot be endogenized into state governance capacity. These dynamics exist independent of state governance capacity, but can certainly be ameliorated by state governance interventions.

102 The discussion below on greed vs. grievance debates appeared in a slightly different form in Thomas Homer-Dixon, Tom Deligiannis, and Dirk Druet, 'The Necessity of Complexity: Taking environmental conflict research beyond mechanism'. Paper presented at 2007 International Studies Association Annual Conference, Chicago, 28 February 2007.

103 Goldstone, 'Demography, Environment, and Security', pp. 88–9.

104 P. Collier and A. Hoeffler, 'Greed and Grievance in Civil War', *Oxford Economic Papers* 56, 2004, 564; Peluso and Watts, 'Violent Environments', p. 22.

105 Ibid., pp. 587–89. See also Paul Collier, 'Doing well out of war: An economic perspective'. Paper prepared for Conference on Economic Agendas in Civil Wars, London, 26–27 April 1999. Online: http://siteresources.worldbank.org/INTKNOWLEDGEFOR CHANGE/Resources/491519-1199818447826/28137.pdf (accessed August 2012). In the context of environment-conflict research, Marc Levy's Willy Sutton analogy offered an early criticism of Homer-Dixon's work that specifically offered greed as an alternative explanation for many of the environment-conflict links in Homer-Dixon's work. See Levy, 'Is the Environment a National Security Issue?', pp. 56–57.

106 Collier and Hoeffler, 'Greed and Grievance in Civil War'; P. Collier and A. Hoeffler, 'Resource Rents, Governance, and Conflict', *The Journal of Conflict Resolution* 49, 2005, 625–33. Recent research by Collier, Hoeffler, and Roehner has refined their conclusions further to stress "the primacy of feasibility over motivation" in order to explain civil war. They write that: "The feasibility hypothesis proposes that where rebellion is feasible it will occur: motivation is indeterminate, being supplied by whatever agenda happens to be adopted by the first social entrepreneur to occupy the viable niche, or itself endogenous to the opportunities thereby opened for illegal income." These conclusions tilt their findings into the camp of conflict scholars who elevate the role of "opportunity structure" as key to explaining civil war. P. Collier, A. Hoeffler, and D. Rohner, 'Beyond greed and grievance: feasibility and civil war', *Oxford Economic Papers* 61, 2008, p. 24. Some researchers argue that the presence of a "youth bulge" – a high proportion of the population made up of 19 to 29-year-olds – increases the likelihood of civil conflict, especially when it is combined with high urban growth rates and low employment rates. For a recent discussion, see H. Urdal, 'A Clash of Generations? Youth Bulges and Political Violence', *International Studies Quarterly* 50, 2006, 607–29.

107 P. Le Billon, 'The political ecology of war: natural resources and armed conflicts', *Political Geography* 20, 2001, 561–84. Michael Ross also distinguishes between non-renewable resources that are easily lootable like gemstones and those that are similarly valuable but unlootable, requiring extensive investment and technological investment to harvest, like gold. These characteristics have important implications for the types of conflict that may emerge, as discussed below. M. Ross, 'Oil, Drugs, and Diamonds: The Varying Roles of Natural Resources in Civil War', in K. Ballentine and J. Sherman (eds), *The political economy of armed conflict: beyond greed and grievance*, Boulder, CO: Lynne Rienner, 2003, pp. 47–70.

108 Collier, 'Doing Well Out of War: An Economic Perspective'. Fearon challenges the statistical analysis in Collier and Hoeffler's work. After reworking their data, he proposes narrowing the significance of their findings to cases where oil production is a major component of primary commodity exports. High oil production as a component of primary commodity exports results in a greater risk of civil war, in his view, because of the influence of weak states relative to level of per capita income and the greater "prize" oil production provides for state or secessionist capture. See J. D. Fearon, 'Primary Commodity Exports and Civil War', *Journal of Conflict Resolution* 49 (4), 2005, 483–507.

109 J. D. Sachs and A. M. Warner, 'Natural Resources and Economic Development: The Curse of Natural Resources', *European Economic Review* 45, 2000, 827–38.

110   R. M. Auty, 'Natural Resources and Civil Strife: A Two Stage Process', *Geopolitics* 9, 2004, 29–49.

111   Le Billon, 'The political ecology of war'.

112   de Soysa, 'Paradise is a Bazaar?'.

113   J. von Braun and R. Meinzen-Dick, '*Land Grabbing by Foreign Investors in Developing Countries: Risks and Opportunities*', IFPRI Policy Brief 13, Washington DC: International Food Policy Research Institute, April 2009.

114   de Soysa, 'Paradise is a Bazaar?', p. 405.

115   B. Korf, 'Rethinking the Greed–Grievance Nexus: Property Rights and the Political Economy of War in Sri Lanka', *Journal of Peace Research* 42, 2005, 201–17.

116   Ross, 'Oil, Drugs, and Diamonds'.

117   An exception among environment-conflict researchers is V. Boge, 'Mining, Environmental Degradation and War: The Bougainville Case', in M. Suliman (ed.), *Ecology, Politics & Violent Conflict*, New York: Zed Books, 1999, pp. 211–27. See also M. Ross, *Extractive Sectors and the Poor*, Oxfam America, 2001; G. McMahon and F. Remy (eds), *Socioeconomic and Environmental Effects in Latin America, Canada, and Spain*, Washington and Ottawa: World Bank/International Development Research Corporation, 2001; C. Ballard and G. Banks, 'Resource Wars: The Anthropology of Mining', *Annual Review of Anthropology* 32, 2003, 287–313; United States Agency for International Development (USAID), Office of Conflict Management and Mitigation (CMM), *Minerals and Conflict: A Toolkit for Intervention*, Washington DC, 2005.

118   See T. Deligiannis, 'The Evolution of Environment-Conflict Research: Toward a Livelihood Framework', *Global Environmental Politics* 12 (1), February 2012.

3

ENVIRONMENTAL SECURITY AND THE RESOURCE CURSE

Indra de Soysa

Introduction

The idea that increased demands on the natural environment due to economic activities of humans lead to social breakdown and ultimately violent conflict has gained widespread acceptance among popular audiences and within scholarly circles.[1] The growing belief that the natural environment and peace are connected is reflected in the recent Nobel Peace Prizes awarded to such personalities as Wangari Maathai and Albert Gore.[2] The latter also won an Oscar for his documentary "An Inconvenient Truth" that paints a dramatic picture of the security implications emanating from global warming. Contrary to this often called neo-Malthusian[3] view on environmental conflict, this chapter aims to demonstrate that poverty and conflict are part of a natural resource trap, where relative abundance of resources explains socio-economic breakdown. It is argued that armed conflict, whether between or within states, results largely from policy failure associated with "the paradox of plenty".[4] The idea is that it is the relative abundance of natural wealth as opposed to resource scarcity that affects economic and governance outcomes; this is often called the "resource curse".

In this chapter I argue that much of the problem of violence can be solved by policies in the realm of global governance,[5] for it is the developmental and governance traps and the lack of "institutional capital" that leads to persistent poverty, human insecurity, and continued dependence on natural resources.[6] In particular, this chapter will highlight how factors associated with globalization and the transformation of the post-Cold War international order can help to mitigate social and economic breakdown by allowing the build up of better institutional capital through international diffusion and global governance. As an example of a region where the risk of violent conflict has been steadily decreasing, precisely because of these positive developments, this chapter examines the case of Sub Saharan Africa (SSA) in some detail. It should be noted here that the issue of natural resource scarcity/abundance and their correlation to violent conflict has been strongly debated above all else in the context of the African continent. In that debate, several people have argued that parts of Africa face the toughest conditions of resource scarcity, where it coalesces with economic depression and bad governance.[7] Contrarily, others have argued that the risk of African conflict can be attributed to relative abundance of natural wealth, rather than its scarcity.[8] While agreeing that social grievances driven by scarcity are unlikely to cause large scale violence, a third group has argued that what matters most is how politics and state policies channel the "curse" of natural wealth. In particular these scholars are concerned with so-called "Dutch Disease", an economic concept that explains the correlation between natural resource wealth and lower economic growth due to the rise in the real exchange rate of the resource exporter relative to other trading partners, leading to a less-productive economy.[9] Given then the recent decline in violent conflict it must be asked, what does this trend suggest with regards to this debate? This chapter will argue that only those countries that lack natural resources (for instance the countries forming Sub Saharan Africa) have seen improvements in their political climates, whereas those resource rich countries that depend on lootable resources and are victims of bad governance (for instance Algeria, Nigeria, Chad, the DRC, and Sudan) continue to be conflict prone.[10] The chapter thus sides with the resource abundance theorists of the above mentioned debate. The empirical evidence for these assertions is presented below. Before making this argument, I have a brief section on methodology.

Methodology

Much of the environment and conflict literature is dominated by case studies.[11] One problem with this is the selection of the dependent variable,[12] as researchers all too often pick countries where conflict is already apparent only then to trace the processes linking the environment to that conflict. "Peaceful countries" with similar environmental problems are rarely examined, with the consequence that we have still learned little of how such countries have avoided conflict. More fundamentally still, even if there are cases where the environment "caused" conflict, how much of a problem it really is and relative to what other factors cannot be reasonably culled from the stories without knowing something about all cases of conflict. Nor is it possible to discern the relative weight one gives environmental factors over others when designing policies for peace. Methodologically, the theories that underpin environmental factors as causes of relatively large, costly conflicts are also suspect. Suggestions for how poor people may overcome their collective action problems for organizing costly conflict are for the most part unpersuasive. For some, being aggrieved is enough. For others, people in stressed environments fight to survive, but is this a rational strategy that people really engage in? Given the weakness of such a starting point, where issues of collective action and rational calculation are ignored, knowing which variables one brings into the analysis relative to what one leaves out is a real methodological conundrum.

Recent work has come a long way in addressing these concerns. The application of rational choice theory and the use of large-N statistical methods have demonstrated that conflicts are driven largely by opportunity factors rather than grievances and that those countries possessing large natural resources have bigger problems than those countries considered natural-resource poor.[13] This is not to say that good case studies of environment and conflict cannot be done, or that they do not exist. In fact, some careful case studies also effectively challenge the earlier stories about the environment and conflict.[14]

The choice of cases to be studied has yet to be justified better than what is apparent from my reading of the earlier literature.[15] In general, zones of conflict are complicated by stories of grievances of all sorts, so that culling objective facts from representations and the strategic behaviour of people is almost impossible.[16] Thus, the best methodological strategy for addressing the thorny question of researching highly charged environments is to correlate objective conditions relating to the environment with conflict outcomes, holding constant all other factors known to predict the outbreak of violent conflicts in large enough samples of countries, which minimizes selection bias. Such quasi-experimental methods allow us to understand what the aggregate effect of environmental factors is relative to others, as well as provide a sense of the magnitude of the impact of the variables focused on, which is generally useful for policymakers. Moreover, the aggregate results also allow one to choose cases for study more intelligently to probe the mechanisms of why environmental conditions do or do not cause conflict. The state of data collection today is very well developed both on conflict as well as environmental indicators, which includes GIS based geo-referenced datasets.[17] This study relies on some of these aggregate data for examining some trends and correlations so as to test the logic of arguments relating environmental factors to conflict in a general sense. It is by no means a thorough and detailed empirical analysis of this complex issue.

Nurture, not nature!

One view commonly voiced by neo-Malthusian environmental conflict scholars is that poor countries are poor and therefore insecure because nature has been unkind.[18] At the same time, many empirical studies by these same scholars suggest that developed states are resilient to shocks caused by natural disasters as well as short-term economic jolts. Wealth, they argue, is thus good beyond simply allowing higher consumption. Development also may mean that risks are mitigated with the application of capital and technology and superior organization.[19] For instance, a hurricane that hits both Florida and Haiti has far different consequences for the two societies that inhabit these two territorial entities.[20] In short, wealth is an insurance policy that increases human welfare far beyond simple consumption. Thus, acquiring wealth, or increasing the level of development of a society, reduces all forms of vulnerabilities, including vulnerabilities emanating from environmental factors and climate change. The issue is whether or not the lack of nature's gifts reduces the possibility of escaping poverty.

The usual way of gauging the level of development is per capita income. Recently, those neo-Malthusian scholars that have linked poverty to the callousness of nature have suggested that poor countries are trapped in a vicious cycle of poverty, underdevelopment, and insecurity because they lack natural resources to overcome social vulnerabilities. Neo-Malthusians thus connect poverty in developing countries to a lack of environmental resources, with poverty, a degraded planet, and environmental scarcity supposedly miring poor countries in a conflict-poverty trap. Increased global climate change is being represented as a variable that can worsen the situation for poor countries by leading to increased vulnerabilities, which, in turn, may lead to social frictions that inhibit development.[21] According to Homer-Dixon:

all types of environmental depletion or damage [are] forms of scarcity of renewable resources [. . .] climate change increases the scarcity of regular patterns of rainfall and temperature on which farmers rely.

[. . .] scarcities can overwhelm efforts to produce constructive change and can actually reduce a country's ability to deliver reform. Consequently, environmental scarcity sometimes helps to drive society into a self-reinforcing spiral of violence, institutional dysfunction, and social fragmentation.

[. . .] A persistent and serious ingenuity gap raises grievances and erodes the moral and coercive authority of government, which boosts the probability of serious turmoil and violence. This violence further undermines the society's ability to supply ingenuity. If these processes continue unchecked, the country may fragment as the government becomes enfeebled and peripheral regions come under the control of renegade authorities. Countries with a critical ingenuity gap therefore risk becoming trapped in a vicious cycle, in which severe scarcity further undermines their capacity to mitigate or adapt to scarcity.[22]

Armed conflict in the post-Cold War world

While neo-Malthusian analyses have been influential in both the academic and policy communities, recent trends in the global level of organized armed conflict raise serious questions about these arguments. As Figures 3.1 and 3.2 aim to show, quite unlike what neo-Malthusians suggest, both international conflict and civil war are on the wane.[23]

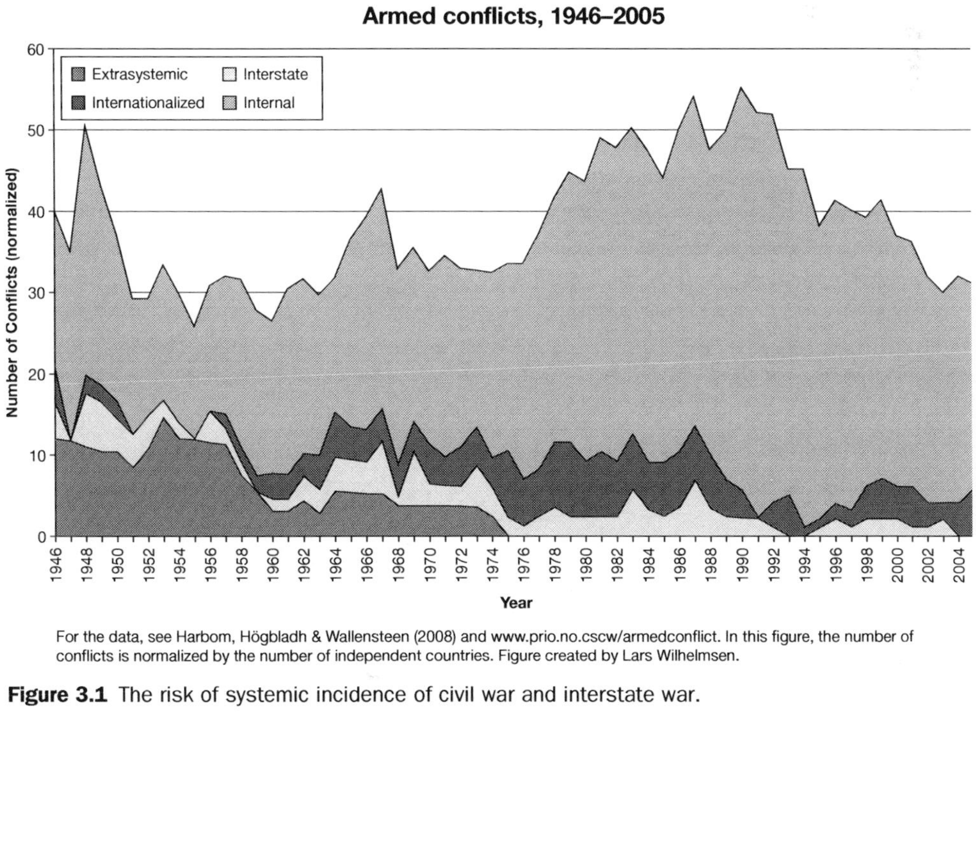

Armed conflicts, 1946–2005

For the data, see Harbom, Högbladh & Wallensteen (2008) and www.prio.no.cscw/armedconflict. In this figure, the number of conflicts is normalized by the number of independent countries. Figure created by Lars Wilhelmsen.

Figure 3.1 The risk of systemic incidence of civil war and interstate war.

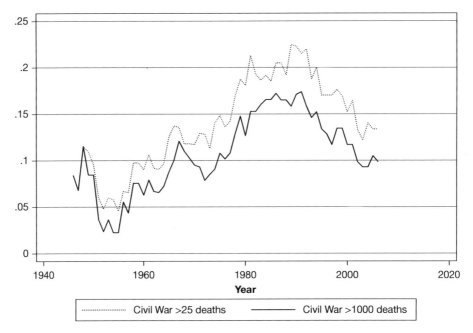

Figure 3.2 The annual average risk of civil war measured at two levels of intensity, 1946–2006.

Source: Uppsala-PRIO dataset on armed conflicts. Figure generated by author.

Since the neo-Malthusian argument about rising temperatures and declining availability of natural resources is applicable at the global level, an overlay of average global temperatures on the civil war figure will show just the opposite – rising temperatures correlate to a reduction in the incidence of civil war! If there is a simple correlation between environmental change and civil conflict, then, does this not mean that the environment is getting better? Of course, the correlation between environmental change, or rising temperatures, and conflict at the global level is only a partial and thoroughly unqualified picture. The implications of climate change should be most severe where poverty, the dependence on rain-fed agriculture, and the dependence on renewable environmental assets are highest among ordinary people for their livelihoods, *ceteris paribus*. It could very well be that only the poorest are vulnerable. Apparently, very local level human insecurity leads to national level crises. Sub Saharan Africa in particular is often singled out as the region most vulnerable to global climate change. Even here, however, reality challenges the neo-Malthusian prognosis. Figure 3.3 disaggregates the global risk of civil war into regions.

As seen there, two regions are generally responsible for pulling down the global risk of civil war, namely Latin America and Sub Saharan Africa. The regions pulling the trend upwards, in turn, are North Africa, the Middle East and Asia where conflicts have increased since the events of 9/11 and the resultant global war on terrorism. The static maps of Figures 3.4 and 3.5 respectively show that most of the existing conflicts in recent years are taking place in Central and South Asia, usually in a long band that stretches from the Caucasus down to South Asia. These are most likely end of empire conflicts following the breakup of the Soviet Union where irredentist and secessionist claims and criminalized violence exist in quasi-ethnic states, from Georgia to Sri Lanka and East

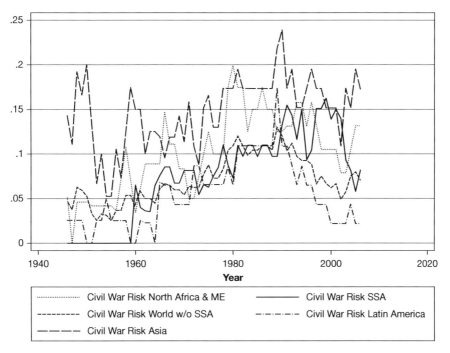

Figure 3.3 The regional risk of civil war with over 25 battle-related deaths.

Source: Uppsala-PRIO dataset on armed conflicts. Figure generated by author.

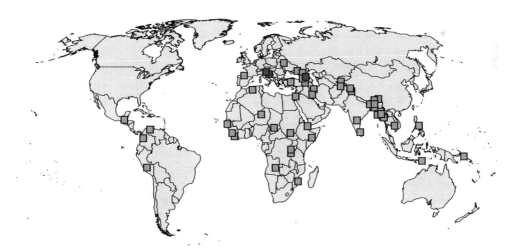

Figure 3.4 The geographic location of conflict with over 25 battle-related deaths in 1992.

Source: UCDP conflict data. Map generated by author using ViewConflicts. The programme is available at www.svt.ntnu.no/geo/forskning/konflikt/viewConflicts/ (accessed August 2012).

Timor in what some have termed the "remnants of war".[24] Contrary to popular opinion, however, Sub Saharan Africa shows a massive decrease in conflicts in the last decade. If climate change was manifesting itself on the ground today, the Sub Saharan African experience simply does not reveal it. Moreover, labelling the war in Darfur in Sudan as

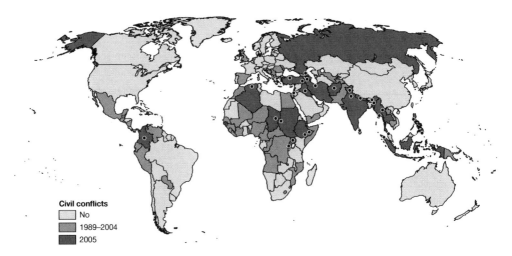

Civil conflicts
- No
- 1989–2004
- 2005

Figure 3.5 The geographic location of conflicts with 25 deaths and above in 2005.

Source: Gleditsch (2008).

an environmental conflict simply absolves the regime in Khartoum of instigating genocide, or at the least of doing nothing to stop it.

The resource curse

I believe that an alternative reading of historical data provides a better explanation of the link between natural resources and violent conflict. This reading stresses two points. First, environmental wealth often is wasted as part of the resource curse, where countries that extract natural wealth suffer economic and political maladies that increase underdevelopment, thereby costing current and future generations the benefits of using resources wisely for diversifying away from dependence on the natural environment. Second, there is little evidence that conclusively suggests natural resources are becoming increasingly scarce globally (with the exception of oil) and that various forms of technology are not displacing them. This idea is akin to cornucopian arguments which hold that human ingenuity is the "ultimate resource" since humans throughout history have adapted to shortage by invention or substitution.[25] Consider, for example, that the livelihoods of people who live in Washington state, the headquarters of Microsoft and home of the richest man on earth, Bill Gates, were once purely dependent on trapping beavers – not so today!

Figure 3.6 shows the trend in the prices of all commodities, energy, edibles, and beverages for roughly the past decade. As seen there, food prices were dropping and have only recently increased above the standard of 100 set for 1995. Moreover, despite worries about the lack of water, the price of beverages has decreased in the past ten years; that is, they got cheaper relative to the price in 1995. In fact, if one were to look at the longer-term trend in prices, the recent upward swing is nothing compared to high prices for all sorts of commodities, particularly food, if we say compare with prices from the 1960s. If the world is suffering growing scarcities, this is failing to register on prices, and this is despite a recent surge in demand for all types of resources in places with strong economic growth such as China, Brazil, and India.

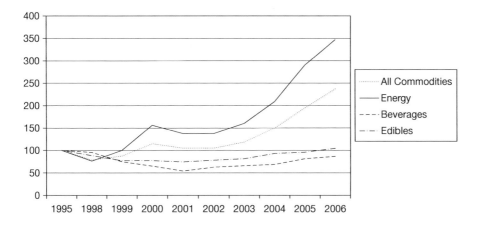

Figure 3.6 Disaggregated commodity price index, 1995–2006.

Source: International Monetary Fund (IMF) website. Figure generated by author.

Moreover, nor does it seem that the poorest countries lack "stocks" of natural wealth relative to the richest countries. The World Bank recently estimated the contribution of natural, human, and physical capital to the total wealth of about 100 countries as part of its "Green Accounting" programme.[26] These estimates allow us to see what kind of capital the poorest countries (defined as the World Bank's "low income" group) lack relative to the highest income countries. As Figure 3.7 shows, what the poor countries

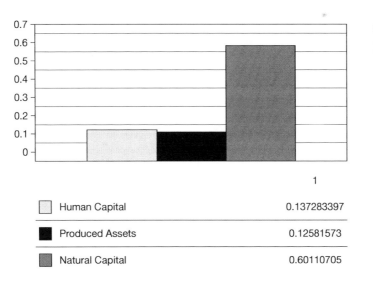

Human Capital	0.137283397
Produced Assets	0.12581573
Natural Capital	0.60110705

Figure 3.7 Poor countries' average per capita share of natural, produced, and human capital relative to the richest countries.

Source: Figure computed by author using capital estimates presented in Kunte, Arundhati, Kirk Hamilton, John Dixon, and Michael Clemens. (1998) Estimating National Wealth: Methodology and Results. Washington, DC: World Bank. (The richest countries were defined as OECD countries and the poorest countries as the Least Developed Countries as designated by the World Bank.)

lack relative to the richest are clearly human and physical capital, not natural capital.[27] Indeed, despite having on average roughly 2 per cent of the income per capita of the richest countries, the poorest countries have more than 60 per cent of the natural capital stock relative to the richest countries. In other words, what, for instance, Angola or the DRC lack relative to, say Portugal or Belgium are not "nature's gifts" in terms of natural wealth, but man-made capital, both physical, human, and perhaps institutional, which, taken together, make up the largest share of general wealth. Indeed as Adam Smith noted several centuries ago, wealth is generated by the productivity of labour.[28]

While grievance-driven factors (environmental scarcity) as explanatory factors for underdevelopment and social peace should thus be viewed with some caution, some scholars have found evidence that suggests that "greed-driven" factors can explain the outbreak and continuation of organized armed violence.[29] This is an assertion shared by theories aiming to explain conflict, which generally concur that extractable wealth in terms of large natural resource supplies (relative abundance) provide the payoff for organizing large scale violence.[30] The key idea is that conflict is not universally harmful, but that a few (can) "do well" out of organizing civil war.[31] Since the provision of justice is a "public good" altruistic individuals rarely spring up to serve justice by bearing all the costs of organizing rebellion. Groups organize for violence because of private gain. Likewise, peace is a public good, which often prevents the majority from organizing for peace, which can be very costly. According to the theory of collective action, larger groups are harder to organize than smaller groups because the payoffs are more concentrated the smaller the size and because free riding can be monitored more effectively.[32] However, given the destruction caused by conflict, people who are productive in society have the necessary incentive to organize for peace. The more any one society stands to lose the easier will be the organization for solving collective action problems. This is the main message in the burgeoning literature on social capital; it perhaps also explains why developed societies are resilient to shocks,[33] as higher levels of development raise the payoff for maintaining peace and strengthen social and state capacities for providing the public force necessary to check socially harmful behaviour.[34]

In all of this, the finding that resource wealth is related to greed-driven conflict is key. Resource wealth not only provides lootable income to private actors, but it leads to semi-private states, or what some call "shadow states", states that are captured by a few private interests.[35] Moreover, the nature of the economic payoff, or viability of the state, provides the rationale for institutional development and strength. A convenient resource stream, such as extractable mineral wealth, leads to withering of institutions around the collection of taxes, thereby weakening the social contract necessary for building a tax base from society. Such processes lead to weak states that will be disinterested in providing the optimal level of public goods, which will ultimately erode social and human capital.[36] In addition, resource wealth allows states to close their economies and practice industrial substitution policies, which some have referred to as "precocious Keynesianism" for state-building along nationalist lines.[37] According to others, it is the "natural resource curse" that seems to be at the heart of some of the social ills facing many poor countries.[38] Taken together these pernicious effects form a powerful cocktail that leads often to state and social disarray, violent civil wars, and continued marginalization.

It is vital to note here that there is nothing automatic about resource wealth that leads to disarray. The development and history of, for example, Norway and Canada are cases in point. Positive examples such as these also mean that the problems faced by weak yet resource rich states may be correctable, for the most part by policy and institutional innovation.[39] In an age of globalization, the technology and capital required for diversification of economies away from dependence on monoculture development are available more easily than ever before. This is promising. Moreover, systemic factors under conditions of globalization are also better today for making resource-led development less pernicious than was possible during the Cold War. Some studies find much evidence to suggest that increased economic openness lowers corruption, and raises the incomes of the poor as well as the political standing of workers.[40] Importantly, the degree of economic openness also seems to condition the effect of natural resources towards better respect for human rights, whereas resource wealth alone tends to increase political repression and politicide.[41] In short, globalization may have an important conditioning effect on the resource curse, because of its influence on economic development.

The conditions of globalization are ostensibly driven by increased trade between states and across regions. Growing trade is further supplemented and complemented by foreign direct investment (FDI).[42] These aspects are generally seen as the hard drivers of globalization.[43] Today, poor countries, which were once against global liberalism, have come to embrace it [44] a point exemplified most clearly – in the economic realm at least – by the economic policies of China, Vietnam, and Cuba.[45] What poor countries lack is capital and technology for improving their trading positions, diversifying away from primary commodity dependence, and creating economic growth. Globalization helps since trade and investment benefit economic growth.[46] Trade dependence and FDI may improve conditions of human rights,[47] social conditions facing labour,[48] environmental conditions, economic sustainability,[49] and even internal peace.[50] In fact, of several variables explaining peace between states, trade, democracy, and joint membership in international organizations are proven to have statistically significant effects on peace, supporting theories of peace going back to the political philosophies of Montesquieu and Immanuel Kant, as well as the Manchester School, all of which subscribed to the assumption that interdependence promotes peace.[51] Increasing interdependence, in turn, is a policy problem that allows room for human agency in building a more peaceful world.

Given all that has been said so far, it should be clear that this chapter contends that conflict and development are largely consequences of policy. Thus, better policy environments locally, as well as globally, have great promise to make a difference. This said, good, or at the very least, better governance does not come about easily; it has to be effected through incentives, and incentives are different in the post-Cold War era when formerly attendant political, ideological, and strategic considerations no longer bear on the political and economic fortunes of poor countries. The good news is that in this era of globalization, the opportunity for building a better environment for development and peace exists. Most countries around the world seek to be more globalized, with the exception of some Sub Saharan African countries, which as some have suggested, despite vast improvements in recent years, still suffer from internal biases against open markets and domestic rent-seeking coalitions that prevent better policies that would move many of these countries away from resource dependence and insecurity.[52]

Reasons for hope

The explanation offered here as to why recent years show a positive trend in African performance is greater liberalization of markets and politics, enabled in large part by the favourable international security and ideological climate after the end of the Cold War. Under these conditions, the "curse" associated with natural resource wealth is likely to be far weaker, largely due to better institutional management of wealth. The case of Sub Saharan Africa is a case in point. This region has historically lagged behind other regions in terms of free markets and encouragement of entrepreneurial talent, although the trend has gone in this direction in recent decades. One report suggests that countries with high natural assets and with a higher level of economic freedom tend to have lower levels of political dissent and repression of human rights.[53] The same might be said for political liberalization. During the Cold War, African leaders who looked "eastwards" implemented communist, one-party dictatorships in the name of socialist development. Likewise, those that at least initially looked "westwards" also tended to be dictatorial and nascent democracy after independence was soon snuffed out. As Figure 3.8 demonstrates, however, this trend has reversed drastically with the end of the Cold War, as ever more African countries become democratic. Correspondingly, Figure 3.9 shows, the same positive trend is apparent with regards to human rights.

Finally, a measure of overall governance developed by the International Country Risk Guide also shows an optimistic picture for the region in recent years, which supplements the findings for political and economic liberalization. As Figure 3.10 illustrates, Africa's

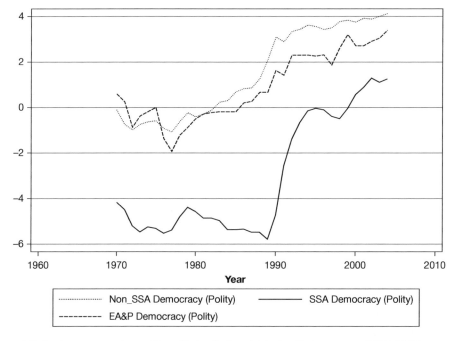

Figure 3.8 Regional comparison of trend in institutional (electoral) democracy, 1970–2006.

Source: Polity IV data.

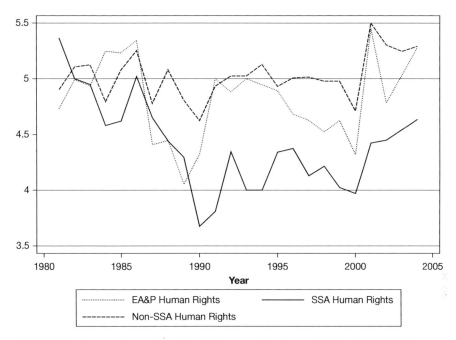

Figure 3.9 Regional comparison of trend in state respect for human rights, 1981–2004.

Source: Figure generated by author. Data available in M. G. Marshall, *Polity IV Project: Political Regime Characteristics and Transitions, 1800–2010*, Online: www.systemicpeace.org/polity/polity4.htm (accessed August 2012).

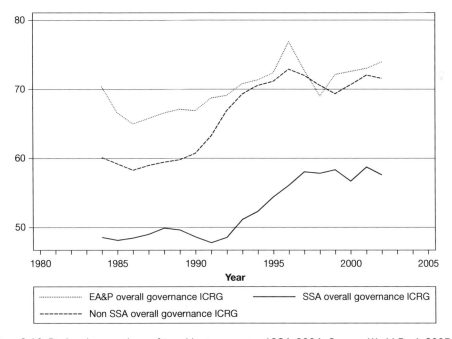

Figure 3.10 Regional comparison of trend in governance, 1984–2004. Source: World Bank 2005.

Source: Figure generated by author. The overall governance measure provided by ICRG takes into account corruption and other risks for international business. The ICRG data are collected for commercial use and are available for purchase at The Political Risk Services Group, Online: www.prsgroup.com/ (accessed August 2012).

performance has been increasing over time, but a massive gap still remains relative to the rest of the World, and to East Asia and the Pacific in particular.

The dimensions just examined in terms of global trends relative to SSA, namely the growth of democracy, the improvements in government respect for human rights, overall economic and political governance, sound economic policies for generating entrepreneurship, and development are all pointing in the right direction. While SSA still lags behind other regions, the direction of change is encouraging and points against arguments that suggest that climatic and environmental conditions can hold vulnerable regions such as SSA back. Of course, much of the phenomena of vulnerability and human security argued in the literature can be highly localized and difficult to pinpoint in aggregated, national level outcomes. Responses to these highly localized vulnerabilities are the preserve however of national states and governments, which is another analytical issue altogether. The fact is that the more responsive governments and states are likelier to develop under conditions of relative scarcity and are not founded purely on the largess of natural wealth.

Conclusion

In conclusion, several parts of the world that enjoy nature's gifts may in fact be "cursed"; the point is, however, that the resource curse may be overcome with conscious effort, which might be well under way in previously problematic areas to a large part due to the implementation of better governance.

Underdevelopment and human insecurity are inextricably linked. Although most of the world's people are increasingly facing better prospects of social and physical security the opportunities afforded by the end of the Cold War and the current age of globalization are not being realized equally by all. The incidence of organized armed violence is declining rapidly since the end of the Cold War, and the risk of organized civil war today is somewhere near what it was at the beginning of the 1960s, despite the fact that there are far more countries and less repression of conflict, as was the case during the Cold War. Globalization promises security and development. Contrary to neo-Malthusian arguments that see a bleak future for security and development based on natural limits to growth and consumption, policy matters. However, natural resource abundance, not its scarcity, hampers both good policymaking and civil peace required for ensuring long-term development and human security. The long-term improvement of the security situation in Sub Saharan Africa flies in the face of the arguments about climate change and the future of development in rain-fed agricultural societies. Either the climate is improving, or it has nothing to do with the weather!

One of the primary issues that increase the risk of conflict seems to be bad governance under conditions of natural resource abundance. The crucial task for global policy would be to direct aid, trade, and investment policies towards diversifying African economies away from extractive activity. The problem, as some pessimistic about the prospects of globalization have claimed, is not that Africa is captured by global markets hungry for resources, but that policies in these countries prevent resource rents from being translated into sustainable development. In fact, resources are being wasted rather than being deployed for productive investment. Angola, Nigeria, and the DRC could be some of the richest places on the planet if wealth were properly managed. Overcoming the "resource curse" might be a worthy challenge for global policy to address. Unfortunately, while governments pay lip service to the issue, very little is done. The private sector initiatives, such as Corporate Social Responsibility (CSR), are a start, but the incentives to shirk (cheat) are so great that only time can judge its success. Besides these initiatives, institutions of global governance, such as the International Financial Institutions (IFIs), could step in to devise programmes that help states manage their wealth better, such as the management of rents from the Chad oil pipeline. These initiatives effected largely through global civil society action may already be having some impact on governance and peace in Sub Saharan Africa.

Poor countries, such as many in Sub Saharan Africa, lack the physical and human capital necessary to join the globalized economies of the world. Many fall behind the rest of the world. Trade and investment could lift these countries out of this trap. However, domestic factors, such as rent-seeking and attitudes of governments and elites against open markets and foreign capital, may still be hindering the rate of convergence with others.[54] The international policy community, and perhaps also the academic communities, could work to increase transparency and awareness around how globalization can benefit these countries. The empowerment of civil society with increased levels of democracy goes some distance towards creating greater transparency and preventing rent-seeking, leading to greater openness of these countries to trade.[55]

The onus is not exclusively on the poor world. Rich countries should also do their bit. Paradoxically, when former colonies are becoming more open to global markets, the rich are cutting them off with tariff and non-tariff barriers.[56] Apparently, jobs are being lost to the South, but consider that the rich world spends over $600 billion per annum on defense, $300 billion on agricultural subsidies, and only $60 billion on aid. One must ask how this allocation improves the security for the rich. Clearly, local and global rent-seeking hinders proper management of wealth, increases inefficient allocation, and perhaps hinders the prospects for development and peace. It is quite right that the "international community" focuses more and more on human security. It might be that future acts of global governance adjust the rich countries' policies that violate the "human security" of the poor, the most odious kind of exploitation. Concerted global governance that addresses these imbalances is now more imperative than ever.[57] The question, however, comes down to political will, but despite goodwill, politicians will not act given imprecise advice. Perhaps consensus should be sought first among epistemic communities who can then bring sharply focused knowledge to bear that raises the costs of self-serving politics. This knowledge is most likely to be generated by paying careful attention to theories supported by systematic, carefully gathered empirical observations of real-world outcomes. The accumulated evidence thus far suggests that natural resource scarcity may be irrelevant for explaining violent conflict and a hindrance to overcoming poverty-related vulnerabilities.

Notes

1   J. Diamond, *Collapse: How Societies Choose to Fail or Survive*, London: Penguin, 2005; T. Homer-Dixon, *The Ingenuity Gap*, New York: Alfred A. Knopf, 2000; M. T. Klare, *Resource Wars: The New Landscape of Global Conflict*, New York: Henry Holt and Company, 2001. For an excellent review of arguments around security concerns due to growing scarcity, see R. A. Matthew, 'Resource Scarcity: Responding to the Security Challenge', *International Peace Institute*, 2008. For an excellent evaluation of the security implications of climate change, see I. Salehyan, 'From Climate Change to Conflict? No Consensus Yet', *Journal of Peace Research* 45, 2008, 315–26.

2   Naturally, the NGO and activist community are particularly taken with this view, with major UN agencies, such as the United Nation's Environment Program (UNEP) and the United Nations Development Program (UNDP), taking the lead.

3   Thomas Malthus famously warned that the increase of population will outstrip society's ability to produce enough food. His theory is widely discredited (see for example, E. Boserup, *The Conditions of Agricultural Growth: The Economics of Agrarian Change Under Population Pressure*, New York: Aldine, 1965; M. Tiffen, M. Mortimore and F. Gichuki, *More People, Less Erosion: Environmental Recovery in Kenya*, London: Wiley, 1994). The neo-Malthusians are those who echo Malthus' warnings about the dangers of a depleting natural resource base and the limits to economic growth and development.

4   T. L. Karl, *The Paradox of Plenty: Oil Booms and Petro-States*, Berkeley: University of California Press, 1997.

5   Environmental degradation and overuse of resources should be addressed by global and local policies on their own right as serious policy issues, but the question of stopping armed violence may require other policies entirely.

6   Please note, by "human security" I mean insecurity to societies caused by physical violence. This sort of insecurity can be affected by global governance and conscious policies both locally and globally. Naturally, human insecurities are caused among other things by scarcity of food and water, but I focus on whether such shortages can cause the type of insecurity that requires global policy intervention as opposed to intervention by local authorities.

7   T. Homer-Dixon, *Environment, Scarcity, and Violence*, Princeton, NJ: Princeton University Press, 1999; J. Markakis, *Resource Conflict in the Horn of Africa*, London: Sage, 1998; M. Suliman, *Ecology, Politics and Violent Conflict*, New York: Zed, 1999.

8   P. Collier, E. Lani , H. Håvard, A. Hoeffler, M. Reynal-Querol and N. Sambanis, *Breaking the Conflict Trap: Civil War and Development Policy*, Oxford: Oxford University Press, 2003.

9   J. D. Sachs and A. Warner, 'The Curse of Natural Resources', *European Economic Review* 45, 2001, 827–38.

10 Of course, conflicts in the horn of Africa, such as in resource poor Somalia, also continue unabated.

11 T. Homer-Dixon and J. Blitt, *Ecoviolence: Links Among Environment, Population, and Security*, Oxford: Rowman & Littlefield, 1998.

12 N. P. Gleditsch, 'Armed Conflict and the Environment: A Critique of the Literature', *Journal of Peace Research* 35, 1998, 381–400.

13 P. Collier, A. Hoeffler and D. Rohner, 'Beyond Greed and Grievance: Feasibility and Civil War', Oxford Economic Papers 2009, 61, 1–27; J. D. Fearon and D. D. Laitin, 'Ethnicity, Insurgency, and Civil War', *American Political Science Review* 97, 2003, 1–16; I. de Soysa, 'Paradise is a Bazaar? Greed, Creed, and Governance in Civil War, 1989–1999', *Journal of Peace Research* 39, 2002, 395–416.

14 N. L. Peluso and M. Watts (eds), *Violent Environments*, London: Cornell University Press, 2001.

15 D. Collier and J. Mahoney, 'Insights and Pitfalls: Selection Bias in Qualitative Research', *World Politics* 49, 1996, 56–91; B. Geddes, 'How the Cases You Choose Affect the Answers You Get: Selection Bias in Comparative Politics', *Political Analysis* 2, 1990, 131–50.

16 T. Kuran, *Private Truths, Public Lies: The Social Consequences of Preference Falsification*, Cambridge, MA: Harvard University Press, 1995; A. Varshney, 'Postmodernism, Civic Engagement, and Ethnic Conflict: A Passage to India', *Comparative Politics* 30, 1997, 1–20.

17 See Centre for the Study of Civil War and Uppsala Conflict Data Program, *Data on Armed Conflict*. Online: www.prio.no/CSCW/Datasets/Armed-Conflict/ (accessed August 2012).

18 Homer-Dixon, *The Ingenuity Gap*; D. H. Meadows, D. L. Meadows and J. Randers, *Beyond the Limits: Confronting Global Collapse, Envisioning a Sustainable Future*, White River Junction, VT: Chelsea Green, 1993.

19 B. Wisner, P. Blaikie, T. Cannon and I. Davis, *At Risk: Natural Hazards, People's Vulnerabiity and Disasters*, London: Routledge, 2004.

20 The devastation of New Orleans by Hurricane Katrina shows that even relatively wealthy places are not immune if governments have been lax and where preparation has been weak. Floridians have learnt this the hard way many times over.

21 Homer-Dixon and Blitt, *Ecoviolence*; T. F. Homer-Dixon, 'On the Threshold: Environmental Changes as Causes of Acute Conflict', *International Security* 16, 1991, 76–116.

22 T. Homer-Dixon, *Environment, Scarcity, and Violence*, Princeton, NJ: Princeton University Press, 1999 pp. 9, 5, 27.

23 Human Security Report, *Human Security Report 2005: War and Peace in the 21st Century*, Oxford: Oxford University Press, 2005; J. Mueller, *The Remnants of War*, Ithaca, NY: Cornell University Press, 2004.

24 Mueller, *The Remnants of War*.

25 J. L. Simon, *The Ultimate Resource II*, Princeton, NJ: Princeton University Press, 1998.

26 See http://go.worldbank.org/EPMTVTZOM0 (accessed August 2012).

27 Please note, the poorest countries are the World Bank's low income category and the richest are the high income category.

28 A. Smith, *The Wealth of Nations (Books I–III)*, London: Penguin Books, 1999 (1776).

29 Collier *et al.*, *Breaking the Conflict Trap*; P. Collier and A. Hoeffler, 'Greed and Grievance in Civil War', *Oxford Economic Papers* 56, 2004, 563–95; de Soysa, 'Paradise is a Bazaar?'; J. D. Fearon and D. D. Laitin, 'Ethnicity, Insurgency, and Civil War', *American Political Science Review* 97, 2003, 1–16; M. Ross, 'What Do We Know About Natural Resources and Civil War?', *Journal of Peace Research* 41, 2004, 337–56.

30 M. Berdal and D. M. Malone (eds), *Greed & Grievance: Economic Agendas in Civil Wars*, Boulder, CO: Lynne Rienner, 2000; P. Collier, 'Rebellion as a Quasi-Criminal Activity', *Journal of Conflict Resolution* 44, 2000, 839–53; Ross, 'What Do We Know About Natural Resources and Civil War?'; de Soysa, 'Paradise is a Bazaar?'.

31 P. Collier, 'Doing Well Out of War' in Berdal and Malone (eds), *Greed & Grievance: Economic Agendas in Civil War*, pp. 91–111.

32 M. Olson, *The Logic of Collective Action*, Cambridge, MA: Harvard University Press, 1965.

33 R. Putnam, *Making Democracy Work: Civic Traditions in Modern Italy*, Princeton, NJ: Princeton University Press, 1993; A. Varshney, 'Ethnic Conflict and Civil Society: India and Beyond', *World Politics* 53, 2001, 362–98.

34   Fearon, *et al.*, 'Ethnicity, Insurgency, and Civil War'; R. H. Bates, *Prosperity and Violence: The Political Economy of Development*, New York: W. W. Norton & Company, 2001.

35   T. L. Karl, *The Paradox of Plenty: Oil Booms and Petro-States*, Berkeley: University of California Press, 1997; W. Reno, 'Shadow States and the Political Economy of Civil Wars', in Berdal and Malone (eds), *Greed & Grievance: Economic Agendas in Civil War*, pp. 43–68.

36   M. Woolcock, L. Pritchett and J. Isham, 'The Social Foundations of Poor Economic Growth in Resource-Rich Countries', in R. M. Auty (ed.), *Natural Resources and Economic Growth*, New York: Oxford University Press, 2001, pp. 76–92.

37   D. Waldner, *State Building and Late Development*, Ithaca, NY: Cornell University Press, 1999.

38   R. M. Auty (ed.), *Resource Abundance and Economic Development*, Oxford: Oxford University Press, 2001; M. Ross, 'The Political Economy of the Resource Curse', *World Politics* 51, 1999, 297–332.

39   R. M. Auty, 'How Natural Resources Affect Economic Development', *Development Policy Review* 18, 2000, 347–64; H. Mehlum, K. M. Halvor and R. Torvik, 'Cursed by Resources or Institutions?', *The World Economy* 10, 2006, 1117–31.

40   D. Dollar and P. Collier, *Globalization, Growth and Poverty: Building an Inclusive World Economy*, Oxford: Oxford University Press, 2001; J. Gerring and T. C. Strom, 'Do Neoliberal Policies Deter Political Corruption?', *International Organization* 59, 2005, 233–54; E. Neumayer and I. de Soysa, 'Globalization and the Right to Free Association and Collective Bargaining: An Empirical Analysis', *World Development* 34, 2006, 31–49; E. Neumayer and I. de Soysa, 'Globalization, Gender Rights and Forced Labour', *The World Economy* 30, 2007, 1510–35.

41   I. de Soysa and H. M. Binningsbø, 'Devil's Excrement or Social Cement? Oil Wealth and Repression, 1980–2004', *International Social Science Journal* 57 (1), 2009, 21–32; E. Kisangani and W. E. Nafziger, 'The Political Economy of State Terror', *Defence and Peace Economics* 18, 2007, 405–14.

42   I focus below on FDI rather than portfolio capital, which is a tiny part of capital flows to the poorest countries because of underdeveloped capital markets.

43   I. de Soysa, *Foreign Direct Investment, Democracy, and Development: Assessing Contours, Correlates and Concomitants of Globalization*, London: Routledge, 2003; R. O. Keohane and J. S. Nye, 'Introduction', in J. S. Nye and J. D. Donahue (eds), *Governance in a Globalizing World*, Cambridge, MA and Washington, DC: Visions of Governance for the 21st Century and Brookings, 2000; B. Simmons, F. Dobbins and G. Garrett, 'Introduction: The International Diffusion of Liberalism', *International Organization* 60, 2004, 81–810.

44   S. D. Krasner, *Structural Conflict: The Third World Against Global Liberalism*, Berkeley: University of California Press, 1985.

45   C. Clapham, 'The Collapse of Socialist Development in the Third World', in B. Gills and S. Qadir (eds), *Regimes in Crisis: The Post-Socialist Era and the Implications for Development*, London: Zed Books, 1995, pp. 4–17.

46   J. Bhagwati, 'Globalization: Who Gains, Who Loses?', in H. Siebert (ed.), *Globalization and Labor*, Tübingen: Mohr Siebeck, 1999; P. Collier and J. W. Gunning, 'Explaining African Economic Performance', *Journal of Economic Literature* 37, 1999, 64–111; I. de Soysa and J. R. Oneal, 'Boon or Bane? Reassessing the Productivity of Foreign Direct Investment', *American Sociological Review* 64, 1999, 766–82; D. Dollar and A. Kraay, 'Trade, Growth, and Poverty', *World Bank, Development Research Group*, 2000; T. N. Srinivasan and J. Baghwati, 'Outward-Orientation and Development: Are Revisionists Right?', *Yale University and Columbia University* 1999.

47 C. Apodaca, 'Global Economic Patterns and Personal Integrity Rights After the Cold War', *International Studies Quarterly* 45, 2001, 587–602; M. Busse, 'Transnational Corporations and Repression of Political Rights and Civil Liberties: An Empirical Analysis', *Kyklos* 57, 2004, 45–66; D. L. Richards, R. D. Gelleny and D. H. Sacko, 'Money With a Mean Streak? Foreign Economic Penetration and Government Respect for Human Rights in Developing Countries', *International Studies Quarterly* 45, 2001, 219–39.

48 G. Garrett, 'Global Markets and National Politics: Collision Course or Virtuous Circle?', *International Organization* 52, 1998, 787–824; E. Neumayer and I. de Soysa, 'Globalization and the Right to Free Association and Collective Bargaining: An Empirical Analysis', *ISS and NTNU* 2004; 'Trade Openness, Foreign Direct Investment and Child Labor', *World Development* 33, 2005, 43–63.

49 G. Garrett, 'Globalization and Government Spending Around the World', Paper read at Annual Meetings of the American Political Science Association in Atlanta, 1–5 September 1999; D. Yu, 'Free Trade is Green, Protection is Not', *Conservation Biology* 8, 1994, 989–96.

50 K. Barbieri and R. Reuveny, 'Economic Globalization and Civil War', *Journal of Politics* 67, 2005, 1228–47; M. Bussmann, H. Scheuthle and G. Schneider, 'Trade Liberalization and Political Instability in Developing Countries', in R. Trapple (ed.), *Programming for Peace: Computer-Aided Methods for International Conflict Resolution and Prevention*, Dordrecht: Kluwer, 2005; de Soysa, 'Paradise is a Bazaar?'; V. Krause and S. Susumu, 'Causes of Civil War in Asia and Sub-Saharan Africa: A Comparison', *Social Science Quarterly* 86, 2005, 160–77; E. Weede, 'On Political Violence and Its Avoidance', *Acta Politica* 39, 2004, 152–78.

51 B. Russett and J. Oneal, *Triangulating Peace: Democracy, Interdependence, and International Organizations*, London: W. W. Norton and Company, 2000.

52 T. J. Moss, V. Ramachandran and M. K. Shah, 'Is Africa's Skepticism of Foreign Capital Justified? Evidence from East African Firm Survey Data', *Center for Global Development*, 2004.

53 De Soysa and Binningsbø, 'Devil's Excrement or Social Cement?'

54 Moss *et al.*, 'Is Africa's Skepticism of Foreign Capital Justified?'

55 H. Milner and K. Kubota, 'Why the Move to Free Trade? Democracy and Trade Policy in the Developing Countries', *International Organization* 59, 2005, 107–43.

56 J. Bhagwati, 'The Golden Age: From a Sceptical South to a Fearful North', *The World Economy* 20, 1997, 256–89.

57 Institutes such as the Center for Global Development measure the commitment of the rich countries to helping the poor. See CGD Online: www.cgdev.org for their index.

4

A POLITICAL ECOLOGY OF ENVIRONMENTAL SECURITY

Michael J. Watts

A form of biopower emerges that addresses the interplay between freedom and danger: *security*. [It] consists of sets of apparatuses that aim to regulate within reality, because the field of intervention is a series of aleatory events that perpetually escape command. . . . Security can be understood as a break with discipline and an intensification of biopolitics . . . [it] is dispersive . . . [its] object-targets are processes of emergence that may become determinate threats. . . . [As] production extends to all of life, all of life must be secured.[1]

Introduction

The famous NASA planet earth ("whole earth") photograph AS17–22727 was taken during the final Apollo 17 mission in 1972. It is a picturing, or rendering, of a certain sort of global nature, global politics and global science all at once – each rooted in the sense of a profound contemporary crisis in which survival and security, for everyone, is at stake. The NASA image was, of course, a planetary image and served planetary purposes. It came to be the lodestar for the United Nations Convention on the Human Environment held in Stockholm in 1972 and resonated deeply with the famous *Limits to Growth* report – penned by a quartet of MIT physicists, cyberneticians and business management theorists – released in the same year. *Limits* represented the apotheosis of a form of catastrophism thinking driven by a deep Malthusianism. What was on offer was a powerful discourse offering the prospect of chaos and collapse rooted in demographically driven scarcity (the five key sub systems calibrated in their World3 computer model were world population, industrialization, pollution, food production and resource depletion).[2] The language of security – despite the predominant Cold War thinking of the time – was almost wholly invisible. If the global modeling exercise of *Limits to Growth* proved to be flawed in all sorts of ways it nevertheless served as an exemplar of "limits modeling" that reappeared, in different guise, in the general atmospheric circulation models (GCMs) three and four decades later as part of the global warming debate. As they gained standing and analytical power, the new wave of global climate change models – without which there would have been no Montreal or Kyoto Protocols or COP 15 – were also draped in the language of crisis and apocalypse, but now thoroughly immersed in the lexicon of security. If 1972–73 witnessed the global food and oil booms, so too is the current moment framed by a run up in food and oil prices to unprecedented levels, and renewed talk of the "scramble for scarce resources" by the chattering classes and policy wonks. Into the mix was added something new: the spectacular collapse of the investment banks and the bursting of the financial speculative bubble. By 2012 the discourse is less Malthusian (but the dead hand of Malthus has certainly not disappeared entirely) and is now redolent with the language of human security, systemic risk, vulnerability, resilience and adaptive capacity. But as Iain Boal notes both versions – now and then – offer dominant, one might say hegemonic, discourses of secularized neo-catastrophism.[3]

Implicit in the science behind the current global climate change debate is a worldview somewhat at odds with the Darwinian orthodoxy of evolutionary gradualism.[4] Climate could, and has, changed historically but for human occupation and livelihood this represented a deep historical time – the very *longue duree*. What is on offer now is something unimaginable until relatively recently: namely abrupt, radical life-threatening shifts framed in the language of uncertainty, unpredictability and contingency. It is an emergent science of planetary disaster demanding an urgent public response – political, policy, civic and business – of an equal magnitude and gravity. A war on global warming, says Iain Boal, must be declared as draconian as the global war on terror: "are we not faced with inhabiting – once again – the rubble of a ruined world?"[5]

In a discursive sense, then, climate change – to take simply one index of environmental security discourse – is a symptom of a planetary emergency. Global warming encompasses, and has direct consequences for, two of the most fundamental human provisioning systems, food and energy, but to these one can add war, conflict and militarism, critical infrastructures, systemic financial risk all of which are now seen to be inseparably and organically linked in a complex of networks of teleconnected effects.[6] This world view mobilizes and enrolls powerful actors around the threat of massive, catastrophic risks and uncertainties. Central to this vision is a construal of the nature of life itself drawing especially upon the molecular and digital sciences – complexity, networks and information are its avatars – which shapes the nature of what is to be governed and how. If life is constituted through complex and continual adaptation and emergence it rests upon a sense of radical uncertainty in which danger and security form an unstable present, what Dillon and Reid call "the emergency of its emergence," or a life "continuously becoming dangerous."[7] Ash Amin sees this as "the condition of calamity," or catastrophism:

> The recurrence, spread, severity and mutability of the world's natural and social hazards are considered as symptomatic of this state (of permanent risk), and its latent conditions are understood to be too volatile or random and non-linear to permit accurate prediction and evasive action. In the apocalyptic imaginary, hazard and risk erupt as unanticipated emergencies, disarming in every manifestation and in every way.[8]

Three things can be said about this state of emergency.[9] First, while there remain important contrasts between different domains of life, there are nonetheless close family resemblances between environmental and biosecurity threats (such bio-terrorism, emerging illnesses and trans-species epidemics, and more generally affinities with a wider panoply of putative threats to human well-being. For Melinda Cooper these commonalities share deeper state-led entanglements which conflated and drew together molecular research and the informational sciences, speculative finance and war.[10] As a particular ontology, catastrophism and a life of radical uncertainty has produced a distinctive culture of risk (and fear) and risk (and fear) management, what Berlant calls an "actuarial imaginary," an assemblage made up of the institutions, technologies, techniques and ethics, the goal of which is to maximize security, profitability and well-being.[11] Across these domains, as Anderson notes, each threat is potentially catastrophic, it is often vague and spectral, and the disaster is imminent or at least foreseeable.[12] For all such disaster talk, the threat is already present, an incubus that can be discerned and visualized through a series of signs, early warnings and simulations.[13]

Second, the threats to life are imminent but indeterminate in a way that prioritizes what Jane Guyer calls the "near future";[14] anticipatory actions become, as a result, the defining qualities of modern liberalism (everyone, Donald Rumsfeld famously said, needs to be proactive and not reactive, less bureaucratic and more like a venture capitalist). Preemption, precaution and preparations are its key deployments – or political technologies – through which a wide range of crises and challenges are to be confronted.[15] Anderson expresses the matter very well: the question is how to protect certain forms of valued life that revolve around a future – uncertain, life-threatening and full of surprises – that diverges from both the past and present. What constitutes governance and action "in the here and now before the full occurrence of a threat or danger"?[16]

The third and related point is that discursively the contemporary crisis is held together by – more properly constituted by – the language of security and by processes of securitization.[17] The hegemony of "security talk" is evident in the extent to which virtually any expression of public concern – energy, terror, finance, environment, food, inequality, globalization, health, pandemics – is now attached to the moniker of security. The extraordinary capaciousness of security as an analytic and as a policy frame reached, in one regard, its final apotheosis in the drawing together of three keywords of contemporary modernity – globalization, security and development – in the 1994 *Human Development Report*.[18] Coeval with the rise of neoliberalism and the *privatization* of everything is the long march of militarized governmentality and the *securitization* of everything.

In this chapter I shall explore environmental security placed on this larger landscape of security and securitization but from the conceptual vantage point of political ecology. Specifically, I want to identify the political ecological toolkit, to identify its distinctive analytics in relation to the other approaches, and to provide some brief examples of what a political ecology might offer. The primary purpose of this chapter is not to review the now vast and complex set of interlocking literatures that operate under the sign of environmental security;[19] neither shall I spend time rehearsing the debates between political ecologists and some of the foundational works in environmental security.[20] My brief is to provide a concise account of how contemporary political ecology might frame and interrogate the overlapping ground of environmental security, human security and global environmental change.

One immediate concern is to acknowledge the challenges posed by the breadth and diversity (and contentiousness) of what now passes as security and environmental security.

Not only has security, narrowly construed as state military security, been extended to new substantive domains (energy, finance, disease and so on) but there has too been a sort of disciplinary and conceptual promiscuity in the ways in which questions are now treated. Environmental security is now resolutely multidisciplinary and conceptually eclectic (Jacques Derrida, Paul Collier, Chuck Holling and Amartya Sen can now occupy the same page and the same conceptual rubric) while at the same time ambitious claims have been made as regards powerful integrative approaches (resiliency, vulnerability, livelihood strategies) to environmental security. These centripetal and centrifugal tendencies make generalization difficult; there is much border crossing and many instances of what Clifford Geertz once called "blurred genres."[21] At risk of mischaracterization of what is a big and baggy field, let me suggest that there are three broad sorts of substantive concern within environmental security:

- causal analysis of the relations between environmental and biophysical change and forms of human conflict (inter-state and civil). This work often turns on the logics of resource scarcity and/or abundance, intervening variables (often the state) and forms of conflict. Work by environmental conflicts theorists associated with the Toronto and Zurich groups[22] and others meet up with research on the logic of civil war and the violent politics of resource-dependency (resource curses and the like) emanating from political science and economics,[23] and with a very different intellectual trajectory and genealogy.
- processes of securitization/desecuritization most associated with the Copenhagen School[24] which explores how some putative ecological threat becomes a matter of emergency politics or security/threat as a result of the actions of powerful securitizing actors. Any form of security is self-referential; it turns on questions of performative speech acts (the securitizing move) and discursive legitimation (complete securitization). The meaning of security in its various forms is a function of what it does. Debates surrounding this post-structural realism and the wider questions of meaning, genealogy and power–knowledge–institutions are central.[25]
- the dialectical intersections of globalization and global environmental change – "double exposure"[26] – and their relation to both analytical (vulnerability, contextual environment, exposure) and normative questions (resiliency, adaptive capacity, precaution).[27] The effect of this work is to place an enhanced sense of security on a larger canvas of global transformations which necessarily links environmental and other securities to the development problems of the global south and simultaneously raises pressing conceptual and methodological questions pertaining to measurement, scale and complex interactive processes.

Running across this trio of substantive concerns are a number of crossing-cutting analytics. One explores the specificities and dynamics of differing security sectors (food and energy say, versus finance and critical infrastructures) the purposes of which is to think about how substantive difference matters analytically. Another is theoretic and conceptual: how to best study causally complex political, social, economic and cultural dynamics of conflict; how to integrate scale and level of analysis under the rubric of globalization; state-centric versus civil society-centric approaches; tensions between normative and analytic narratives; how processes, outcomes and responses interact in "complex systems." And not least there is the growing dominance of resiliency thinking as a way of providing a sort of unified field theory capable of integrating theory and practice. Much of this has in the environmental security arena been associated with the mitigation and adaptation debates over global climate change[28] but resiliency talk is now the primary discourse which covers a multitude of security discourses and arguably provides the master-narrative for thinking about human security at large.[29] Resiliency is always joined to other keywords: vulnerability, exposure, adaptation, adaptive capacity and more generally to the structure of living, self-organizing complex systems.[30]

I want to make two broad sorts of argument. The first is that political ecology in its founding started from a more-or-less Marxist analysis of political economy in which the social relations of production, access to and control over resources, and power relations rooted in the state and capital, and the dynamics of specific historical forms of capitalist accumulation were its central starting points. Its object of critique was, in fact, ecological anthropology and cultural ecology which seemed functionalist, collapsed the social into the ecological and deployed crude notions of adaptation, and Malthusianism ("population pressure"). Both approaches largely ignored the expanding powers of commodification (taken from Karl Polanyi) and the circuits of capital (taken from Karl Marx). While much analysis in environmental security claims to have integrated these insights[31] in the use of concepts like vulnerability, exposure and global processes, it has in fact been smuggled back in many of the earlier living systems and systems theoretic notions so roundly rejected by political ecology: that is to say, some of the historical strengths – a historical and spatially rooted analysis of social relation of production and exchange and the struggles and contestations over access to and control over resources, and an attentiveness to the specificities and materialities of differing ecologies and resources (tropical forest are not Sahelian rangelands, oil is materially different from cotton or coal) – have fallen out of environmental security or at least are radically diluted. The result is that much of what passes as the analysis of individual or household or community vulnerability or insecure livelihoods is not attentive enough to the rhythms of accumulation, the exercise of power and historical trajectories of place and space (which typically emerge from historical and ethnographic research).

The second is that environmental and other securities can be productively located on the larger canvas of biopolitical security derived from Foucault's discussion of biopower, capitalism and liberal rule.[32] Biopower encompasses a variety of forms of governing associated with the modern technologies of risk and threat management. Foucault asks the question: what is the principle of calculation for the manufacture of liberal freedom?[33] The principle he says is *security*. In our epoch, biosecurity, or more properly biopolitical security, deploys a vast array of technologies of risk. From the perspective of contemporary neoliberalism and its technologies of rule, there is a close affinity between resiliency thinking and a vision of how a spontaneous market order will be built from and out of an individual and community self-making and self-regulation through means of calculation and commodification shaped by their own peculiar exposure to the necessary and unavoidable contingencies of life. As Anderson properly notes "through neoliberal logics of governing, the contingency of life has become a source of threat and opportunity, danger and profit".[34] The strength of political ecology I argue is that it attempts to link its fundamental focus on regimes of accumulation (in our time the dynamics of neoliberalism) with forms of contemporary governing and the operations of liberal governmentality.

What might a critical political ecology consist of?

Political ecology emerged in the 1970s – at a moment of what might be called a peasants studies boom and a reconsideration of Marxist theory – as a critique of two disciplinary approaches to the study of human relations with the environment: namely, ecological anthropology and cultural ecology and natural hazards research.[35] Both privileged the role of culture (broadly defined) as a set of means by which human population adapted to the ecosystems (or living systems) of which they were part. Both approaches saw manifestations of culture – religion, ritual, local perceptions and knowledge – as functionally adequate for ecosystem stability (Rappaport's famous argument that Tembaga-Maring ritual served a sort of thermostat to regulate human and pig populations in highland Papua New Guinea is a paradigmatic case[36]). Political ecology conversely – drawing inspiration from the study of peasant societies undergoing important economic and cultural transformations, and from Marxian political economy focused especially on the role of the state in post-colonial development – was especially shaped by the agrarian question: that is to say, (to quote Karl Kautsky[37]) the forms in which capital was taking hold of and transforming agriculture. Much of this work was focused on the global south in largely agrarian (and peasant) societies – the emphasis was typically on community based field study –in the throes of what Karl Polanyi called the "great transformation."[38] Central to political ecology was political economy: rather than examining the functional adequacy of culture or social structure, political ecology started with the relation of producers to the market, the commodification of land and labor, the forms of surplus extraction and the prismatic forms of social differentiation with peasant communities, the breakdown of the moral economy, emerging forms of class structure and the changing relations of production.[39] Rather than seeing environmental questions through the prism of society *and* nature or human action and biophysical effect, political ecology drawing on Marxist ideas of the labor process and notions of first and second nature, saw nature and society as dialectically constituted.[40] Equally, environment was not some pre-given but was an object that could be construed in different ways by different communities and classes (this was an echo of Rappaport's cognized and perceived environmental models).

Political ecology questioned the emphasis on adaptation, on what appeared often to be a heavy dependence on systems theory and homeostatic processes, and the idealism of well-adapted societies living in harmony with their environment. Rather, it turned the torch toward the dynamic commercialization of agrarian societies, how communities were being torn asunder and social structures and culture retooled and redeployed under the conditions of development and the post-colonial state.[41] Peasant and other communities were being differentiated in complex ways. By focusing on patterns of accumulation, access to and control over resources, and changing class structure, political ecology could demonstrate that some individuals and households were rendered marginal (to their resource base) and made vulnerable to anticipated and unanticipated environmental processes in new ways. Small farmers might be degrading their environment because they had no choice; forests were destroyed as a desperate strategy to establish property rights in areas where the rule of law was lacking; peasants worked harder and longer, often degrading their land, in order to ensure social reproduction in the face of price squeezes. In short, this political ecology had Marx as its reference point and might be framed by what I would call *regimes of accumulation*: how the political economy of social agents (peasants, pastoralists, landlords and workers) were shaped and limited in their capacity to manage their resource and regulate their environment by the social relations of production and exchange.

In the 1990s and thereafter this early political ecology was broadened in two ways: to put it crudely, to Marx and regimes of accumulation were added Foucault and *regimes of truth*,[42] and Gramsci (and Foucault) and *regimes of rule/hegemony*.[43] This shift was partly a result of changing intellectual fashion (the growing influence of forms of post-structuralism), partly a function of cross-fertilization with others' fields (science studies, race theory, environmental history, green justice), partly a function of the interest of deploying political ecology in industrial and advanced capitalist settings (rather than the world of peasants in the global south), and not least because of the blind spots within the Marxian optic. One early development was driven by the realization that political ecology was relatively silent on the forms and dynamics of political contention surrounding the environment. Environmental movements, the role of civil society, and later armed struggle (militant struggles over forests or oil) pushed political ecology to expand and deepen its understandings of the operations of power. Not surprisingly the knowledge–power–institutions nexus, drawing especially from post-structural and discourse analysis, was taken up quickly. Careful examinations were made of forms of environmental expertise, how institutions like the World Bank were "greened," how conventional models of environmental degradation (tragedy of the commons) constructed referent objects in particular ways with consequences for policy, and especially a focus on forms of green governance (for example understanding the effects of decentralized governance on forest regulation or common property institutions, and the politics of differing management regimes).

Following Foucault I have dubbed this line of reasoning as regimes of truth. Inevitably there was a sort of spill over to broader questions of rule and to the relations between the Marxian concerns with capital and class (regimes of accumulation) and Gramscian focus on hegemony, coercion and consent; that is to say, why some ideas and practices became dominant, and how subalterns or oppressed groups might build counter-hegemonic or positional practices and centers of power. A key question here naturally is the relation between identity and the environment and particular sorts of green subjectivities. Arun Agrawal in his book *Environmentality*, observed the tendency over time of communities who self-govern resources to change their attitudes about the environment, and slowly internalize responsibility for governing nature.[44] In his case research in India, communities that long opposed colonial and government control of forests – going so far as to set them on fire rather than see them dominated by an external sovereign state – eventually come to fully accept the mantle of protecting forests on behalf of the state, changing the way they think about forests, but also their own views about themselves. Agrawal refers to this transformation as a case of "environmentality" where people's identities, activities and attitudes come to internalize previously external norms or mandates.

If political ecology has sought to keep these three regimes (or perspectives) in a sort of dynamic tension, it is also true that it contained certain predispositions of a methodological sort. One was its predilection for local and historically rooted ethnographic and field based research (whether the objects of ethnographic study were farming communities in Ecuador, the World Bank environment program, or the Wise Use movement in the US southwest). In part this reflected a growing sensitivity to power, territoriality and green governance (the scale question) and to the fact that political ecology was not solely concerned with how the environment was socially constructed but took very seriously the site specificities and material particularities of local environments. An attentiveness to the material and biophysical qualities of resources and ecosystems and how their properties matter analytically has been a hallmark of the field.

A political ecology capable of maintaining fidelity to Marx, Foucault and Gramsci is no small order of course, and there remains much to be done to further and deepen such a project around for example the subject of environmental security. There are nevertheless a number of works that capture what such an endeavor might entail. Li has explored these questions through the lens of environmental rule in Indonesia, exploring the power of a liberal "will to improve," understood as a two century long project to secure the welfare of populations, but rooted in a historically complex situation of government practice, operating within the jagged rhythms of capitalist accumulation.[45] Li is especially concerned with the ways government programmers draw boundaries around, and "render technical," aspects of landscape, conservation and livelihood. Simultaneously, she demonstrates how these practices have limits, imposed by the contradictions between improvement and sovereign power, and between the rationalities and practices of government and their ability to actually regulate dynamic social relations. These open up the terrain of "contestation and debate between people with different interests and claims".[46] Focusing explicitly on politics surrounding a national park, she has shown how technical conservation efforts served to screen out marginal households among recipient communities, a process which produced limited development benefits and encouraged community radicalization. In one case a Free Farmers Forum emerges from a century of failed improvement; in another highland villagers reject the park and efforts by the Nature Conservancy. In all of this, local politics turns on the contradictions of a form of rule – trusteeship – in which agents with power are ultimately unprepared to relinquish their authority, however much it is draped in the rhetoric and discourse of participation and empowerment.

Jake Kosek's book *Understories* provides another illustration of how forests (in this case the US southwest) are classified, organized and ruled in a way that is intended to produce particular sorts of subjects (including Smokey the Bear) and property relations.[47] Yet, at the very moment that forests are declining as local sources of revenue and employment, they become the basis for powerful (yet different) sorts of insurgent consciousness and practice among both Hispanic and white rancher communities. Kosek's new work on the bee crisis points to all of these issues at work, and to the deep complexity of science, political economy, politics and context. He shows how the current state of the honeybee is undeniably dismal, experiencing considerable decline in populations even before recent reports of "colony collapse disorder."[48] Global environmental changes have been devastating, whether the intensification of industrial agriculture, toxic pollution, climate change, loss of habitat or the spread of disease and parasites. Few researchers, however, pose the more fundamental question: how has the modern bee come into existence in a way that has made it vulnerable to new threats? In fact, the largest funding for bee research and bio-engineering during the Bush administration was by military intelligence and weapons research agencies who hope to harness and develop bees' abilities as part of the "war on terror."

It needs to be reiterated that political ecology from its very inception spoke directly to those questions that currently animate the environmental security/global environmental change field: livelihoods, poverty, inequality, marginalization, vulnerability, exposure. These questions are certainly central to what is arguably the most pressing environmental security question now, namely global climate change. But two things need to be made clear: first, it was not the existence of poverty or inequity as such that distinguished political ecology as much as its Marxist toolkit and its focus on the social relations of production and exploitation. In much of the environmental security literature what appears is a shopping list of factors and the constant invocation of the most ambiguous if meaningless word in the English language: context. The sort of hard-edged political economy has dropped out completely. Second, political ecology was from the get-go deeply critical of the epistemological and analytical legacies of folding the study of environmental change into a systems-theoretic biologically derived theory. Adaptation, adjustment and adaptive capacity were deeply problematic. This raises then how and in what ways the current environmental security approach to global environmental change, which is replete with the language of adaptation, has smuggled back in what political ecology rejected. Pelling has properly noted that adaptation is deployed in a number of ways and some (the IGPCC work for example) seems little different from the work on natural hazards during the 1960s.[49] But the vulnerability literature and the development of so-called social resiliency (moving resiliency from ecology to social science) is, as Bassett has suggested, largely bereft of critical social science.[50] So much of the socio-ecological systems (SES) and resiliency work as it accounts for social and institutional dynamics is utterly anodyne and bland; what I have called regimes of accumulation is utterly absent;[51] the same might be said of the studies of political dynamics and institutional change and adaptive governance[52] and globalization.[53] One can endorse the view that so-called transformational adaptation[54] and human security framing of vulnerability have affinities with political ecology;[55] there are clearly points of tension and difference. At the very least, there needs to be a more serious engagement with critical social science and differing theoretical traditions which generate quite different conceptual apparatuses, meanings and framings and therefore policy prescriptions.

A political ecology of environmental security: risk, neoliberalism and the actuarial imaginary

> I think the multiplication of the enterprise form within the social body is what is at stake in neoliberal policy.... The stake of all neoliberal analyses is the replacement every time of *homo economicus* as partner of exchange with

a *homo economicus* as entrepreneur of himself, being for himself his own capital, being for himself his own producer. . . . The individual's life itself – with his relationships to his private property, for example, with his family, his household, insurance and retirement – must make him into a sort of permanent and multiple enterprise.[56]

The influential Copenhagen School, especially the work of Wæver,[57] have traced the way in which a core logic of security has expanded its referent objects through securitization and has, as a result, widened its circumference from the modern concept of security (modeled in accordance with *raison d'état* and necessity) consolidated in the 1940s. Questions of threat and survival become the basis for assigning absolute political priority without conventional constraint.[58] This approach, which is at heart discursive, draws upon a strange (and on its face highly contradictory) group of theoretical bedfellows: Austin, Derrida, Schmitt and Waltz among them. It is however odd that the work of Foucault does not figure more centrally in this project. After all, it was Foucault who made the point that what distinguished modern rule (over the last four centuries) was a new form of biopower that links freedom and danger, namely *security*. Historically, there have been for Foucault two great *dispositifs* of security. One turns on sovereign territoriality and is customarily thought of as the geopolitics of security. The other as we have seen turns on the problematic of life and populations, this is the biopolitics of security. Both have coexisted, and do coexist, but as Dillon notes, once life and survival is made the principle of formation around which the problematization of security, fear and danger revolves, then the politics of security are transformed.[59]

Biopolitics in the Foucauldian sense points of course to both a rather different way of thinking about environmental security and about nature or life more generally, both historically and contemporaneously. Nature – or perhaps more appropriately the world of living things – has always provided the ground and substance for economics and politics and in this sense the relations between environment and governance or forms of rule have a long history.[60] But it has also provided, as Michel Foucault shows, a way of linking security and the apparatuses and rationalities of modern government in the context of the growth of a capitalist economy.[61] Biopower refers to the strategic coordination across a number of forces that make up life or living beings, and represents a distinctively new raft of modern governmental techniques and technologies of power. Biopower entails the rendering of life or living people as an object of regulation, and it rests upon forms of intervention aimed at optimizing life against particular threats ("making life live").

It goes without saying that what passes as life or a living thing (and what is necessary to support it) is a historical question[62] – life is, as it were, a project rather than an accomplishment. Life as a living system and as an emergent system characterized by self-organization – a vision central to our contemporary sense of the life sciences – is central to the constitution of the biopolitical securitization.[63] Biopolitics, in other words, changes in relation to the forms in which life processes are, as Dillon has shown brilliantly in his work,[64] made transparent to knowledge. If life as a biological phenomenon is the object of biopolitical security technologies then to the same degree the technologies – now dominated by the technologies of risk and the wider actuarial imaginary – are constituted by, and regulated through, the historical forms of circulation of life (what Foucault calls *être biologique*). Currently the modalities through which life reproduces itself are trans-actional, complex, non-linear and combinatorial.[65] Dillon (2007) makes the important point that it is the idea of radical contingency which stands at the heart of this view of life.

Foucault originally drew a distinction between two political technologies of biopower: discipline (an anatomopolitics focused on the body and deployed in institutions like the prison) and biopolitics (especially a state biopolitics applied to humans as living species). In *Security, Territory, Population* Foucault explored how early biopolitics revolved around the aleatory properties and processes of populations which were normalized through forms of classification, ordering, testing.[66] In this way force is brought to bear less through direct discipline than in "regulating overall conditions of life and naming threats to the balance or equilibrium of life".[67]

Just as populations are aleatory, however, so contingency is said now to be generically constitutive of life as biological existence. As contemporary life scientists offer a view of life as incalculable, non-algorithmic and beyond our capacity to predict, so the biosphere is, in this account, constructed by the emergence and persistent co-evolution of autonomous agents. The science of complexity as much as the life sciences *per se* focus now on the emergent properties of living systems. Adaptation within these living systems is unpredictable, self-organizing and self-engendering. Against this backdrop of the molecular and digital revolutions, biopolitics cannot be achieved solely through the avalanche of statistics; neither is the homeostatic goal of biopower possible or even desirable. As Anderson puts it, a productive life can only be secured in relation to threat and danger that resides within life, and securing life must involve a creative relation to the contingent (the free, self-governing subject must be open to and embrace the radical uncertainty of life).[68] Modern life is, says Dillon, governed by laws of emergence and what he calls a distinctive moral and behavioral economy of existence:

> Biopolitically speaking, at the beginning of the 21st century, biological being as emergent being is enjoined to secure itself through securing its future by experimental participation in the engendering and unleashing of its own emergent potential. While allied to other ways of taming chance, risk technologies are also now deeply implicated in this novel biopoliticised securing of the life of emergent entities.[69]

In epistemological terms, contingency is the defining quality of human life which biopolitics is committed to securing. As a consequence, biopolitical security can only be meaningfully performed *through* contingency not apart *from* it, which is to say that biopolitical security will be conducted through shaping our exposure to, and the creative exploitation of, contingent events and processes in nature and from the "independent actions and interventions of biological being itself".[70] As we shall see, these forms of intervention are complemented in our time by other forms associated with high neoliberalism which are consistent with the vision of biological being as emergent, namely the vision of an economy as a living, self-organizing and self-correcting system.[71]

Radical contingency as a condition of existence and as a form of prescription suggests a new form of biopower that links freedom and danger, namely security. Modern biopolitics of security cannot be formed by safety and protection as such but by cultivating the very principles of adaptive emergence. Contingency and transformation are the means to survive and be safe, or more properly qualitative change in the nature of the living thing itself is the very condition of possibility of security. It is this notion that aleatory events are now perpetually beyond any sense of control – that the objects of security are seemingly an infinite horizon of emergent threats, a sort of paranoid world of permanent emergency – which leads Anderson to suggest that security represents now something quite new: as he puts it "security can be understood as a break with discipline and an *intensification* of biopolitics".[72] Security poses the prospect of interventions extending to all of life in order to make life live for the market. The question is what technologies of risk management constitute such a security strategy? And what sort of economy provides the "basic grid of intelligibility" for this contemporary historical form of the bio-securitization of politics? If as Foucault suggests, neoliberalism is "a whole way of being and thinking," a "grid of economic and sociological analysis," what is its relation to the governing of contingency?[73]

One part of this story – the relation between contingency and liberalism – is well understood.[74] Risk calculation and uncertainty in the nineteenth century represented not simply something ultimately incalculable but a *dispositif de sécurité* focused on foresight, contract, prudence and enterprise. In our time risk operationalizes the biopolitics of emergent life through the commodification of contingency, which is taken to be constitutive of what it is to be a living, emergent, transactional being.[75] Risk is by definition about the probability of the unpredictable: it speaks to danger but also market opportunity and source of profit by making contingency a fungible commodity. Risk may invoke a device – insurance or underwriting – for offsetting the consequences of injury but it has come to encompass a much larger social and economic ground, indeed it underwrites many orders of governance. It is, in Gramscian terms, the hegemonic practice governing "the conduct of conduct," it is now an ontology, a way of being. Risks multiply and circulate; they are monetized and securitized in a veritable avalanche of forms and norms. Risk technologies which are in the business of underwriting exposure to contingency are central mechanisms to self-governing – that is to say self-making and self-regulating subjects saturated in, and bound by, a world of calculation and commodification determined by their peculiar profile of exposure to the necessary and unavoidable contingencies of life. Put rather differently, risk is about the governance of threat in an uncertain near future: it speaks to anticipatory action in a contingent world. Anderson notes that this anticipatory governance – the management of risks – is rooted in the calculative and performative actions of preparedness, preemption and precaution;[76] that is, stopping events before they happen or reach points of irreversibility and/or halting the effects of potential threats and disruptions, all in the name of care and safety. But risk as universal account has come to mean that risk can itself become a risk, potentially producing its own negation.

Risk has become an assemblage of universal account. If security is a form of biopower it leads inexorably to Foucault's inquiry on the relations between government and the transformations in the circuits of capital. Naturally this takes us deep into the heart of neoliberalism – and Foucault's astonishingly prescient lectures on German ordo-liberalism and its aftermath[77] – and the twin processes of financialization and securitization associated with the rise of what is called money-manager capitalism. What lies behind the rise of finance capital as a dominant force within capitalism (led by the US restructuring of the

1970s) is neoliberalism as a whole way of life – or as Hayek said of seeing liberalism as "a living thought," as a "utopia." It necessitates generalizing the enterprise forms to the entire social fabric, the economization of the entire social field;[78] competition is the measure or norm for all of life while the enterprise form is the model for living a life (i.e. *homo economicus*). If the neoliberal economy as a self-organizing and self-correcting living system composed of individuals whose life is "a sort of permanent and multiple enterprise",[79] it is rendered actionable, says Anderson, by "environmental technologies."[80] These technologies address not so much capabilities through discipline or the management of populations as the future-oriented operating environment, the rules of the game rather than the players. That is to say markets and competition as a way of grasping the future or the event before it occurs. This is what is contained in the transformation of risk from a management device to a universal system of account and an entire order of governance. The "actuarial imaginary" constitutes an assemblage made up of the institutions, technologies, techniques and ethics the goal of which is to maximize (future) security, profitability and well-being. But if security is to be located within the circumference of a normalized prudentialism resting on the individualized and dispersed risk regimes of the neo-liberal order, there is much to suggest, to return to Foucault, that life has not so much been integrated into forms of governance as it has escaped from them.[81] The governance of risk itself seems to be in crisis in the face of a triple crunch: the meltdown of the financial sector so central to neo-liberal accumulation strategies over the last three decades, depleting strategic resources and global climate change that imperils both accumulation and civilization itself through a "slow death."

Climate change and resiliency

In my previous work I have tried to explore the intersections of security and neoliberalism in the Sahel region of Africa which has again (recall that the Sahel was the site of much Malthusian thinking about drought-famine in the 1970s) become a laboratory for the study of climate (now global climate change) and security (the slow death of food insecurity and famine).[82] The new biosecurity here is the language of adaptation to climate change and the resiliency of socioecological systems, what Rose in another setting has called "government through community."[83] The origins of the resiliency work lay in the 1970s with the work of Holling, which attempted to locate the equilibrium-centered work of systems ecology on the larger landscape of the biosphere as a self-organizing and nonlinear complex system. Complexity science – the hallmark of contemporary systems ecology – represents a meeting point of several multidisciplinary strands of science, including computational theory, non-equilibrium thermodynamics, evolutionary theory and earth systems science.[84] At the heart of Holling's early work was how systems retain cohesiveness under stress or radical perturbations (such as climate variation). Resilience determined the persistence of relations in a system. In his later work, he explored the implications for management of ecosystems by emphasizing less stability than the unpredictable and unknowable nature of complex system interdependencies, which implied (in policy terms) a need to "keep options open, the need to view events in a regional rather than a local context, and the need to emphasize heterogeneity".[85] Through his Resilience Alliance and later the Resilience Center, resilience and adaptive systems thinking were pushed far beyond ecology to encompass a co-evolutionary theory of societies and ecosystems as a single science ("panarchy").[86] Holling extended his view of resiliency by suggesting that all living systems evolved through disequilibrium, that instability was the source of creativity: crisis tendencies were constitutive of complex adaptive systems. Indeed, resiliency is now so central to the notion of environmentally sustainable development – the cornerstone of the major multilateral development and international environmental NGOs – that the complex adaptive systems framework (including the sorts of measures of standardization and accounting for assessing ecosystem resilience) has been taken up by the likes of the World Bank and UN HABITAT in such diverse arenas as sustainable urbanism, ecosystem services, and climate adaptation and mitigation.[87] A key policy document, *Roots of Resilience*– bearing the imprimatur of UNEP, the World Bank and the World Resources Institute – reads:

> Resilience is the capacity to adapt and to thrive in the face of challenge. This report contends that when the poor successfully (and sustainably) scale-up ecosystem-based enterprises, their resilience can increase in three dimensions. They can become more economically resilient – better able to face economic risks. They – and their communities – can become more socially resilient – better able to work together for mutual benefit. And the ecosystems they live in can become more biologically resilient – more productive and stable.[88]

Resiliency is the form of governmentality appropriate to *any* form of perturbation and uncertainty: dealing with extreme weather events is not merely analogous to coping with recurrent financial shocks. It provides the means through which economics and social resilience is to be achieved.

At the time that Holling was laying out his first ideas (and in the midst of the Sahelian famine in Africa), Friedrich Hayek delivered his Nobel Prize speech, which, as Cooper and Walker brilliantly show, has an elective affinity with Holling's ideas.[89] Hayek was moving toward his mature theorization of capitalism as an exemplar of the biological sciences: the extended market order is "perfectly natural . . . like biological phenomena, evolved in the course of natural selection".[90] In his Nobel lecture, he returned to the epistemology of limited knowledge and uncertain future, a position which led him to explicitly reject and denounce the Club of Rome *Limits to Growth* report. It was to biological systems and complex, adaptive, and nonlinear dynamics that he turned to provide the guide for his "spontaneous market order" of capitalism. Both endorse the view of limited knowledge, unpredictable environments and order through survival.

The notion of adaptive capacity and government through communities does of course rest on a substantial body of research which demonstrates how rural communities in Africa (and elsewhere) adapt to climate change through mobility, storage, diversification, communal pooling and exchange by drawing on social networks and their access to resources.[91] Yet what is on offer instead is a bland and bloodless shopping list of "conditions" for adaptive governance, including "policy will," "coordination of stakeholders," "science," "common goals" and "creativity." *Roots of Resilience* offers a full-blown theory of development ("growing the wealth of the poor") in the same way that socioecological governance and adaptive, flexible climate risk management offers a form of neoliberal governmentality for climate change in Africa. The new post-Washington Consensus looks like it is even more consistent with Hayek's mature vision of the market order than its earlier raw and crude neoliberal version.[92] *Roots of Resilience* proposes to scale up "nature based income and culturing resilience," which require ownership, capacity and connection. Ecosystem-based enterprises, rooted in community resource management, will entail local–state and private–civic partnerships and enterprise networking. Markets in ecosystem services, and delegation of responsibility to communities and households as self-organizing productive units, will constitute the basis for survival in biophysical, political, economic and financial worlds defined by turbulence, risk and unpredictability. Some will be resilient, but others will be too resilient or not resilient enough.

We are in the world of catastrophic events, thresholds of survival, and maladaptation. Ecological resiliency is the calculative metric for a brave new world of turbulent capitalism and the global economic order and a new ecology of rule. The Sahel's "bottom billion" provide a laboratory in which the poor will be tested as the impacts of change manifest. Resiliency has become a litmus test of the right to survive in the global order of things.[93] To return to Foucault and his notion of an expanded sense of biosecurity, resiliency is an apparatus of security that will determine the process of "letting die." Africa, once again, is the testing ground for a vision of security and care in which life is nothing more than permanent readiness and flexible adaptiveness. As such, it is a deeply Hayekian project – an expression of the neoliberal thought collective – in which the idea of a spontaneous market order has become, ironically, a form of sustainable development. Building resilient peasants and resilient communities in West African Sahel turns on an amalgam of institutions and practices geared toward individuals armed with traditional knowledge but rooted, to return to Dillon in a distinctive moral and behavioral economy of existence.[94] Adaptation in these systems is unpredictable, self-organizing and self-engendering – consistent with life itself – but consistent with a sort of Hayekian economic subject and a spontaneous economic order.[95] The challenges of adapting to the radical uncertainties and perturbations of global climate change, produce a new sense of *homo economicus*; the African peasant as Foucault says becomes "an entrepreneur of himself," a hedge-fund manager for his own impoverished life.[96]

Conclusion

From the vantage point of political ecology, then, a judicious mix of Marx and Foucault would place environmental security on the larger canvas of modern forms of biopower. This builds upon Foucault's own account – his deep history – of *dispositifs* of security in which security (either geopolitical or biopolitical) is linked to modern forms of rule and govermentality. In doing so, environmental security is situated with respect to the *longue durée* of forms of life and it sees the securitization process as described by the Copenhagen School as part of a wider logic of neoliberal rule. In addition, while Foucault was only tangentially concerned with the emergence of capitalism as such and the circuits of merchant, industrial and finance capital, his lectures on biopower point to the need to think about the norms and forms of neoliberal capitalism, not simply as discourses but as actually existing forms or varieties of capitalism. Security, risk management, resiliency provide the contemporary hegemonic language and practice in which particular forms of life constitute the basis of neoliberal rule and governance. While the rise and dominance of a certain sort of national security state has been key to this neoliberal logic – Masco's work analyzing biosecurity and American empire makes this point[97] – it also extends beyond it (hence Foucault's comments on civil society and neoliberalism). At its most ambitious political ecology – in contradistinction to the quarter of theoreticians on whom the Copenhagen School relies – tries to weave together Marx, Foucault and Gramsci as *its* conceptual triumvirate.

Political ecology is – not unlike environmental security – a dynamic, evolving and expansive approach to thinking about the political economy of the environment. The thrust of my remarks is to try and retain some of the founding conceptual architecture – and its critical Marxist orientation – while incorporating ideas and concepts from social and political theory. In this way, the points of tension and difference between political ecology and other approaches is hopefully clear – and to that extent what a political ecology of environmental security offers. This account is necessarily theoretical and abstract and political ecologies historic strength and distinctiveness has always resided, as Matthew et al. note, in its "detailed and contextual approach"[98] – often employing the sorts of extended cases study method rich in historical and geographical specificity of the sort outlined by Michael Burawoy.[99]

One can, nevertheless, point to several examples of the political ecology of environmental security. Peluso and Vandergeest's pathbreaking work on forests and counter-insurgency;[100] Bohle and Funfgeld"s analysis of the political ecology of violence in Sri Lanka;[101] and finally my own work on energy security in the Gulf of Mexico[102] are suggestive of how for example questions of accumulation and dispossession, nation-state building, and liberal governmentality might be powerfully conjoined in the analysis of biosecurity.

Notes

1 B. Anderson. 2010. Affect and biopower. *Transactions of the Institute of British Geographers* 37, 28–43: 34, my emphasis.
2 The debate continues. Donnella Meadows, Jørgen Randers and Dennis Meadows updated and expanded the original version in *Beyond the Limits* in 1993 – a twenty-year update on the original material. The most recent updated version was published in 2004 by Chelsea Green Publishing Company and Earthscan under the name *Limits to Growth: The 30-Year Update*. In 2008 Graham Turner at the Commonwealth Scientific and Industrial Research Organization (CSIRO) in Australia published a Working Paper entitled "A Comparison of 'The Limits to Growth' with Thirty Years of Reality" which examined the past thirty years of reality with the predictions made in 1972. He found that changes in industrial production, food production and pollution are all in line with the book's predictions of "economic and societal collapse in the 21st century" (see www.csiro.au/files/files/plje.pdf, accessed August 2012).
3 I. Boal. 2009. *Globe, Capital, Climate*. Berkeley, CA: RETORT: 3.

4 S. Weart. 2004. *The Discovery of Global Warming*. Cambridge, MA: Harvard University Press; Boal, *Globe, Capital, Climate*.

5 Boal, *Globe, Capital, Climate*: 5.

6 OECD. 2003. *Emerging Risks in the 21st Century*. Paris: Organization for Economic Cooperation and Development. Online: www.oecd.org/dataoecd/20/23/37944611.pdf (accessed August 2012); World Economic Forum. 2012. *Global Risks*. World Economic Forum: Davos. Online: www.weforum.org/reports/global-risks-2012-seventh-edition (accessed August 2012).

7 M. Dillon and J. Reid, J. 2009. *The Liberal Way of War*. London: Routledge: 85.

8 A. Amin. 2012. *Land of Strangers*. London: Polity: 138.

9 G. Agamben. 1998. *Homo Sacer: Sovereign Power and Bare Life*. Stanford, CA: Stanford University Press.

10 M. Cooper. 2008. *Life as Surplus*. Seattle, WA: University of Washington Press.

11 L. Berlant. 2007. Slow death (sovereignty, obesity, lateral agency). *Critical Inquiry* 33/7, 762.

12 B. Anderson. 2010. Preemption, precaution, preparedness. *Progress in Human Geography* 34/6, 777–98: 779–80.

13 Ibid.

14 J. Guyer. 2009. Prophecy and the near future. *American Ethnologist* 34/3, 409–21.

15 Anderson, Preemption, precaution, preparedness; Amin, *Land of Strangers*.

16 Anderson, Preemption, precaution, preparedness: 780)

17 R. Floyd. 2010. *Security and the Environment: Securitisation Theory and US Environmental Security Policy*. Cambridge: Cambridge University Press.

18 R. Matthew, J. Barnett, B. McDonald and K. O'Brien (eds). 2010. *Global Environmental Change and Human Security*. Cambridge, MA: MIT Press.

19 S. Dalby. 2009. *Security and Environmental Change*. London: Polity; J. Barnett. 2001. *The Meaning of Environmental Security*. London: Zed Press; R. Leichenko and K. O'Brien. 2008. *Environmental Change and Globalization*. Cambridge: Cambridge University Press.

20 See, for example, N. Peluso and M. Watts (eds). 2005. *Violent Environments*. Ithaca, NY: Cornell University Press.

21 C. Geertz. 1980. Blurred genres. *American Scholar* 29, 165–82.

22 See Deligiannis, chapter 2 in this volume; see J. Barnett and N. Adger. 2007. Security and climate change: towards an improved understanding. *Political Geography*, 6, 1–21 for a review.

23 See de Soysa, chapter 3 in this volume; P. Collier. 2008. *War, Guns and Votes*. New York: Harper; S. Kalyvas. 2006. *The Logic of Violence in Civil War*. Cambridge: Cambridge University Press.

24 B. Buzan and O. Wæver. 2009. Macrosecuritisation and security constellations: reconsidering scale in securitisation theory. *Review of International Studies* 35, 253–76.

25 Barnett, *The Meaning of Environmental Security*; Floyd, *Security and the Environment*.

26 Leichenko and O'Brien, *Environmental Change and Globalization*.

27 See Barnett, et al. (eds). 2010. *Global Environmental Change and Human Security*; N. Adger. 2006. Vulnerability. *Global Environmental Change* 16, 268–81; E. Schipper and I. Burton (eds). 2009. *Adaptation to Climate Change*. London: Earthscan.

28 Schipper and Burton, *Adaptation to Climate Change*.

29 M. Duffield. 2011. Total war as environmental terror. *South Atlantic Quarterly* 110/3, 757–69.

30 See B. Smit and J. Wandel. 2006. Adaptation, adaptive capacity and vulnerability. *Global Environmental Change* 16, 282–92; C. Folke. 2006. Resilience. *Global Environmental*

Change 16, 253–67; R. Nelson, W. Adger and K. Brown. 2007. Adaptation to environmental change. *Annual Review of Environment and Resources* 32, 395–419.

31 See Adger, Vulnerability; M. Fussel. 2006. Vulnerability. *Global Environmental Change* 16, 155–67.

32 M. Foucault. 2008. *The Birth of Biopolitics*. London: Palgrave; see M. Dean. 2005. *Governing Societies*. London: Open University Press.

33 Foucault, *The Birth of Biopolitics*: 65.

34 Anderson, Affect and biopower.

35 In the same way that environmental security has broadened its theoretical horizons, so too has political ecology over the last three decades. I provide a very simplified gloss here. These changing currents can be easily seen by glancing at: R. Peet and M. Watts (eds). 1985. *Liberation Ecologies*. London: Routledge; R. Peet and M. Watts (eds). 1996. *Liberation Ecologies*. London: Routledge; and R. Peet, M. Watts and P. Robbins (eds). 2011. *Global Political Ecology*. London: Routledge – political ecology readers that cover the period from the 1980s to the present.

36 R. Rappaport. 1967. *Pigs for the Ancestors*. New Haven, CT: Yale University Press.

37 K. Kautsky. 1977. *The Agrarian Question*. London: Zed Books.

38 K. Polanyi. 1947. *The Great Transformation*. Boston, MA: Beacon.

39 For fuller accounts of the history and development of the field see P. Robbins. 2004. *Political Ecology: A Critical Introduction*. New York: Blackwell; R. Neumann. 2004. *Making Political Ecology*. London: Hodder.

40 N. Smith. 1977. *Uneven Development*. New York: Routledge. Long before the emergence of resiliency theory and socio-ecological systems, political ecology started from the notion of the reciprocal human appropriation and transformation of nature, that necessarily involved the humanization of nature and the naturalization of human.

41 M. Watts. 1983/2012. *Silent Violence*. Berkeley: University of California Press; M. Davis. 2000. *Late Victorian Holocausts: El Nino Famines and the Making of the Third World*. London: Verso.

42 T. Forsyth. 2003. *Critical Political Ecology*. London: Routledge; T. Li. 2007. *The Will to Improve*. Durham, NC: Duke University Press.

43 A. Gramsci. 1971. *Prison Notebooks*. London: Lawrence & Wishart; D. Moore. 2005. *Suffering for Territory*. Durham, NC: Duke University Press; S. Malette. 2009. Foucault for the next century: eco-governmentality, in S. Binkley and J. Capetillo (eds), *A Foucault for the Twenty First Century*. Cambridge: Cambridge University Press, 222–39.

44 A. Agrawal. 2005. Environmentality: Technologies of Government and the Making of Subjects. Durham, NC: Duke University Press.

45 Li. *The Will to Improve*.

46 Ibid.: 270.

47 J. Kosek. 2006. *Understories*. Durham, NC: Duke University Press.

48 J. Kosek. 2011. The nature of the beats, in M. Watts, R. Peet and P. Robbins (eds), *Global Political Ecology*. London: Routledge, 227–53.

49 M. Pelling. 2011. *Adapting to Climate Change*. London: Routledge.

50 T. Bassett. 2012. Déjà vu or something new? Unpublished manuscript, Department of Geography, University of Illinois.

51 See M. Cote and A. Nightingale. 2011. Resilience thinking meets social theory. *Progress in Human Geography* 43, 1–15.

52 C. Folke, K. Hahn, P. Oslo and J. Norberg. 2005. Adaptive governance of socio-ecological systems. *Annual Review of Environment and Resources* 30, 441–73.

53 O. Young, O. et al. 2006. The globalization of socio-ecological systems. *Global Environmental Change* 16, 304–16.

54 M. Pelling. 2011. *Adapting to Climate Change*. London: Routledge.

55 Leichenko and O'Brien, *Environmental Change and Globalization*.

56 Foucault, *The Birth of Biopolitics*.

57 See B. Buzan, O. Wæver and J. de Wilde. 1998. *Security: A New Framework for analysis*. Boulder, CO: Lynne Rienner; B. Buzan and O. Wæver, O. 2009. Macrosecuritisation and security constellations: reconsidering scale in securitisation theory. *Review of International Studies* 35, 253–76.

58 Floyd, 2010. *Security and the Environment*: 39.

59 M. Dillon. 2008. Underwriting Security. *Security Dialogue* 39/2–3 (April), 309–32.

60 M. Schabas. 2007. *The Natural Origins of Economics*. Chicago: University of Chicago Press.

61 M. Foucault. 2007. *Security, Territory, Population*. London: Palgrave; Foucault, *The Birth of Biopolitics*.

62 F. Jacob. 1989. *The Logic of Life: A History of Heredity*. London: Penguin Books; F. Jacob. 1989. *The Possible and the Actual*. London: Penguin Books; S. Kaufman. 2000. *Investigations*. New York: Oxford University Press.

63 See M. Dillon. 2007. Governing through contingency: the security of biopolitical Governance. *Political Geography*, 26, 41–47; M. Dillon, 2008. Underwriting security. *Security Dialogue* 39/2–3 (April), 309–32.

64 This section draws extensively on Dillon, Governing through contingency; Underwriting security; see also Anderson, Affect and biopower; Preemption, precaution, preparedness.

65 See Kaufman, Stuart. 2000. *Investigations*. New York: Oxford University Press.

66 Foucault, *Security, Territory, Population*.

67 Anderson, Affect and biopower: 32.

68 Ibid.: 33.

69 Dillon, Underwriting security: 314.

70 Ibid.: 315.

71 Hayek, *The Fatal Conceit: The Errors of Socialism – Collected Works of F.A. Hayek*, vol. 1. London: Routledge; M. Cooper. 2008. *Life as Surplus*. Seattle: University of Washington Press.

72 Anderson, Affect and biopower: 34, my emphasis.

73 Foucault, *The Birth of Biopolitics*.

74 P. O'Malley. 2000. Uncertain subjects: risk, liberalism and contract. *Economy and Society* 29, 460–84; P. O'Malley. 2006. *Governing Risks*. Aldershot: Ashgate.

75 Dillon, Underwriting security. 2005.

76 Anderson, Preemption, precaution, preparedness.

77 Foucault, *The Birth of Biopolitics*.

78 Ibid.: 242.

79 Ibid.: 241.

80 Anderson, Affect and biopower: 38.

81 Foucault, *The Birth of Biopolitics*: 143.

82 M. Watts. 2011. Ecologies of rule, in C. Calhoun and G. Derlugian (eds), *The Deepening Crisis*. New York: New York University Press, 67–92.

83 N. Rose. 1999. *Powers of Freedom*. Cambridge: Cambridge University Press.

84 C. Folke. 2006. Resilience. *Global Environmental Change* 16, 253–67.

85 C.S. Holling. 2001. Understanding the complexity of economic, ecological and social systems. *Ecosystems* 4, 390–405.

86 L. Gunderson and C.S. Holling (eds). 2002. *Panarchy: Understanding Transformations in Human and Natural Systems*. Washington, DC: Island Press. J. Guyer. 2009. Prophecy and the near future. *American Ethnologist* 34/3, 409–21.

87 M. Leach (ed.). 2008. *Reframing Resilience*. STEPS Center, IDS, University of Sussex; A. V. Bahadur, M. Ibrahim and T. Tanner. 2011. *The Resilience Renaissance? Unpacking*

of Resilience for Tackling Climate Change and Disasters. Institute of Development Studies, University of Sussex, Brighton. Online: http://community.eldis.org/.59e0d267/ resilience-renaissance.pdf (accessed August 2012); F. Lentzos and N. Rose 2009. Governing insecurity. *Economy and Society* 38/2, 230–54.

88 WRI. 2008. *Roots of Resilience.* Washington, DC: World resources Institute/UNEP/ World Bank: 123.

89 M. Cooper and J. Walker. 2011. Genealogies of resilience. *Security Dialogue* 14 (2), 143–60.

90 F. Hayek. 1988. *The Fatal Conceit: The Errors of Socialism – Collected Works of F.A. Hayek,* vol. 1. London: Routledge, cited in M. Cooper and J. Walker, Genealogies of resilience.

91 Adger, N., Lorenzoni, I. and O'Brien, K. (eds). 2009. *Adapting to Climate Change.* London: Cambridge University Press.

92 "[T]he extended order is perfectly natural, in the sense that it has itself, like similar biological phenomena, evolved naturally in the course of natural selection' (Hayek, *The Fatal Conceit*: 244).

93 M. Cooper, 2010. Turbulent worlds: financial markets and environmental crisis. *Theory, Culture & Society* 27/2–3, 167–90.

94 M. Dillon, Underwriting Security. *Security Dialogue* 39/2–3 (April), 309–32.

95 P. O'Malley, Resilient subjects. *Economy and Society* 39/4, 488–509.

96 Foucault, *The Birth of Biopolitics*: 241.

97 J. Masco. 2011. Pre-empting biosecurity. Unpublished manuscript, Department of Anthropology, University of Chicago.

98 R. Matthew et al. (eds). *Global Environmental Change and Human Security*: 13.

99 M. Burawoy. 2008. *The Extended Case Method.* Berkeley: University of California Press.

100 N. Peluso and P. Vandergeest. 2011. The political ecology of war and forests. *Annals of the Association of American Geographers* 101/3, 587–608.

101 H.-G. Bohle and H. Funfgeld. 2007. The political ecology of violence in eastern Sri Lanka. *Development and Change*, 38/4, 665–87.

102 M. Watts. 1983/2012 *Silent Violence.* Berkeley: University of California Press; and see N. Hildyard, L. Lohmann and S. Sexton. 2012. *Energy Security: For What? For Whom?* Sturminster Newton: Corner House.

5

FROM CONFLICT TO COOPERATION? ENVIRONMENTAL COOPERATION AS A TOOL FOR PEACE-BUILDING

Achim Maas and Alexander Carius with Anja Wittich

Introduction[1]

Recent debates on potential impacts of climate change have catapulted environmental security to the top of the agenda of many political bodies.[2] The discourse, with its Malthusian connotation and alarmist tone, is filled with vivid images of potential water wars, food crises and legions of environmental refugees fleeing from disaster. The United Nations (UN) Security Council has debated the implications of climate change on international security, while the European Union (EU) developed an action plan to marshal all instruments at its disposal to combat climate change. High-level political figures such as UN Secretary-General Ban Ki Moon retrospectively associated environmental change with events such as the violence in Darfur,[3] thus linking climate change with what many believe to be an incidence of genocide.[4] In comparison, the ways in which environmental change – or environmental affairs in general – may contribute to peace is far less discussed. In recent years a solid if small body of literature on this topic has developed.[5] In addition, a number of governmental and non-governmental organizations have attempted to use environmental topics as entry points for peace- and confidence-building measures. The focus of their work is often the joint management of environmental affairs and natural resources with a view to overcoming direct, indirect and potential causes of conflict related to the environment and natural resources. In addition, the dialogue between different parties to a conflict may also support confidence-building and reconciliation measures. Occasionally, the latter may even be the primary intention. The underlying idea is that when people meet and jointly work on common problems, they recognize that they share needs and interests, making cooperation the more rational choice than conflict. Hence, it is imperative to create strategic social space in which conflicting parties can meet in an adequate setting to unlearn stereotypes about their respective "Other".[6] In this sense, environmental issues are but one topic among many that can facilitate trust and dialogue.

So far, empirical evidence analysing the connection between environment and cooperation offers mixed results. To some extent this is due to a lack of systematic research on the role of environment and natural resources across different regions, conflict types as well as specific (geo-) political and wider socio-economic contexts. Furthermore, assessment methodologies to either anticipate or retrospectively assess the effectiveness of environmental peace-building are either rarely used or underdeveloped. Finally, the ways in which environmental peace-building has been operationalized by different organizations varies greatly in scope and the concrete results are in many cases difficult to assess. This includes avoiding exacerbating difficult situations.[7] Thus, assuming the environment is a topic suitable for peace-building, a number of not fully answered key questions remain, particularly regarding conditions, constraints and consequences.

Building on prior conceptual work and experiences[8] as well as research findings in the context of the "Initiative for Peacebuilding" (IfP)[9] carried out in 2007–09, this chapter aims to highlight opportunities and pitfalls of environmental peace-building. In order to do this the chapter focuses on environmental peace-building in the South Caucasus, covering the political events in the region until late 2009.[10] The South Caucasus is interesting for a number of reasons. First, there are several organizations working within the framework of environmental peace-building, many of them openly label their programmes as such despite the sensitivity this may carry. Second, they cover many aspects of environmental peace-building and provide good examples as well as lessons to be learnt. Third, the region is home to a diverse range of types and stages of conflicts.

Environment and peace-building

A recurrent issue in the debates on the link between environment, peace and security is the absence of an agreed-upon definition of any of the terms. Consequently linking them – as has been done in the case of "environmental security" – does not necessarily improve the understanding of linkages between environment, peace and (violent) conflict. Instead the result may be a vague term formed of two other insufficiently defined and often contested terms, which may be even vaguer than its constituent parts.[11]

"Environmental peace-building" is no exception and many other terms – environmental peace-making, ecological peace-building, environmental diplomacy – have emerged over the years. Environmental peace-building is neither a coherent theoretical school nor a concrete and distinct set of practical activities. Instead, it should be considered as an umbrella term that covers a wide range of aspects, which are united by their focus on the relationships between environment, conflict and peace. It has been applied to a variety of environmental topics, ranging from preserving eco-system services to pollution control, biodiversity conservation, and resource scarcity as well as the use of natural resources.[12]

Peace may also include a variety of issues, among others overcoming direct, structural and even cultural violence,[13] ranging from addressing direct causes of conflicts to societal transformation. However, conceptual thinking about the potentials of environmental peace-building is highly academic unless it is grounded in a concrete local and practical context. Hence, instead of labelling environmental peace-building a school of thought or theory, the debate about environmental peace-building can be more aptly considered as a discourse in which a wide range of people from academics to politicians, environmentalists and others participate. Within this discourse, three major dimensions can be identified:

1. Addressing environmental causes of conflict

While environmental issues such as land degradation, water scarcity or the exploitation of natural resources may often be linked to violent conflict, they are rarely the only or even the major causes of armed conflict. Instead, they often constitute indirect or structural sources of conflict, eroding societal resilience and capacities for non-violent conflict resolution and transformation.[14] Furthermore, where wealth sharing from resource exploitation is inequitable between segments of society, this may fuel inter-group grievances. The latter may also develop in response to events such as disasters, which tend to affect populations unequally – thus environmental issues can escalate latent social frictions or even contribute to creating new ones. Activities under the umbrella of environmental peace-building would then include identifying and addressing these actual or potential causes of conflict. In some cases, this may be relatively straightforward, for example by introducing new technologies or management techniques to improve efficiency and so reduce relative resource scarcity. In more complex cases, such as inequitable resource management, a broader and more sustained approach is necessary as the implied changes in the current socio-economic structures will most often run counter to the interests of those groups benefiting from the inequity. In other cases, however, the respective authorities may lack the capacity and resources to cope with a given situation or may simply be unaware of the fact. In these cases, the reverse to the above mentioned may be true. This is not to say, however, that institutions are unable to cope with environment-related social friction and conflict potential, but rather that they gave rise to them in the first place. What is then needed is broader societal transformation.

2. Environmental cooperation as a platform for dialogue

Given that environmental issues are often lower on the political agenda than other issue areas and thus are less visible, environmental issues may provide a good entry point for dialogue and cooperation – even between parties to a conflict where environmental variables do not play a role in a given conflict. In such cases it is the objective to use environmental issues such as transboundary water pollution to create a social space in which representatives of conflict parties can meet, discuss issues and cooperate with a view to developing (or creating) common solutions. In contrast to the first dimension, however, it is assumed here that environment is not a direct cause of conflict in itself, but rather a medium that may facilitate dialogue processes. The broader aims are thereby threefold: first, to serve as a confidence-building measure by providing a platform for continuous people-to-people communication, thus overcoming stereotypes. Second, to serve as a symbol of successful and peaceful cooperation between conflicting parties, thus proving that peaceful co-existence and non-violent conflict resolution is possible. In this regard, environmental cooperation may also serve as conflict resolution capacity building for conflict parties by allowing them to "engage" on a less conflict-ridden topic. And third, to support changes in mind-sets by developing awareness of the need to find common solutions to common problems, regardless of whether they are environmental or political. The most important aspect from the view of peace-building is the sustained process that brings people together consistently. Thus ideally, the more "technical" cooperation developed around environmental issues would spread to other, more political arenas.[15]

3. Sustainable development as contribution to durable peace

Although sustainable development has become something of a catch-all term it is a key concept for environmental peace-building and conflict transformation. This stems from the fact that sustainable development implies a rational, efficient and equitable use of natural resources and treatment of the environment in order to avoid degrading the foundations of society. This in turn requires a certain level of transparency, good governance, rule of law and the capacity of local authorities to combat corruption, illegal resource exploitation, organized crime and related issues. It further includes resolving disputes over resources, for example questions of land tenure, in a non-violent way. Building up the capacity of local authorities, or even whole states, in the field of environment and resource governance may strengthen states as a whole. Furthermore, improving environmental awareness and strengthening civil society organizations may add to democratization efforts and show ways to address other issues in non-violent ways.[16] Particularly against the background of climate change, environmental peace-building thus also receives a conflict prevention aspect as it strengthens state capacity to cope with adverse environmental change. In contrast to the first and second dimension, this third dimension is the most long-term and structural approach, as it is not only working on immediate issues as in the first dimension or changing mind-sets as in the second dimension, it is about fundamentally changing socio-economic processes to make them more resilient to events that may, in more fragile societies, facilitate the outbreak of violence.[17]

The three dimensions not only partly overlap, but they are also inextricably linked. Thus, successfully overcoming the causes of conflict includes preventing their recurrence, most often resulting in changing existing institutions, habits and structures. On the other hand, where people are in dire humanitarian need, issues of sustainable development or elaborate institutionalized cooperation may be inadequate to address the immediate suffering (see Figure 5.1).

It is interesting to note that none of the three dimensions is particularly innovative with regard to peace-building. Within peace studies, for instance, exists a broad body of literature examining how causes of conflict can be overcome that focuses on the role of dialogue, mediation and cooperation in reconciliation and conflict transformation, as well as on the relevance of good governance, emancipation, transparency and strengthened civil society as contributions to peace (-building). The discourse on environmental peace-building may thus be considered as one particular aspect of peace research, with a strong, though broadly understood, focus on environment issues.

Although environmental peace-building provides an instrumental lens for addressing environmental conflict, it is important to note that this does not necessarily mean that practitioners will be working on conflict directly. More often than not the linkage between environment and conflict is also quite indirect. Often, therefore, activities labelled as environmental peace-building are those that are conducted in the context of an actual

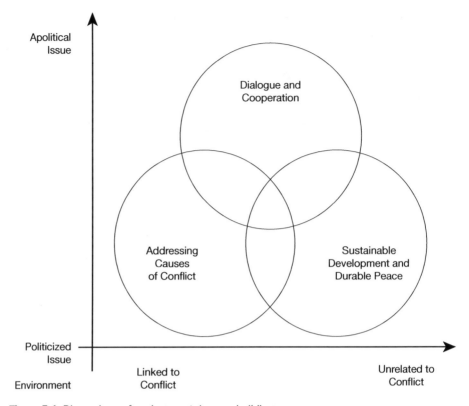

Figure 5.1 Dimensions of environmental peace-building.

or potential conflict with a view to transforming this context in such a way as to inhibit escalations and facilitate de-escalation and conflict transformation. Thus, from a practical standpoint, environmental peace-building can be labelled as a tool or measure that can initiate or augment peace processes; it cannot replace them.

Whether or not environmental peace-building is the right tool in any given case is very much dependent on the situation. So far, if to varying degrees, it has been applied by several organizations in many areas of the world. These activities can often be located in the second and third dimensions outlined above and focus on the creation of (trans-boundary) protected areas, the cooperative use of resources and information sharing. Table 5.1 provides a selection of projects and programmes from Africa, Latin America and Asia.

These projects and programmes differ hugely in scope and activities and it is beyond this chapter to outline their respective qualities. Moreover, systematic cross-regional research has only been carried out on specific approaches such as the role of protected areas, while comparative analysis on different aspects such as types of conflict, conflict phase, regimes and so on is unavailable.[18] To highlight the different aspects, opportunities and pitfalls of environmental peace-building, we will now focus on environmental peace-building in the South Caucasus.

Table 5.1 **Selected environmental peace-building projects and programmes**

Project/Programme Title	Type	Executing and Funding Institutions	Countries
Transboundary Biosphere Reserve	Protected area	German Federal Nature Conservation Agency	Russia, Mongolia, China, Kazakhstan
Ai-Ais Richtersveld Transfrontier Park	Protected area	Ministries of the Environment of Namibia and South Africa	Namibia, South Africa
ECOPAS / W Parc ECOPAS Project (Ecosystéme protégés en Afrique sahélienne)	Protected area	Benin, Burkina Faso, Niger, European Union	Benin, Burkina Faso, Niger
Great Limpopo Transfrontier Park	Protected area	South Africa, Mozambique, Zimbabwe	South Africa, Mozambique, Zimbabwe
International Gorilla Conservation Programme	Protected area	Democratic Republic of Congo (DRC), Uganda, Rwanda, African Wildlife Foundation, Flora & Fauna International, World Wide Fund for Nature (East Africa)	DRC, Uganda, Rwanda
Kgalagadi Transfrontier Park	Protected area	Department of Wildlife and National Parks, USAID, Peace Park Foundation, SAN Parks	Botswana, South Africa
National park Thayatal (Podyji)	Protected area	Czech Republic, National Park Management Austria	Czech Republic, Austria
Selous Conservation Programme	Protected area	German Technical Cooperation (GTZ), Tanzanian Wildlife Division	Tanzania, Mozambique
Selous-Niassa Wildlife Corridor Research Project	Protected area	GTZ	Tanzania, Mozambique
Trifinio Plan	Protected area/ cooperative resource use	Organization of American States, Inter-American Institute of Cooperation for Agriculture	El Salvador, Guatemala
OKACOM (Permanent Okavango River Basin Water Commission)	Protected area/ cooperative resource use	Botswana, Namibia, Angola	Botswana, Namibia, Angola
Nile Basin Initiative	Cooperative resource use	World Bank, UN Development Programme, Canadian International Development Agency, etc.	Egypt, Ethiopia, Kenya, DRC, Rwanda, Somalia, Sudan, Tanzania, Uganda
Good Water Makes Good Neighbors	Cooperative resource use	Friends of the Earth Middle East, EU, US government Wye River Program	Israel, Palestinian Territories, Jordan
SADC Protocol on Shared Watercourse Systems	Cooperative resource use	Southern African Development Community (SADC)	Angola, Botswana, Lesotho, Malawi, Mozambique, Namibia, South Africa, Swaziland, Tanzania, Zambia, Zimbabwe
Regional Water Data Banks Project	Information sharing	EU, France, The Netherlands, USAID	Israel, Palestinian Territories, Jordan

Source: Adapted from Carius 2006: 5f.

The South Caucasus and environmental peace-building: experiences, opportunities and pitfalls[19]

The South Caucasus is an environmentally, economically and politically vibrant region. It includes Armenia, Azerbaijan and Georgia and overlaps with parts of Iran, Russia and Turkey. Home to a wide range of unique species, the South Caucasus is a biodiversity rich region, including species such as lynx and bisons. It is also home to many natural resources including oil, gold and copper. Economic development has strongly increased over the past years and the region is becoming an increasingly important transport corridor. Environmental and economic priorities, however, remain far from reconciled with the current emphasis on economic growth and natural resources exploitation resulting in environmental degradation.[20] Many environmental problems, such as pollution in the Kura-Araks river basin, would benefit from transboundary cooperation and a regional approach.

The South Caucasus is home to a wide range of manifest and latent conflicts.[21] To begin with, several secessionist conflicts are active within the region, Abkhazia and South Ossetia vying for independence from Georgia, and Nagorny Karabakh from Azerbaijan being the most obvious. All three have achieved de facto independence but remain internationally non-recognized entities (NRE). The conflict in Chechnya has cooled down in recent times.[22] The peace processes remain in a political stalemate, except for phases of escalation and de-escalation over the past two decades, resulting in their being labelled as essentially "frozen conflicts" in the past. The most severe escalation so far happened in early August 2008, when Georgia attempted to militarily recapture South Ossetia, resulting in a military confrontation with Russia.[23]

These "internal" conflicts are interlinked with a number of interstate conflicts, particularly between Armenia and Azerbaijan over the status of Nagorny Karabakh and the territories of Azerbaijan currently occupied by Armenia. In addition, Russia has started to formalize its relations with Abkhazia and South Ossetia which escalated its ongoing disputes with Georgia proper to a new level,[24] particularly after their political disputes turned into violent conflict when Russia not only supported South Ossetia in combating Georgian forces, but also started bombarding targets in Georgia proper.[25]

Being at the fulcrum of Asia, Europe and the Middle East, the South Caucasus is a site where the interests of many regional and global powers meet and occasionally clash. A crystallizing point, among others, is Georgia's aspiration to join NATO, which is supported by the USA and many European powers, while adamantly opposed by Russia. This web of confrontational relations also includes those between Turkey and Iran, two other major regional powers vying to maintain or increase their sphere of influence. Turkey has traditionally supported Azerbaijan and Georgia in their claims for territorial integrity, while Iran tends towards supporting Armenia, not least because of its large Azeri minority and due to disputes with Azerbaijan over exploiting the hydrocarbon resources in the Caspian Sea.[26] To add yet another layer of complexity, neither Armenia and Azerbaijan nor Georgia are consolidated stable states. The recent presidential elections in Armenia and Georgia were accompanied by riots and states of emergency. The 2005 elections in Azerbaijan failed to meet international standards.[27] All three countries suffer from corruption, which also impacts on the work of international organizations (IO) and non-governmental organizations (NGO). Besides the direct impact on conducting projects, the distrust of the population is aroused if IOs and NGOs cooperate with corrupt officials or elites, placing their integrity in doubt.[28]

This mix of internal, international and global dynamics continues to harbour a significant escalation potential of spawning a new war. Yet there is a copious number of actors present in the South Caucasus who support confidence and peace-building measures that avoid escalations and prepare the ground for conflict transformation.[29] The wide range of approaches also includes environmental peace-building.

A number of – mostly governmental – organizations, initiatives and programmes directly link issues of environment, peace and security. One of the most prominent is the so-called *Environment and Security Initiative* (ENVSEC), which is made up of several international organizations[30] and works closely with the governments of Armenia, Azerbaijan and Georgia on environmental risks and threats with a view to improving cooperation and coordination between the countries. Main areas of work include environmental degradation and access to natural resources in conflict regions, the management of cross-border environmental concerns and population growth and rapid growth in capitals. Another such initiative is the *South Caucasus River Monitoring* (SCRM) project, a long-running project supported by NATO and the OSCE dealing with the water quality of the Kura-Araks river basin.[31] A third is the *Caucasus Initiative* (CI) of the German Federal Government, with the explicit goal of fostering peace and stability in the South Caucasus through cooperation between the recognized countries in different thematic fields, including the environment. Within the context of the CI, the Trans-boundary Joint Secretariat (TJS) was established with the objective to support governmental authorities in the establishment of trans-boundary nature conservation parks. A fourth is USAID's *South Caucasus Water Program* on managing regional water issues, which has been promoted by USAID as being one of the preconditions for peace and security in the region. Similarly, the *European Neighbourhood Policy* (ENP) aims at creating a ring of stability and security in the immediate European neighbourhood by supporting intra-regional development and cooperation in various policy fields, including environment. In addition, a number of organizations such as the *Swiss Development Cooperation* (SDC) also engage in projects across conflict divides, particularly in Georgia, with a view to increasing people-to-people contacts. None of these organizations or initiatives has either the explicit goal or mandate of resolving any of the conflicts. Their emphasis is rather on preparing a climate that is more conducive to political negotiations by tackling environmental risks and threats to human security, while at the same time aiming to promote dialogue and (peaceful) cooperation. Finally, a number of international organizations attempt to bring different parties together, including conflicting parties such as Armenia and Azerbaijan, to work on environmental issues – without a direct reference or objective to promote peace and stability in the region. These are, for example the Regional Environment Centre for the Caucasus (REC)[32] or the Caucasus Biodiversity Council (CBC).

Although these various initiatives pursue all of the three above-mentioned dimensions of environmental peace-building, the majority focus on dialogue and cooperation coupled with sustainable development as opposed to addressing the causes of conflict.

In part this is because environmental issues or resource use cannot be considered a direct cause of violence or a major driver of conflict in the South Caucasus. The ENVSEC initiative has identified a number of potential environmental risks in the South Caucasus, which are mostly issues of human security concerning access to resources. None of these risks has been directly linked to the outbreaks of political conflict outlined above; however, they are believed to exacerbate existing tensions.[33] Also, several conflict parties have raised the illegal exploitation of natural resources within the NRE as an issue, in particular Azerbaijan. However, work on issues of environmental degradation in the NRE is severely limited due to a number of constraints. Furthermore, "illegal" exploitation is a question of definition as anything conducted within the NRE could be legitimately considered as illegal given its unresolved status. Still, a number of projects have been conducted in the NRE, such as the EU funded rehabilitation programme in Abkhazia and SO, which are focusing on improving livelihoods and alleviating human insecurity. While they suffer from a set of obstacles – among others the insecure situation, travel restrictions, occasional escalations in the political climate and so on – these projects still take place, showing that basic activities are possible.

All the different approaches in the South Caucasus have in common that they work along the internationally recognized political geography. Initiatives actively seeking to promote cooperation and dialogue, such as the CI, exclude the NRE and only focus on the three recognized states. Even organizations such as the CBC, which define the Caucasus ecologically and not politically, exclude the NRE. While this could still serve as a confidence-building measure at least between Armenia and Azerbaijan, one of the major causes of this friction – the status of Nagorny Karabakh – is excluded. Similarly, within the scope of ENVSEC, the environmental situation in the NRE was identified as a major cause of concern – but the NRE are excluded from the deliberations of ENVSEC, making them an object, not a subject of ENVSEC. The reason for this is in part the unresolved status of the NRE, should Abkhazia, South Ossetia and Nagorny Karabakh be engaged in a similar manner to the recognized states, this could be interpreted as acknowledging them on the same level as independent states, thus justifying their claims and implicitly taking sides in favour of independence. This is, of course, opposed by Georgia and Azerbaijan, as it would devalue their aspirations to re-integrate their break-away regions, even though both countries may be ready to grant them a high level of autonomy. Similarly, international organizations, in particular the UN and OSCE, have enshrined the principle of territorial integrity in their charters, which they perceive as a natural bias against the NRE. Consequently, the confidence-building aspect on a political level is in many cases non-existent between the relevant parties to a conflict as they are treated unequally. However, even full cooperation would not necessarily mean that the three NRE would be integrally involved; cooperation often also implies interdependence and thus dependence, which is contrary to the aspirations of the NRE.

This is not to say that cooperation is completely impossible, only that it may often take a parallel approach, with shuttle diplomacy between both sides helping parties to agree on "common projects" that are then implemented separately and in parallel, without direct contact. In other words, cooperation is vertical; external/international parties cooperate individually with the states and the NRE while the latter two are only connected via externals. Horizontal cooperation, that is, directly across conflict divides between conflicting parties, does not take place. Furthermore, apolitical fields are not necessarily neutral, but may be seized upon by conflicting parties to support their claims – such as the government of Azerbaijan using the 2006 forest fires in Nagorny Karabakh to show how "careless" the de facto authorities are.

Another example is the case of the REC, of which both Azerbaijan and Armenia are members, but whose disputes severely inhibit its work. Azerbaijan often rejects any proposal that implies direct cooperation with Armenia outright and regardless of the topic, which makes labelling a highly important and difficult issue. In short, due to its larger symbolism, dialogue and cooperation in the field of the environment is not apolitical.

Beyond symbolism, environmental issues are dominated by a number of – often externally – financed NGOs, scientists and experts in national ministries and donor organizations. The direct target of dialogue and cooperation is thus limited to a small expert community. This is a good start; it would be naïve to think that entire segments of society could be engaged with only single activities. Still, we should question whether this very select group is the most adequate primary target group given the low relevance of environment in the South Caucasus (see below). However, even in areas where there is no major conflict – such as between Georgia and Azerbaijan – the existing international cooperation formats are not satisfying, as Georgia increasingly tends to re-orientate itself towards the Black Sea and the West, disassociating itself from both Armenia and Azerbaijan.

With regards to the third dimension of environmental peace-building (sustainable development as a contribution to durable peace) it is first of all important to realize that environmental issues are currently not a political priority in the South Caucasus. While comprehensive environmental legislation has been passed in Armenia, Azerbaijan and Georgia alike all three ministries of the environment (MoE) are politically comparatively weak organizations. All three governments currently exploit natural resources with a view towards economic growth. Indeed, the respective governments – except for the MoEs – perceive nature protection and environmental conservation primarily as obstacles to economic development. The notion of using protected areas economically, such as for eco-tourism, is only slowly emerging. The deficits in implementing existing environmental legislation as well as the low level of enforcement of environmental obligations are indicative of this low priority setting. Furthermore, all three governments ensure their survival by raising standards of living for their constituency, focusing on short-term measures and lacking a strategic vision.

The situation in the NRE is similar, though the humanitarian situation in these areas is far worse than in the recognized states. While, on the one hand, this accelerates environmental degradation and thus in the long term worsens the humanitarian situation, there is little room to manoeuvre against the manifest background of people's suffering. Environmental projects often carry no direct and recognizable short-term benefits for people and may therefore lack ownership. Consequently, the interest and readiness of authorities to cooperate is not genuine, but rather purely donor driven. Concurrently, capacity building and strengthening environmental agencies may have currently only a very limited reach. In fact, the multiplicity of donors, projects and programmes even occasionally exceeds the absorption capacity of the respective MoE. Sustainable natural resource management is still an alien concept. Environmental projects initiated by donors may therefore not be viable due to the lack of interest and ownership. In the case of the NREs, projects face the additional obstacle that they are, at least officially, still the territory of Georgia (Abkhazia, SO) and Azerbaijan (NK) respectively and working there requires the permission of the governments in addition to that of the de facto authorities. Georgia and Azerbaijan, however, have no interest in building capacities in the NRE. Hence, a major source of funding for implementing environmental projects is unavailable to them, barring the building of relevant capacities or working on environmental issues.

Beyond these issues, three more factors need to be taken into account. First, the larger geopolitical conflict dynamics are not covered and also cannot be covered with a (sub-) regional approach. Still, they need to be kept in mind as escalations on the global level may trickle down to lower levels and hamper, if not erase, progress made in confidence building. Second, the conflicting parties, particularly the recognized states and third parties from outside the region, have a very different time horizon. For the recognized states, the priority is conflict resolution in the short term, as the respective governments legitimize themselves, particularly during election campaigns, using claims of resolving the conflicts favourably for their constituency. One of the results is the readiness of conflicting parties to trade longer-term progress in conflict transformation for short-term successes. Furthermore, events outside the South Caucasus such as the recognition of Kosovo may be perceived as creating additional time pressures, as Georgia and Azerbaijan proper may consider time is playing against them.[34] CBM, which is the focus of environmental peace-building activities, is a long-term approach focusing more on conflict transformation as a precondition for conflict resolution. The priority for third parties is therefore rather to sustain processes over the medium to long term to allow the development of a climate more conducive to peace. Hence, while the ultimate aims of conflicting parties and extra-regional third parties may be similar, they are contradicting themselves in their priorities and time horizons. Finally, what should not be underestimated are the interests of powerful groups in continued conflict. In the South Caucasus the status quo of neither peace nor war has created a mix of shadow, war and coping economies which offers ample opportunity for black market activities, abuses of power, corruption and related issues.[35] Hence, even if there would be genuine interest and even successes in cooperation in the field of environment, spill-over effects to other political levels may be actively prevented by power holders.

Conclusions for the environment and peace-building in the South Caucasus

For environmental peace-building more generally the following lessons can be learnt from the example of the South Caucasus. First, state-driven cooperation and dialogue may be structurally inadequate to work on conflicts where statehood, particularly recognition and sovereignty, are of high symbolic value for all conflicting parties. While actual cooperation on a technical level is possible, institutions and structures always also symbolize relationships. If they run contrary to what the participating parties are ready to accept and value, actions are likely to be stalled.

Second, although the environment as such may not be a priority issue, it becomes politicized by one of the participating parties, either intentionally or unintentionally, as soon as it is somehow related to a conflict. Hence there is no true apolitical or neutral issue and even organizations not working on peace and conflict issues suffer the (often adverse) consequences.

Third, environmental awareness and an authentic interest are preconditions for environmental peace-building. While third parties may highlight and focus on long-term issues and structural causes of conflict such as environmental degradation, the views and perspectives of the parties to a conflict are very much focused on short-term developments and the actual conflicts.

Fourth, states and organizations that have initiated environmental peace-building activities are hesitant to accept the consequences this may entail, that is, meaningfully engaging all sides of the conflict. Failing to do so empowers one side of the conflict over the other, when in fact a neutral or apolitical topic requires equality or – more unlikely – the acceptance of inequality.

Fifth and finally, large numbers of third parties, especially donor organizations, conducting projects and programmes with parties to a conflict individually do not necessarily have a cumulative effect particularly where coordination is missing. It may even exceed the absorption capacities of conflicting parties. The example of the South Caucasus shows that environmental peace-building as such is not an approach that can be transplanted to every type of conflict or region in the world. The perspectives of conflicting parties on the environment as well as the context very much determine to what extent environmental peace-building may succeed or fail. This is particularly the case where environmental peace-building, intentionally or unintentionally, symbolically touches upon the interests of the conflicting parties. Conversely, where the environment is not a priority issue for them and their attention is focused on other issues such as conflict resolution or economic development, environmental peace-building may simply fall flat.

The example of the South Caucasus, and the "lessons learnt", enables us to amend the three theoretical dimensions of environmental peace-building outlined earlier in the chapter. With regards to the first dimension we can now say that: in many cases the causes of conflict can only be addressed where conflicting parties allow for this to occur. Although a particular interest and/or feeling of ownership are not necessary requirements for peace-building, the former might help with alleviating suffering and human insecurity.

With a view to the second dimension we can conclude that for cooperation to be fruitful, and in order to learn about and understand the motivations and interest of others, (conflicting) parties need to enter into a dialogue as equals. This is unlikely to happen when respective objectives are highly incompatible – especially where this incompatibility is existential (as in case of NRE versus the respective states they are seceding from).

And, finally in view of the third dimension we can now say that: sustainable development is a viable precondition for a durable peace only where interest and necessary commitment from the conflicting parties exists. If this is not the case, or where this runs contrary to their priorities and interests, the chances that activity will have the intended long-term transformative effect are remote.

In short, our case study shows that theoretical discourse on environmental peace-building needs to incorporate a number of additional issues. Above all else, it needs to include conflicting parties' views on the conflict itself and not only – as has hitherto been the case – their view on the environment. Depending on the answer to these questions, an assessment can be made to the extent that the environment could be an entry point for dialogue. Generally speaking, the less compatible the aims, the more likely it is that environment – and indeed any other topic – becomes politicized and linked to the conflict. Thus, Figure 5.1 needs to be complemented and modified – see Figure 5.2.

If the aims between the conflicting parties are incompatible, and moreover when environmental issues are unimportant topics for the parties (upper left corner), dialogue and cooperation can hardly be an option. On the contrary, in such a case parties could link the environment to conflict and politicize it without any authentic interest in truly resolving the issue. Conversely, where the environment is an important topic for conflicting parties it may become a politicized issue but working on it, even in the face of incompatible goals in other areas, could be possible due to the interests of the parties (lower left corner). It should not be expected however, that technical cooperation will automatically produce any spill over on the political level.

In the South Caucasus, the incompatibility of goals manifests itself in the question of recognition and sovereignty – a question that is still unresolved, even though the peace processes started two decades ago. Institutions and structures that incorporate this feature but leave it unaddressed are therefore unlikely to reach the intended results, as the relevant parties to a conflict are either not included or may simply refrain from participating. Consequently, new structures, forms of cooperation and dialogue are

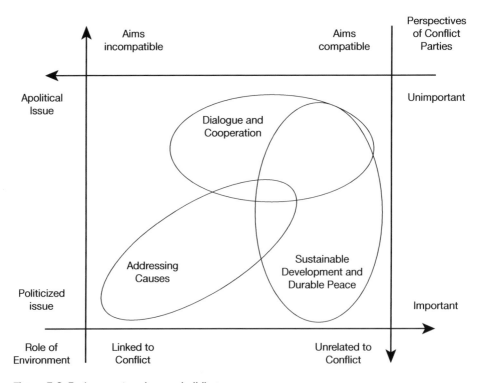

Figure 5.2 Environment and peace-building.

needed. Finding a way to purposefully include the NRE is of major relevance, if confidence-building and conflict transformation is the aim.

One relatively successful example from the economic realm is the *Caucasus Business Development Network* (CBDN), which incorporates non-state representatives and organizations from Abkhazia, Armenia, Azerbaijan, Georgia, Nagorny Karabakh, South Ossetia and Turkey. While CBDN is subject to a number of obstacles, such as travel restrictions, it has initiated a couple of joint projects. Whether a similar format is possible in the environmental realm, however, is a different question given the general low priority and awareness of environmental affairs in the South Caucasus. An entry point may perhaps be found at the nexus where environmental affairs and priority areas such as economic development can be fruitfully combined.

Environmental peace-building as an analytical method

As mentioned above, environmental peace-building in general is strongly related to other approaches in peace research and conflict management. Its added value is the focus on one specific thematic area: the role of the environment, including natural resources, in mitigating and transforming violent conflicts. It provides a distinct perspective on conflict, one that may highlight factors not sufficiently scrutinized yet. The key is to go beyond a mere understanding of how environmental issues interact with conflict, but in a sense use this understanding as a tool for peace-building and conflict management – thus revealing untapped potentials. This method is useful in areas where conflicts over resources exist; however, even in other instances environmental peace-building can provide an innovative solution for protracted conflicts, where other forms of mediation and peace-building were less successful.

The starting point for analysis from an environmental peace-building perspective is the theoretical assumption that the environment can contribute to peace. Of course, taking such a deductive approach is not without its problems, as it may predetermine results. Also, the examples of the South Caucasus show that this assumption is not always true, as environmental topics may not resonate well with conflicting parties and other stakeholders. In extreme cases, the attempt to create platforms for dialogue may even provide a new arena where parties can carry on their conflict. In order to determine whether environmental peace-building may be appropriate in any given case it is useful to start with a thorough conflict analysis, focusing on the role of the environment and natural resources.[36] This emphasis could take the form of explicating a set of hypotheses on the role of the environment in the conflict as well as its potential for contributing to its transformation. A grounded theory approach may be useful where no information on the role of the environment is available, as it allows for capturing the uniqueness of a particular (conflict) situation.[37] Aside from this, the following set of guiding questions may be a useful starting point for analysis:

- Which people or groups of people are directly dependent on natural resources and on their immediate environment (e.g. farmers or miners)?
- How is the environment impacted by conflict?

- How are the environment and natural resources framed by the parties to a conflict? To what extent do they appear in their rhetoric?
- To what extent are livelihoods related to natural resources – for example food, water and energy – satisfied or the necessary resources available to satisfy them?

The analysis will specifically address the first dimension outlined above and help identify whether the environment or natural resources are a cause of conflict between conflicting parties. This can point to further analysis and to developing peace-building strategies that address the causes of conflict, for example looking at how a specific resource contributes to conflict. However, a warning must be given against simplifying the situation: First, "removing" one cause of conflict is unlikely to change the overall situation, as accumulated grievances may have transcended original causes of conflict. Second, there may be additional factors that allowed a cause of conflict to emerge in the first place; cronyism and corruption for example can lead to a waste of resources and environmental degradation, which can further result in grievances and potentially violent conflicts.[38] Hence, identifying underlying linkages and the role of institutions in environmental and resource governance is of key importance. This is also closely linked to the third dimension of environmental peace-building: sustainable development as a foundation for a durable peace. The focus of this dimension is assessing how environmental institutions could contribute to responsible governance and/or state-building efforts. Priority should be given to areas where environment and natural resources could contribute to recovery efforts, such as job creation and income generation. The following set of questions may help in this regard:

- Who controls environmental affairs and access to natural resources? Who has access and who is actively or passively excluded?
- Are there (power) asymmetries between different groups or conflicting parties? If yes, what are they?
- What are the exploitation and/or consumption patterns?
- Is environmental and resource governance perceived as fair and equitable?
- Do natural resources serve special economic functions, such as being the primary (or the only) export goods?
- Who is benefiting from the status quo? Who is interested in changing it?

A more complex issue is the question of whether dialogue and cooperation over environmental issues can serve as confidence-building measures. Beyond conflict analysis, a needs assessment can determine to what extent the environment can play a role in bridging conflict divides with dialogue and cooperation efforts. However, it is necessary to avoid imposing an outside view: particularly in times of conflict, questions of sustainable development may appear misplaced to those suffering and struggling for their daily livelihoods. Also, there exist very different perspectives on the environment: everything labelled "environment" can resonate with a specific connotation and when of low priority may also be of low impact on confidence-building, as stakeholders may have few reservations about abandoning environmental cooperation. Creating awareness for environmental issues and needs may thus be a precondition for allowing cooperation in the first place. Understanding that there are common needs across conflict divides could be considered progress in this regard, and the analyst conducting field research may thus contribute positively: providing food for thought and revealing different opportunities for a situation that seems intractable. Still, this has to be contextualized within the specific perspective of the respective stakeholder or party to a conflict.

Another key issue is the target group: who are the actors between whom confidence should be built? The answer to this question depends on the level of analysis chosen at the outset. A focus on individuals and organizations working on environmental issues may seem an obvious choice, but they may not be ideally suited to serve as agents of change capable of influencing attitudes, perspectives and their society as a whole.[39] Focusing on cross-community cooperation in frontier areas could be more viable; it not only helps to produce regions of relative stability and confidence, but also regions that can serve as symbols. The inherent attractiveness (or coerciveness) of a third party may provide incentives for other actors to cooperate, but if they feel no ownership and have no interest in cooperation then it is only a superficial action.[40] Hence, the third party may be the only link between two parties. Moreover, any form of dialogue, confidence-building and thus ultimately peace-building and reconciliation depends on the readiness of all sides to let the first step happen. Where there is not even a minimal willingness or interest, environmental peace-building may contribute very little in terms of confidence-building and conflict transformation.

The last crucial issue looks at the actual structure or format that dialogue and cooperation may take. How can a viable structure be established where people-to-people contacts can be sustained, so that over time stable expectations and trust can unfold? And how can it be made sustainable, meaning that the ownership and interests of those involved is high enough to continue the efforts even when external support has ceased? What obstacles exist, for example is it physically possible for parties to engage in dialogue? This question also includes how such efforts are best communicated to serve as successful examples for peaceful co-existence. This is not without risks for those engaged in this activity, as cooperation that leads to a changed status quo may create an incentive for stakeholders benefiting from it to obstruct such efforts. Additionally, those participating in a cooperation effort may sometimes have no interest in any publicity for their efforts.[41] Thus, assessing the sensitivity of a proposed activity from a "do no harm"[42] perspective is vital to avoid unintended damage. This needs to include a far-reaching perspective, as the field of stakeholders may change after, for example, the return of refugees or internally displaced persons.

While the above-mentioned approach is comprehensive, analysis from an environmental peace-building perspective could also take an evaluative form. For instance, a specific activity in the environmental realm – for example, a water management project across conflict divides – is evaluated according to the role it plays, or may play, in peace-building efforts.[43] Depending on the results, an ongoing project that contains no peace-building component may be refined accordingly. However, the focus should not be too narrowly on "purely" environmental topics, in order to avoid simplifying the complexities of peace and conflict. Instead, analytical findings should always be compared against the overall conflict picture, and ways to integrate environment with other areas such as the economy should be identified. Where possible the findings should always be checked against the perspective of those living in the region and suffering from the conflict, using tools such as stakeholder workshops and focus group discussions.

In summary, though far from offering a cohesive theory or research field, environmental peace-building can provide different avenues for thinking about protracted conflicts. Of course, the risk of securitizing the environment by explicating the links between environment and peace remains.[44] At best this approach could provide a tool or mechanism for those engaged in conflict transformation; at the very least the analytical process itself may contribute to new thinking. Given the nature of violent conflict, a warning should be given against exaggerated expectations. Further research will be necessary to understand the role that environmental peace-building can play – and the roles it cannot – within differing contexts of conflict, regimes and stakeholder constellations.

Notes

1   The authors would like to thank Irina Comardicea for many helpful comments and editorial support.
2   For an overview, see A. Carius, D. Tänzler and A. Maas, *Climate Change and Security – Challenges for German Development Cooperation*, Eschborn: GTZ, 2008.
3   B. K. Moon (2008) 'A Climate Culprit in Darfur'. Online: www.washingtonpost.com/wp-dyn/content/article/2007/06/15/AR2007061501857.html (accessed August 2012).
4   To what extent the outbreak of war and violence in Darfur and Sub-Saharan Africa is related to environmental change is contested. See, *inter alia*, C. S. Hendrix and S. M. Glaser, 'Trends and triggers: Climate, climate change and civil conflict in Sub-Saharan Africa', *Political Geography* 26, 2007, 695–715; S. A. Mason, A. Schnabel, A. Müller, R. Alluri and C. Schmid, *Linking Environment and Conflict Prevention: The Role of the UN*, Zurich and Bern: ETH Zurich and Swisspeace, 2008, pp. 30f.
5   For a general overview, see S. H. Ali (ed.), *Peace Parks. Conservation and Conflict Resolution*, Cambridge and London: MIT Press, 2007; A. Carius, *Environmental Cooperation as an Instrument for Crisis Prevention. Conditions for Success and Constraints*, Berlin: Adelphi Consult, 2006; K. Conca and G. D. Dabelko (eds), *Environmental Peacemaking*, Washington, DC: Johns Hopkins University Press, 2002.
6   J. P. Lederach, *The Moral Imagination. The Art and Soul of Peace-building*, New York: Oxford University Press, 2005.
7   See A. Hammil and C. Besançon (2003) 'Promoting Conflict Sensitivity in Transboundary Protected Areas: A Role for Peace and Conflict Impact Assessments'. Online: www.iisd.org/pdf/2006/security_promoting.pdf (accessed August 2012); M. B. Anderson, *Do No Harm. How Aid can support Peace – or War*, Boulder, CO and London: Lynne Rienner, 1999.
8   See among others, Carius, *Environmental Cooperation*; K. Conca, A. Carius and G. D. Dabelko, 'Building Peace Through Environmental Cooperation', in M. Renner, H. French and E. Assadourian (eds), *State of the World Report 2005*, World Watch Institute, 2005, 144–57; A. Carius, G. D. Dabelko and A. T. Wolf, 'Water, Conflict, and Cooperation', *Environmental Change and Security Program Report* 10, 2004, 60–66.
9   IfP is an EU-funded consortium of ten civil society organizations and think tanks conducting evidence-based research on a variety of topics related to peace-building around the world. See *Initiative for Peacebuilding* (www.initiativeforpeacebuilding.eu) for more information.

10   Please note, we highlight specific aspects only and do not attempt to fully capture the complexity of the region and its conflicts – this would go far beyond the scope of this chapter and rightfully deserves its own books. See among others, S. E. Cornell, N. L. P. Swanström, A. Tabyshalieva and G. Rcheishvili, *A Strategic Conflict Analysis for the South Caucasus. With a Focus on Georgia*, Washington, DC: Central Asia-Caucasus Institute, 2005; T. de Waal, *Black Garden. Armenia and Azerbaijan through Peace and War*, New York and London: New York University Press, 2003; P. Champain, D. Klein and N. Miramanova (eds), *From War Economies to Peace Economies in the South Caucasus*, London: International Alert, 2004.

11   Cf. S. Dalby, *Environmental Security*, Minneapolis and London: University of Minnesota Press, 2002; B. Buzan, O. Wæver and J. de Wilde, *Security. A New Framework for Analysis*, Boulder, CO, and London: Lynne Rienner, 1998. Within the scope of this chapter, we will focus on acute/violent conflict.

12   See Conca and Dabelko, *Environmental Peacemaking*.

13   J. Galtung, *Peace by Peaceful Means: Peace and Conflict, Development and Civilisation*, Thousand Oaks, CA: SAGE, 1996.

14   For an overview on types of environment-induced conflicts, see S. Mason, A. Schnabel, A. Müller, R. Alluri and C. Schmid, 'Linking Environment and Conflict Prevention: The Role of the United Nations' Bern: Swisspeace, 2008.

15   On the possible role and value of dialogue projects for conflict resolution and transformation, see N. Ropers, 'From Resolution to Transformation: The Role of Dialogue Projects', 2003. Online: www.berghof-handbook.net/uploads/download/ropers_handbook.pdf (accessed August 2012).

16   On the role and impacts of democratization, see among others E. Matthews and G. Mock, 'More Democracy, Better Environment?', 2003. Online: http://earthtrends.wri.org/pdf_library/feature/gov_fea_dem.pdf (accessed December 2008).

17   See also E. Weinthal, *State Making and Environmental Cooperation. Linking Domestic and International Politics in Central Asia*, Cambridge and London: MIT Press, 2002.

18   See instead, among others, Ali, *Peace Parks*; Carius, *Environmental Cooperation*; and Conca *et al. Environmental Peacemaking*. See also chapter 10 by Ali and Watzin in this volume.

19   The following chapter is largely based on research conducted within the framework of IfP in 2008, including three field research phases in Spring, Summer and Fall 2008.

20   P. Agrasot, N. Tavitian and R. Kefferpütz (eds), *Greening the Black Sea Synergy*, 2008. Online: http://assets.panda.org/downloads/black_sea_full_report.pdf (accessed August 2012).

21   For a selection of sources, see supra note 10.

22   See A. Wittich, A. Maas and A. Carius, 'Peace, Security and the Environment in the Black Sea Region', in Agrasot *et al.*, *Greening the Black Sea Synergy*, pp. 39–43.

23   Hostilities ceased by the time of writing, but the political ramifications of these events are still difficult to assess. As of mid-August 2008, the status quo ante seemed to have become re-established.

24   International Crisis Group, *Georgia and Russia: Clashing over Abkhazia*, Europe Report No. 193. Brussels: International Crisis Group, 2008.

25   While Georgian president Saakashvili declared that Georgia and Russia were in state of war on 9 August 2008 (C. Georgia, 'Georgia in State of War with Russia – Saakashvili', 2008. Online: www.civil.ge/eng/article.php?id=18988&search=), neither Georgia nor Russia officially declared war on one another.

26   Cornell *et al.*, *A Strategic Conflict Analysis for the South Caucasus*.

27   International Crisis Group, *Azerbaijan's 2005 Elections: Lost Opportunity*. Europe Briefing No. 40. Brussels: International Crisis Group, 2005.

28   N. Mirimanova and D. Klein, *Corruption and Conflict in the South Caucasus*, London: International Alert, 2006.

29   See W. Kaufmann, 'Die Rolle von Nichtregierungsorganisationen bei der Bearbeitung von Konflikten im Südkaukasus' in A. Klein and Silke Roth (eds), *NGOs im Spannungsfeld von Krisenprävention und Sicherheitspolitik*, Wiesbaden: VS Verlag, 2007, pp. 299–312.

30   The Organisation for Security and Co-operation in Europe, the UN Development Program, the UN Environment Program, the UN Economic Commission for Europe and the Regional Environmental Centre Caucasus plus the North Atlantic Treaty Organisation as associated member.

31   See A. Wittich and A. Maas, *Regional Cooperation in the South Caucasus. Lessons for Peacebuilding, from Economy and Environment*. Brussels: Initiative for Peacebuilding, 2009.

32   The Regional Environmental Centre for the Caucasus. Online: www.rec-caucasus.org/ (accessed August 2012).

33   See UNDP, UNEP and OSCE, *Environment and Security. Transforming risks into cooperation. The case of the Southern Caucasus*. Geneva/Vienna/Bratislava: UNEP/ OSCE/UNDP, 2004. See also Wittich *et al.*, 'Peace, Security and the Environment'.

34   See T. de Waal, 'Soft annexation of Abkhazia is the greatest legacy of Putin to his successor', *Caucasian Review of International Affairs* 2 (3), 2008. Online: www.cria-online.org/Journal/4/Interview_with_Thomas_de_Waal.pdf (accessed August 2012).

35   E. Bell with D. Klein, *Peace-building in the South Caucasus: What can the EU contribute?*, London: International Alert, 2006; Mirimanova and Klein, *Corruption and Conflict*.

36   There are various forms of conflict analysis methodology. *The Berghof Handbook for Conflict Transformation*, available online, provides a comprehensive overview in this regard. See Berghof Foundation, online: www.berghof-handbook.net/ (accessed August 2012).

37   For an overview, see N. Pidgeon and K. Henwood, 'Grounded Theory', in M. Hardy and A. Bryman (eds), *Handbook of Data Analysis*, London/Thousand Oaks, CA/New Delhi: Sage, 2004, pp. 625–48.

38   Kenya provides an instructive example in this regard. See C. Kahl, *States, Scarcity, and Civil Strife in the Developing World*, Princeton, NJ and Oxford: Princeton University Press, 2006.

39   For the concept of agent of change, see Lederach, *The Moral Imagination*.

40   Cf. R. J. Fisher, *Methods of Third-Party Intervention*, 2001. Online: www.berghof-handbook.net/uploads/download/fisher_hb.pdf (accessed August 2012).

41   Cf. Annika Kramer, *Regional Water Cooperation and Peace-building in the Middle East*, 2008. Online: www.initiativeforpeacebuilding.eu/pdf/Regional_Water_Cooperation_ and_Peacebuilding_in_the_Middle_East.pdf (accessed February 2009).

42   On the "do no harm" approach, see M. B. Anderson, *Do No Harm. How Aid Contributes to Peace – or War*, London: Lynne Rienner, 1999.

43   For an example from the Middle East, see Kramer, *Water Cooperation*.

44   Cf. R. Floyd and R. A. Matthew, chapter 1 in this volume.

6

ENVIRONMENTAL DIMENSIONS OF HUMAN SECURITY

Simon Dalby

Introduction

At the end of the Cold War security was rethought in numerous ways and the policy debates suggested that states were no longer the only referent objects in need of securing. Indeed given the threats many states posed to their own populations and those of surrounding states in some cases it was clear that states, and in particular their nuclear arsenals were the threat that needed to be tackled. Disasters at Chernobyl and Bhopal in the mid 1980s had pointed to the technological hazards of industrial systems, adding to the arguments concerning nuclear winter as a potential danger that made nonsense of claims that warfare between major industrial states was a viable policy option. The larger human predicament in terms of environmental dangers and the pressing necessities to alleviate poverty through development were simultaneously being addressed by discussions of sustainable development.[1]

Political discourse also returned to some of the original themes in the debates about the establishment of the United Nations dealing with human rights and such matters as genocide conventions and collective defence. Wars in Iraq and the Balkans and state collapse in Somalia suggested that new forms of intervention were needed and that the safety of populations was now a higher priority than the perpetuation of state sovereignty. Likewise the origins of threats were broadened to include matters that rendered people insecure even if no direct aggression was involved in causing their plight. But in these cases, and the environmental case in particular, it was far from clear how traditional military notions of security had any bearing. In Gwyn Prins' pithy formulation the Top Guns were apparently powerless to deal with Toxic Whales.[2]

Much of this discussion coalesced under the rubric of "human security" in the mid 1990s. But the term itself is not easy to define with any precision. It clearly refers to the safety of people not states, places obligations on international institutions to deal with what became known as complex humanitarian emergencies, and bundles these together with aspirations to a liberal order where economic progress and human rights are combined as a policy desideratum as well as some sort of analytical lens.[3] The tensions and difficulties in this discussion are considerable as later sections of this chapter will outline, but the aspirations to human security for all are widely articulated. Most recently the discussions of human security have been explicitly connected with matters of global warming, and in the aftermath of Hurricane Katrina in particular, questions have been raised about the apparently enhanced vulnerabilities of people due to climate change. Fears that increasingly severe storms and weather disruptions including floods and droughts will both kill people, and when they migrate to avoid catastrophe, cause political instabilities, have linked human, environmental and traditional concerns with national security once again.[4] But far from all of the discussion about human security has been linked, even indirectly, to matters of environment.

This chapter starts with the widely cited formulation of human security in the 1994 edition of the United Nations Development Report.[5] This ambitious agenda has shaped much of the discussion since but the ambiguities as to how environment is conceptualized in terms of human security are clear right from the start here. Subsequently Michael Renner, a key thinker in the Worldwatch Institute's engagements with security and environment, extended the formulation and linked it to discussions of globalization. Human security has also focused on the matter of natural hazards and more recent work has suggested both that social networks are key to surviving disasters, and that human security should be explicitly extended to link up with disaster research in a focus on making people free from hazard impacts.[6] Finally the policy framework of human security has raised crucial questions of international responsibilities in the case of humanitarian emergencies and the question of how and where the international community has the "responsibility to protect" people in danger. This has recently been extended to matters of environmental emergencies and the impacts of climate change; the chapter closes with a discussion of how human security now relates to the emergent frameworks of "sustainable security" and the need to think about these in the new geological context of the Anthropocene.

Human security and the United Nations Development Program

In 1994 the United Nations Development Program (UNDP) encapsulated the new agenda in a discussion of human security. The UNDP authors argued that:

> The concept of security has for too long been interpreted narrowly: as security of territory from external aggression, or as protection of national interest in foreign policy or as global security from the threat of a nuclear holocaust. It has been related more to nation-states than to people.[7]

This suggestion challenged the implicit assumption in most security studies that states were the most important entity and that war is the primary danger facing humanity. Once the nuclear standoff between the superpowers was over, as it clearly was by the early 1990s, other matters could intrude on discussions of survival, state policy and long term security planning.

The concept of human security had, they asserted, at least four essential characteristics. First, it is a universal concern relevant to people everywhere. Second, the components of security are interdependent. Third, human security is easier to ensure through early prevention. Fourth, and perhaps the crucial innovation in this formulation, is the shift of the referent object of security from states to people. More explicitly this formulation defines human security as:

first, safety from such chronic threats as hunger, disease and repression. And second, it means protection from sudden and hurtful disruptions in the patterns of daily life – whether in homes, in jobs or in communities. Such threats can exist at all levels of national income and development.[8]

The argument suggests that the concept of security must change away from Cold War and realist preoccupations with territorial security to focus on people's security, and from armaments towards a reformulation in terms of sustainable human development. Demilitarization is obviously part of this agenda, but human welfare broadly conceived is the overall thrust of the concept. As such there are numerous threats to human security, although specific threats are likely to be locale dependent. Nonetheless, according to the UNDP threats to human security come under seven general categories: economic, food, health, environmental, personal, community and political.

While many on the UNDP list of threats to human security are local threats, global threats to human security are said to include at least six categories, caused more by the actions of millions of people rather than deliberate aggression by specific states. As such they would thus not be considered security threats under most narrow formulations of security studies. These six are: 1) unchecked population growth, 2) disparities in economic opportunities, 3) excessive international migration, 4) environmental degradation, 5) drug production and trafficking, and 6) international terrorism.[9] These categories and the formulation of human security underlay a growing number of assessments of the state of world politics in the 1990s and attempts to encapsulate political agendas for reform of the international system.[10] While these themes were temporarily removed from the international agenda in the immediate aftermath of events of September 11, 2001, they reappeared later in the decade and have been recently readdressed in part as a result of the rejuvenated discussion of climate change and its implications.

More specifically when the UNDP authors turned their attention to environmental matters they looked both to national concerns and environmental threats within countries, including in developing countries where clean water supply, loss of forests and desertification are noted as particular problems.[11] In industrial states air pollution in big cities and the loss of forests and agricultural land due to pollution are noted. Salinization of farmland and severe floods and droughts due to land use changes are noted, as is the fact that people who move onto marginal land are more vulnerable to natural hazards. Cyclones, which hit Bangladesh regularly, get a special mention. The discussion of global environmental threats includes acid rain in Europe, chlorofluorocarbons and the ozone layer depletion, greenhouse gas emissions, loss of biological diversity, accelerated coastal marine pollution and declines in fish catches as well as coral reef loss and the fragmentation of valuable habitats.[12]

But while this comprehensive agenda of policy aspirations, and its attempt to reinterpret the priorities for international politics, covered many of the issues discussed in the 1990s, it offered neither obvious priorities for policy nor an effective analytical lens through which research could be conducted on the parameters of human insecurity. Policy aspiration and academic analysis merge here frequently in various ways but the sheer breadth of matters that related environmental matters to human insecurities precluded any clear research agenda or policy focus. Nonetheless the link between human security and environment had been clearly made in this document and it was to influence many researchers subsequently. The fact that this formulation of human security came from the UNDP, with its focus on development, emphasizes the discussion of the vulnerability of the most marginal and the poor; those for whom development supposedly promises a better life. Human security is formulated by the UNDP precisely in terms of providing the conditions that make development possible.

Globalization and environmental insecurity

Ironically it was precisely these poor and marginal peoples who were the object of some of the early 1990s literature that formulated matters in Malthusian terms of population growth as the primary threat to the national security of at least the United States. While the most high profile articulation of this argument came in Robert Kaplan's infamous *Atlantic Monthly* article on "the coming anarchy" published in 1994, the same year as the UNDP formulation of human security, it was far from the only source of alarm.[13] The contrast between the Malthusian approach, which effectively blames the poor as the authors of their own misfortune, and the more sophisticated thinking that starts with simple empirical questions concerning the causes of their insecurity, emphasizes the importance of the politics implicit in formulations of human security.[14] This is the key methodological point in the whole debate and has shaped most of the discussion.

As Michael Thompson's scathing critique of Malthusian environmental security thinking clearly showed, the "ignorant fecund peasant" model of overpopulation leading to environmental crisis that was frequently popular with Northern environmentalists, is wildly inaccurate. This is so not least because of the gross over-generalizations about peasant behaviour, but more importantly because of the overlooked intricacies of rural political ecologies, as well as global trade interconnections that frequently disrupt marketing arrangements and cause disastrous price fluctuations.[15] Local farmers actually usually know a lot about their land, water supplies, animals and what will grow where and when, but in the last couple of centuries they have frequently been removed from their land, made the scapegoats for their own misfortunes and blamed for their ignorance of modern ways. Nonetheless such misconceptions continue to shape government policy, not least in Asia where Thompson's critique was specifically directed.[16] Now marginalized herders and farmers are part of the new discourse on climate security, although the more thoughtful commentators do understand that their vulnerabilities are mostly not of their own making.[17]

In contrast to these simplistic neo-Malthusian understandings in the 1990s, Michael Renner of the Worldwatch Institute in Washington used the United Nations' concept of human security to investigate the global connections between environmental problems and social conflict.[18] Pointing to the growing inequities in the global political economy he suggested that these are related in various ways to both degradation and conflict. Renner asserted that:

> It is becoming clear that humanity is facing a triple security crisis: societies everywhere have to contend with the effects of environmental decline, the

repercussions of social inequities and stress, and the dangers arising out of an unchecked arms proliferation that is a direct legacy of the cold war period.[19]

Human security worries, he continued "are now magnified by the unprecedented scale of environmental degradation, by the presence of immense poverty in the midst of extraordinary wealth, and by the fact that social, economic, and environmental challenges are no longer limited to particular communities and nations".[20] Political violence is often widespread in places where war is not officially occurring.

But political fragmentation, fuelled by the proliferation of weapons, is only one half of Renner's story of human insecurity. The other half is the accelerating processes of globalization that are interconnecting the world's economies and cultures in ways that often operate to undercut traditional economies and challenge the sustainability of agricultural and survival practices. Wars in the South are now frequently about who will control the revenue from the extraction of resources for sale on the global market; a form of shadow globalization that has gradually come to be recognized as a matter of insecurity due to the connections of the global economy rather than any intrinsic local scarcities.[21] The processes of accelerating globalization explain much more than the geopolitically inadequate formulations of Malthusianism, which continue to assume that difficulties are driven by local autonomous processes of over-population.[22]

Renner's argument, picking up on contemporaneous 1990s critiques of the operation of the global economy and matters of international debt and development, suggested that the forces setting populations in motion was a crucial dimension of human insecurity. Looking to the global interconnections puts human security into the circuits of the global economy, and reminds readers that famines are about poverty much more than about food shortages. The historical patterns of commodification, overexploitation and domestication of resources in the expansion of commercial economies are not new.[23] Neither are the population disruptions and displacements that often result. These displaced people may in turn damage marginal lands unsuited to agriculture, or swell the ranks of the poor in urban shanty towns caught up in processes that Thomas Homer-Dixon has called "ecological marginalization" in his framework of the links between environmental scarcities and political conflict.[24] But, as it turns out, only rarely do these populations cause political violence that can be reasonably attributed to environmental degradation. The interconnections between globalization and environmental degradation are crucial to Renner's argument, but the processes in motion are also those carried out by political and military elites. Modernization is about state making. The enclosures and displacement of rural populations that are part of the state backed and enforced expansion of commercial agriculture cause the displacement of many "environmental refugees".[25]

On the largest scale, these people become the object of Malthusian security narratives on the crises of migration, and invoke alarm as migrants apparently threaten the integrity of the nation to which they flee. Renner's point was that rural land reform and the construction of less unequal land holding arrangements is often the key to improving environmental conditions and other facets of life in rural areas. But such initiatives may run directly up against elites whose power and wealth rest on the unequal landholdings, and whose businesses dominate the expanding commercial sector of the agricultural economy of developing states. As Colin Kahl's research has subsequently illustrated, where this is engaged in agricultural production for export, sometimes partly as a result of internationally imposed economic conditionalities and structural adjustment policies, the links between globalization and peasant struggles are especially direct.[26]

This has become especially pressing in the first years of the new century as states and corporations are involved in gaining access to agricultural lands, in Africa and elsewhere as strategies that anticipate future food shortages. Securing supplies for distant urban populations is leading to practices of land-grabbing that accentuate rural inequities and displacements. This is especially ironic in terms of environmental security where the land is being used to grow biofuels, justified as an ecologically sensible alternative to fossil fuels that are causing climate change.[27] Getting the geography of these insecurities clear is key to contextualizing both human security and the debate about the relationships of environment and conflict.[28] Without this clarity on who is insecure, and specifically where, numerous generalizations can easily follow about resources and conflict but they suffer analytically from this lack of clear contextualization. This in turn may lead to policy suggestions that are inappropriate for dealing with the sources of insecurity.

Frameworks for analysis

The contrast between a focus on local environments as determining social life, and the potential for conflict arising in particular places as a result of large-scale economic and political changes, parallels the shift in focus from national security to a much more encompassing formulation of security in the UNDP framework. On the one hand human security does look to the smaller scale and the practical conditions of life for people, not just states. On the other hand it looks to the global economy, and political processes which flow across boundaries to render people insecure in many places. It's about governance broadly understood, but also about the simple but crucial fact that people are sometimes insecure not as a result of the deliberate actions of another state, but as the accidental consequences of economic and environmental changes which happen to have unfortunate effects in particular places. Most recently the growing alarm about the effects of climate change has added an important additional link between the local and the global causes of insecurity.

Modernity and the expansion of the global economy is a disruptive force, and one in part based on the appropriation of rural natural resources to supply commodities to the metropoles. Many parts of the world were incorporated into the global economy in the nineteenth century expansion of European empires and the patterns of poverty and dependence established in this period still have consequences generations later because of the property arrangements and trading patterns.[29] This is especially clear in Richard Matthew's research work in Pakistan conducted and published immediately prior to the events of September 11, 2001. In trying to unravel the human security consequences of the environmental situation in the North Western regions of Pakistan Matthew found it necessary both to look back to the legacy of the British Empire in the region as well as the contemporary patterns of poverty and tribal rivalries.[30] The imperial legacy, stretching back to the wars of the nineteenth century, and to British attempts to control the region through a combination of logging projects and inefficient irrigation schemes, has shaped the ecological context within which contemporary problems of underdevelopment and conflict have to be understood.

Across the rest of the subcontinent multiple patterns of insecurity are related to environmental matters in complex ways.[31] Widespread poverty in India coexists with a burgeoning middle class consumer society and a state that has a nuclear arsenal at the apex of a complex military establishment. Bangladesh is a densely populated country with a huge subsistence agriculture population vulnerable to storms and flooding as well as to droughts that are at least potentially aggravated by dams built upstream on the major rivers in India. Sri Lanka's civil war was a complicated struggle in which entitlement to agricultural land and water was an integral part of the rural struggle. Nepal's marginal state has been aggravated by a lengthy insurrection. All this may be complicated in the future by climate change and the melting of the glaciers in the Himalayas which will change the flows of the rivers supplying water to the food growing areas for a substantial part of humanity.

None of this is amenable to resolution by traditional military means, although military institutions can be of considerable assistance in disaster relief and such things as building storm shelters; clearly human security is something that is not provided for by tanks, guns and fighter-planes, much less nuclear weapons. Human security requires a shift of focus from states and armies to the conditions that make people insecure in particular places. Clearly the environmental dimensions of human insecurity are more a matter of vulnerability to the disruptions of daily life than a matter grasped through the traditional focus on national security. In so far as climate change disrupts directly by intensifying such things as the monsoon rains in Asia, which caused massive flooding in Pakistan in 2010, then global processes are key to insecurities in particular places.

Global environmental change and human security

Precisely these kinds of vulnerabilities are at the heart of the research programme undertaken by the Global Environmental Change and Human Security project (GECHS).[32] This attempts to step back from the detailed within-state analysis of human security and link the security discussion explicitly into the global environmental change literature. The implicit assumption in much of the literature on climate change in particular and more generally in discussions of environmental change more broadly, is that these changes will threaten societies with both long term stresses and immediate hazards, and possibly political violence too.[33] Both long term environmental change with its consequences on agriculture, forestry and rural land use, and the immediate hazards as a result of storms, floods, droughts and wild fires apparently will threaten the human security of people in many places in coming decades. Here human security discussions link up with disaster research and the larger discussions of human vulnerabilities of various kinds.

But it is not so obvious how these links can be made, nor how research can be conducted to examine such vulnerabilities. Especially in commercial economies, and in the case of farming in developed states, climate change may be only a minor concern relative to technological changes, commodity prices and short term meteorological fluctuations.[34] While earlier work on disasters and natural hazards emphasized the immediate disruptions and loss of life caused by discrete events, more nuanced understandings recently have suggested that a more complex approach is needed. In particular work on famines has long suggested that hazards turn into disasters as a result of complex social phenomena; shortages of food are compounded by a lack of social entitlements that make particular people vulnerable to starvation.[35] This analysis of disasters suggests very clearly that access to social networks and the availability of cash to purchase essentials is key to survival. The poor, and especially women and children, suffer precisely when they do not have these social networks in place to fall back on. Hazards frequently disrupt networks and do so especially when evacuations are necessary in the face of flooding or other disruptions.

In so far as global environmental change is increasing the frequency and severity of hazards, and not only that of the large events that garner global media attention, then these insecurities are likely to be aggravated. But they are also now inflicted on a population that has become more urban and, in the case of many of the burgeoning slums of the Southern megacities, is living on marginal land in situations that are exposed to floods, landslides and other hazards.[36] Slums frequently do not have basic infrastructure requirements, nor are building structures robust enough to withstand storms; hence vulnerability is also a directly physical phenomenon for these urban residents. Vulnerability has to be understood as a complex social and ecological situation that is entwined with the social and economic entitlements available to particular people in specific circumstances.

In addition to vulnerability understood in these terms it is also helpful to understand coping capacities as part of what needs to be examined.[37] How systems adapt to stressors, or how they reconstruct themselves after a major disruption, is key to understanding the long term outcomes of disasters. Resilience refers to the ability to bounce back after a disruption; most ecosystems have the ability to rebound provided most of the building blocks are still available. They may not of course replicate exactly the prior situation, but can adapt and respond to disruptions. Forests regrow after fires if seeds survive the conflagration. But of course if they are completely removed as might happen in a major landslide, or as a result of a volcanic eruption, then resilience is overwhelmed and eventually entirely new ecologies will result. Dynamic and adaptable systems are much more likely to rebound quickly; those without such flexibilities are likely to either take much longer to bounce back or fail to do so entirely. Hence resilience is key to understanding disaster outcomes; the speed of possible reconstruction too then is an important factor in vulnerability.

While there are numerous dimensions to these phenomena it is clear that those with access to property, credit and cash are much more likely to recover quickly after a disaster. The story from New Orleans in the aftermath of Hurricane Katrina in 2005 suggests the rapid recovery of institutions such as universities and sports franchises, while the poor tenants evacuated from the ninth ward of the inundated city in particular were mostly incapable of rebuilding their lives.[38] Renters are obviously dependent on their landlords to either rehabilitate or reconstruct buildings. The availability of public housing for the poor adds an explicitly political dimension to their plight; if the city and state choose not to reconstruct damaged housing then those without access to the private housing markets are simply out of luck. Similar dynamics work in other cities, but the informal sectors of vulnerable Southern cities may in some ways be able to adapt better than those dependent on state institutions in such places as New Orleans. Indeed one of the notable factors in the Mumbai floods, a month prior to Katrina, is the speed and flexibility of how the informal networks in the slums of Mumbai responded to the crisis, although they alone were not adequate to cope with the inundation.[39]

Policy frameworks

Such considerations raise a host of issues for scholars of security policy. While there clearly is a role for a military dimension in disaster relief, the traditional understandings of national security and the priority given by states to military preparations have little to offer by way of guidance as to policy and preparations for dealing with increased hazards and disasters. If traditional policy responses emphasize migrants and disaster victims as a threat to national security, as an external disruption that must be contained, then the fate of those subject to environmental disruptions may simply become worse.[40] If however, disaster politics works to encourage international cooperation and humanitarian assistance then new habits of collaboration may change security policies quite dramatically. The point is that states, and especially non-governmental organizations, are frequently motivated to assist across national borders when disaster strikes. How they do so matters and the disaster research clearly suggests that enhancing resilience is important, as is post-disaster reconstruction. Financial arrangements that try to insure against catastrophic disruptions are frequently of little use to marginalized peoples.[41] Who decides on what is reconstructed is key; the ability of dynamic commercial sectors and political elites to take advantage of the destruction brought by disasters, and to do so at the cost of the more vulnerable parts of society, is a long established pattern.[42]

Broadening security to include such considerations, and the United Nations formulation of human security in the 1994 United Nations Development Programme does just this, requires a much more complicated formulation for security than present in traditional security studies. This broadened agenda requires consideration of security for whom, from what threats, and provided how in what specific circumstances. Vulnerability analysis makes it clear that simple invocations of society are not enough to formulate policy; clearly gender, age, social circumstances and wealth all matter in how insecurity happens. In trying to respond to this problem Úrsula Oswald Spring has suggested a conceptualization in terms of "human, gender and environmental security". While HUGE is certainly a very apt acronym for this agenda, the necessity to try to think all the facets of human insecurity together is now unavoidable.

Oswald's formulation of gender extends it beyond a simple male and female dichotomy to suggest that gender is key to a more complicated sociological imagination that includes children, elders, indigenous peoples and other minorities as agents in the human dimensions of environmental challenges, the need for peacebuilding in numerous places and in attempts to tackle equity issues.[43] Gender understood in these terms connects with security understood in terms of livelihood, food, health, education, public safety and cultural diversity. The critical dimension challenges hierarchies and violence that have made so many vulnerable by how gender roles have shaped societies, and suggests very clearly that human security has to focus on these people rather than on maintaining the social order that simply maintains the control and safety of rich men running states.

HUGE thus also looks to other social organizations, equity and development, and such things as ethical investment strategies, and broad public participation in political decisions to tackle discrimination and violence, to secure humanity. Consequently "environmental security" concerns are incorporated because a healthy environment and strategies of resilience building for vulnerable groups and especially for women should reduce the impacts of hazards. To do so, especially in areas that are hazard prone, this approach focuses on both the potential for financial and technical innovations which enhance women's own resilience from the bottom up and the provision of state institutions that can warn of impending hazards, organize evacuations and facilitate relief and reconstruction efforts subsequently. Advance planning to deal with disaster in isolated regions might help greatly to prevent famine or violent conflict in the aftermath. But to do all this requires both an understanding of the complexity of human networks and support systems and a political commitment to aid all citizens facing hazards. It also requires recognition of the increasingly interconnected social and ecological systems that are the contemporary human context.

A similar set of concerns structures Hans Günter Brauch's formulation of matters in terms of Human and Environmental Security and Peace (HESP), explicitly linking together the environmental and structural factors that render people insecure with an orientation to international institutions and peacebuilding.[44] This is both an attempt to develop a research agenda and orient it to the particular circumstances that make people insecure, as well as an attempt to promote a policy agenda that takes this orientation seriously. Much conceptual work needs to be done to link environmental change with specific regional outcomes, and the key themes of sustainable development and how it can be linked to sustainable peacebuilding simultaneously with a reconceptualized theme of security.

Building on earlier formulations of human security, in particular in terms of the themes of the freedom from fear, and freedom from want, promoted by the Human Security Network and the Commission on Human Security respectively, Brauch has suggested a "freedom from hazards" policy approach. This is loosely similar to the focus in HUGE emphasizing the importance of building resilience and dealing promptly with disaster impacts. Consistent with the original UNDP orientation, its focus is on prevention and anticipating dangers rather than on reaction after the event. As such it links directly to international disaster prevention initiatives and emergency prevention policy measures. But it does so with an explicit recognition of the growing dangers of global environmental change and the need to think about infrastructure provision and international cooperative efforts. The great difficulty with this mode of thinking is simply the numerous things that have to be brought together to analyse contemporary human security in the age of globalization.[45]

These links between globalization and environmental change have been discussed in terms of "double exposures", emphasizing that the GECHS framework is about both environmental and global economic phenomena.[46] Looking at both together shows how people are vulnerable because of both, but in different ways in different places depending on how these two related phenomena play out in particular contexts. But globalization is all about economic development which is largely powered by fossil fuels which are causing greenhouse gas increases in the atmosphere, and causing deforestation and land use changes as well as the urbanization which makes people vulnerable in new circumstances. These are all parts of the same process, and they raise the biggest questions of the obligations of governments and international organizations to provide security for people, not just for states in these new times. Once again human security challenges the simple assumption that states provide security.

Humanitarian intervention and the responsibility to protect

This insistence on thinking about the international dimensions of human security, and the obligations on the part of states and the international community to protect citizens of all states from security threats, links up with the discussion of human security in terms of the appropriate response to humanitarian emergencies and what has become known, following the title of the 2001 report of the International Commission on Intervention and State Sovereignty (ICISS), as *The Responsibility to Protect*.[47] This formulation suggests that states have the obligation to provide for the safety of their citizens, and should they very obviously fail in this duty the international community has the obligation to intervene to provide the necessary assistance.

While highly controversial in a system where sovereignty has long been the principle on which the United Nations system at least notionally operates, the theme of the necessity to intervene in a crisis where a state is incapable, is part of the larger human security agenda, and it reflects the larger shift toward cosmopolitan sensibilities that globalization has initiated. The arguments for intervention are heavily circumscribed to deal with the obvious objections to the violation of national sovereignty. Both pressing urgency and serious consequences are necessary to justify an intervention; the principal is clearly an emergency measure to be invoked in times of complex humanitarian emergency, rather than as a matter of the routine operation of international relations. But given the current circumstances of environmental change, of which climate change is only perhaps the most obvious manifestation, the question of environmental obligations across borders has once again become unavoidable.

Rarely are the political dilemmas implicit in such a formulation more clearly delineated than in a 2007 essay by political theorist Robyn Eckersley.[48] Pondering the circumstances in which the criteria of the "Responsibility to Protect" might apply to environmental matters, she concludes that only in some unusual circumstances might an ecological intervention be possible and justified. Specifically her argument suggests that the harm caused by the extinction of a major species such as the great apes in Rwanda, or the imminent melt-down of a nuclear reactor, in a replay of the Chernobyl accident of 1986, might meet the criteria of immediate emergency circumstances and great harm if no intervention is attempted. But is this an adequate formulation of environment in which human security might now be understood? Where Eckersley draws attention to fairly narrow terrestrial environmental phenomena and poses the question of interventions to deal with these matters, the larger environmental "emergency" that we collectively face is not obviously amenable to interventions of the sort she discusses.

Taking ecology seriously as a science requires a larger and more encompassing view of what might be in need of ecological defence. In the last couple of decades science has made dramatic strides in understanding the biosphere and the dynamics of planetary systems.[49] Earth system sciences have made clear that humanity and the rest of the biosphere are interlinked much more closely than the normal assumptions of life in territorial states suggests. This science has also suggested that the most important drivers of the biosphere are in many ways not terrestrial, but the atmosphere and the oceanic system that between them determine the climatological conditions for land based species. The most important mechanisms shaping our biosphere are oceanic and atmospheric systems that we have already altered to such a degree that it is now commonly asserted that we live in a new geological period, which has been named "the Anthropocene". These changes have been accelerating in the last half-century literally driven by the fossil fuelled global economy.[50]

All of which raises the important point that the international system of states, granted responsibility for ensuring protection to its peoples, can be judged to have fairly systematically failed to act in a prudent manner to head off the worst imminent effects of these changes. The Kyoto protocol, for all its faults, is an international agreement under which some states have obligations to reduce their carbon emissions. Failure to do so is fairly directly leading to changes which will, in the foreseeable future, have consequences for the territorial integrity of many states and even the physical survival of a few. While this may, as Eckersley argues, be a more gradual process than a nuclear reactor melt-down, or perhaps, although this is really unlikely, a more reversible process than the elimination of a species closely similar to humans, as in her Rwanda example, the sheer scale of the changes in the biosphere and its fundamental challenge to the survival of low lying Atoll states in the Indian and Pacific oceans is surely a much more compelling case for emergency action to prevent their inundation and elimination as member states of the United Nations and as peoples? While their vulnerabilities are complicated and so far at least the stresses on the Islands have not led to violent conflict, their human populations are insecure in a number of interconnected ways.[51]

The Atoll states in the Pacific and Indian oceans, and the low lying littoral states, only most obviously perhaps Bangladesh, have no military options to intervene in this threat to their physical survival, now knowingly exacerbated by the affluent states, the annex I countries under the Kyoto Protocol, who disregard their international commitments. What then is to be done, and by whom? How might a "responsibility to protect" be acted upon by the poor and marginal states in response? Those directly subject to sea level rise, the possibility of more severe hurricanes and other possible hydro-meteorological hazards, surely have a compelling ethical argument for recourse to "ecological defence" in the face of the profligate use of fossil fuels in developed states which indirectly, albeit probably unintentionally, endangers their populations and territory.

Despite such arguments, climate change and such matters are still considered a matter mainly for state action, and in so far as powerful states have in many cases signally failed to live up to their obligations, the big political question raised by climate change in particular is what other options there might be for ethical action on this matter in light of the need to act to protect the human security of island populations, and many others rendered vulnerable in various places as a result of climate change. International affairs is no longer only a matter of territorial states and military force, and discussion of ethical action and human security needs to consider other actors including corporations, citizens and all manner of other communities not constrained by state boundaries.

Suggesting that climate change is not an emergency, because it is not immediate, as Eckersley does, avoids confronting the consequences of extravagant consumption and once again points the finger of accusation at poorer and marginal states as in need of interventions, rather than looking directly at the sources of the biospheric disruptions in the Anthropocene and linking matters of justice to resource extractions and luxury consumption.[52] It also fails to accept the extent to which climate changes are already contributing to insecurity in many places.[53] The inadequacies of the existing state system to deal with these matters once again emphasizes the importance of the broader human security agenda, but also suggests that ethical action in the larger cause of ecological defence will have to overcome the limits of "intervention" as defined in the ICISS framework. Nonetheless, even in recent evaluations of the likely security consequences of climate the focus on national security rather than people or human security remains entrenched in scholarly analysis.[54]

Reformulations: sustainable security

Recent attempts to grapple with these complicated interconnected security matters have suggested to thinkers on both sides of the Atlantic that some larger formulation of security in terms of "sustainable security" is needed if security is to be about addressing the looming disruptions that climate change is bringing to societies in many places. Environmental security is no longer a matter "out there" in the rural hinterlands, but as residents of many cities, and notably Bangkok and Manila in late 2011, know to their cost, a part of contemporary urban life now when extreme events inundate cities. The Oxford Research Group's Sustainable Security project now links the causes of insecurity to four main drivers. These are climate change, competition over natural resources, global militarization and the marginalization of populations as a result of contemporary changes.[55] The focus here is on the interconnected nature of the dangers that make people insecure. Consistent with the 1994 UNDP formulation of human security as primarily a matter of prevention, rather than violent action in the face of emergencies, the necessity of thinking through the connections between ecological, economic and social changes is foregrounded in their analysis. The alternative may be, as Oxford Research Group member Paul Rogers puts it, a matter of military force "keeping the violent peace" and policing the disruptions that contemporary changes are setting in motion.[56] But human security for the majority of the world's peoples this certainly is not!

Military interventions to aid in the provision of human security have not been successful in some high profile cases, especially the intervention in Somalia in the early 1990s. There, assisting in the provision of humanitarian aid, and food in particular, did seem to improve matters, but the attempts to deal with the civil war by attempting to capture Mohamed Aidid in particular aggravated rather than resolved matters.[57] Clearly drought was part of the disaster and as in so many wars in the past food became a weapon and a resource to be fought over, nonetheless the larger failure of governance is key to explaining why so many people died in the 1990s and more recently too. Intervening too late allowed the situation to spiral out of control; prevention wasn't tried despite clear warnings of imminent difficulties. Military interventions designed to remove particular regimes from power are clearly not enough; in many cases they simply make things worse as the troubling legacy of the "war on terror" makes clear.

In Washington a major think-tank has produced a loosely similar sustainable security analysis in charting the interconnected nature of contemporary globalization and in the process made it clear that development is in many ways at the heart of contemporary insecurities. Migration, poverty and dislocation are not matters that can be dealt with by force, although military agencies clearly have a role to play in crisis situations.[58] A security strategy that fails to deal with the processes already in motion that are making people insecure is one doomed to failure. Even the American military is beginning to grapple with some of the implications of such thinking.[59] Clearly a paradigm shift is needed in how security is thought; the violence of the war on terror has been largely counter-productive, and using war-fighting strategies for dealing with such things as poverty, drugs and climate change is not making Americans or anyone else more secure. The human security agenda, spelled out in the early 1990s is, it seems, finally gaining traction as a new mode of thinking about governance even within the Washington security establishment where traditional notions of national security are still dominant.

Human security in the Anthropocene

But if these reformulations of security in terms of sustainability are to follow through in terms of completing the paradigm shift in security thinking, they also need to incorporate the change of perspective that has informed ecological sciences in the last few decades, and is gradually changing understandings of environment, and the priorities of environmental movements too in the last few years.[60] Earth system sciences have made it increasingly clear that humanity is now a major force shaping the surface of the earth, drastically changing the species mix of the planet, acidifying the oceans and, by turning rocks into air so enthusiastically by the combustion of fossil fuels, changing the composition of the planet's atmosphere. In short we now live in a new geological age, the Anthropocene, one shaped by industrial humanity. This provides the context within which we now need to think about security and humanity's place in the larger scheme of things. We need a new "political geoecology" for the Anthropocene, where security involves quite literally thinking about how we are making the planetary future, in whose interests, and how it might be appropriately shared so as to render most people free from the most obvious dangers to their security.[61]

The sheer scale of human activity has changed the context of humanity in the last couple of centuries, and this changing context, the new Anthropocene era, is the appropriate framework within which human security must now be understood.[62] Combining HUGE and HESP with a consideration of the sources of endangerment, and the inadequacy of existing institutions to cope with the vulnerabilities now obvious in the Anthropocene, starkly presents the dilemmas that face both analysts and policy makers trying to tackle matters of environmental security. But what is clear from the analysis in this chapter is that the traditional focus on states as protectors is no longer an adequate geopolitical specification for what needs to be done; the formulations of human security make this point unavoidable, the recognition that equity and impacts do not follow state boundaries reinforces this point,[63] but they do not resolve the dilemmas that need urgent attention.

On the other hand, remembering the key point in the UNDP's initial formulation is especially important: human security is mostly about prevention. It's about anticipating dangers and acting to head them off before downward spirals happen. This is even more important in the context of climate change and the long term increasingly interconnected future of humanity. Both the immediate contexts in which people live, and the larger environmental processes that are now increasing the hazards that make people vulnerable in these new urban contexts are partly caused by humanity's actions. This is very troubling for people still thinking in terms of an external environment as the context for human activities, but the new ecological circumstances of the Anthropocene in which we all live make it clear that we are collectively shaping our vulnerabilities, both directly in terms of where people live, how they are fed, and how sanitation and other necessities are provided, and indirectly through our changing the biosphere, precisely by the economic activities and the changes to ecosystems that we have set in motion. Thinking about these sources of environmental dangers, and the possibilities of international actions to "intervene" to promote human security, makes the political dilemmas especially acute.

The good news about environment and human security is that there is actually considerable potential for international cooperation on environmental matters and resource management to facilitate peacemaking and peacebuilding.[64] Cooperation in many places has shown itself to be conducive to improving lives, providing jobs and ameliorating the worst environmental disasters. Once again thinking about security as building for the future, rather than preparing to fight when the worst happens, turns out to be the successful strategy. But this requires a very different framework for analysis. It requires thinking about environmental matters as key to providing security rather than as a threatening external matter that might cause conflict.[65] Here the human security framework is key to thinking about the international cooperation needed to deal with building a post-carbon fuel economy just as much as it is key to the smaller scale wise use of river water resources or forest projects which alleviate erosion and facilitate land management. Clearly it is necessary to start building cities and economies in ways that simultaneously increase resilience while reducing the total artificial throughput in the biosphere. This is key to a notion of human security that takes sustainability seriously. All this requires understanding humanity as an active force shaping ecosystems at a variety of scales, not a passive victim of a capricious nature. It requires thinking very hard about what we are making, and how, and what we are building, in terms of institutions, buildings, infrastructure, economies and landscapes, and hence what kind of a world we are trying to secure for future generations of humanity.

Notes

1   World Commission on Environment and Development, *Our Common Future*, Oxford: Oxford University Press, 1987.

2 G. Prins and R. Stamp, *Top Guns and Toxic Whales: The Environment and Global Security*, London: Earthscan, 1991.

3 F. O. Hampson, J. Daudelin, J. B. Hay, T. Martin and H. Reid, *Madness in the Multitude: Human Security and World Disorder*, Toronto: Oxford University Press, 2002.

4 J. W. Busby, 'Who Cares about the Weather? Climate Change and U.S. National Security', *Security Studies* 17, 2008, 468–504.

5 United Nations Development Program, *Human Development Report*, New York: Oxford University Press, 1994.

6 H. G. Brauch, 'Conceptualising the Environmental Dimension of Human Security in the UN', in M. Goucha and J. Crowley (eds), *Rethinking Human Security,* Chichester: Wiley Blackwell, 2008, pp. 19–48.

7 *Human Development Report*, 1994, p. 22.

8 *Human Development Report*, 1994, p. 23.

9 *Human Development Report*, 1994, p. 34.

10 See for instance The Commission on Global Governance, *Our Global Neighbourhood*, Oxford: Oxford University Press, 1995; The Independent Commission on Population and Quality of Life, *Caring for the Future*, Oxford: Oxford University Press, 1996.

11 *Human Development Report*, 1994, pp. 28–30.

12 *Human Development Report*, 1994, pp. 35–36.

13 R. D. Kaplan, 'The Coming Anarchy', *The Atlantic Monthly*, 273 (2), 1994, pp. 44–76.

14 J. Barnett, *The Meaning of Environmental Security*, London: Zed Books, 2001.

15 M. Thompson, 'The New World Disorder: Is Environmental Security the Cure?', *Mountain Research and Development* 18, 1998, pp. 117–22.

16 P. M. Blaikie and J. S. Muldavin, 'Upstream, Downstream, China, India: The Politics of Environment in the Himalayan Region', *Annals of the Association of American Geographers* 94 (3), 2004, pp. 520–48.

17 C. Parenti, *Tropic of Chaos: Climate Change and the New Geography of Violence*, New York: Nation Books, 2011.

18 M. Renner, *Fighting for Survival: Environmental Decline, Social Conflict and the New Age of Insecurity*, New York: Norton, 1996.

19 Renner, *Fighting for Survival*, p. 17.

20 Renner, *Fighting for Survival*, p. 18.

21 P. Le Billon, *Wars of Plunder: Conflicts, Profits and the Politics of Resources*, London: Hurst, 2012.

22 This key argument runs through the literature on political ecology. See N. Peluso and M. Watts (eds), *Violent Environments*, Ithaca, NY: Cornell University Press, 2001; A. Salleh, *Eco-Sufficiency and Global Justice: Women Write Political Ecology*, London: Pluto, 2009.

23 On the history of this see M. Davis, *Late Victorian Holocausts: El Niño Famines and the Making of the Third World*, London: Verso, 2001.

24 See T. Homer-Dixon, *Environment, Scarcity and Violence*, Princeton, NJ: Princeton University Press, 1999.

25 M. Gadgil and R. Guha, *Ecology and Equity: The Use and Abuse of Nature in Contemporary India*, London: Routledge, 1995.

26 C. Kahl, *States, Scarcity, and Civil Strife in the Developing World*, Princeton, NJ: Princeton University Press, 2006.

27 P. Matondi, K. Havnevik and A. Beyene (eds), *Biofuels, Land Grabbing and Food Security in Africa*, London: Zed Books, 2011.

28 S. Dalby, *Environmental Security*, Minneapolis: University of Minnesota Press, 2002.

29 A. Baviscar (ed.), *Contested Grounds: Essays on Nature, Culture and Power*, Delhi: Oxford University Press, 2008.

30   R. Matthew, 'Environmental Stress and the Human Security in Northern Pakistan', *Environmental Change and Security Project Report* 7, 2001, 17–31.

31   A. Najam (ed.), *Environment, Development and Human Security: Perspectives from South Asia*, Lanham, MD: University Press of America, 2003.

32   R. Matthew, J. Barnett, B. McDonald and K. L. O'Brian (eds), *Global Environmental Change and Human Security*, Cambridge MA: MIT Press, 2010.

33   J. Barnett and W. Neil Adger, 'Climate Change, Human Security and Violent Conflict', *Political Geography* 26, 2007, 639–55.

34   M. Brklacich, 'Advancing our Understanding of the Vulnerability of Farming to Climate Change', *Die Erde* 137, 2006, 181–98.

35   M. Watts and H. Bohle 'The Space of Vulnerability: The Causal Structure of Hunger and Famine', *Progress in Human Geography* 17 (1), 1993, 43–67.

36   M. Davis, *Planet of Slums*, London: Verso, 2006.

37   Vulnerability is a very complicated matter, see N. Adger, 'Vulnerability', *Global Environmental Change* 16, 2006, 268–81; and H. M. Fussel, 'Vulnerability: A generally applicable conceptual framework for climate change research', *Global Environmental Change* 17, 2007, 155–67.

38   R. Green, L. K. Bates and A. Smyth, 'Impediments to Recovery in New Orleans' Upper and Lower Ninth Ward: one year after Hurricane Katrina', *Disasters* 31 (4), 2007, 311_35.

39   R. B. Bhagat, M. Guha and A. Chattopadhyay, 'Mumbai after 26/7 Deluge: Issues and Concerns in Urban Planning', *Population and Environment* 27, 2006, 337–49.

40   P. J. Smith, 'Climate Change, Mass Migration and the Military Response', *Orbis* 51 (4), 2007, 617–33.

41   K. Grove, 'Insuring "Our Common Future"? Dangerous Climate Change and the Biopolitics of Environmental Security', *Geopolitics* 15 (3), 2010, 536–63.

42   N. Klein, *The Shock Doctrine: The Rise of Disaster Capitalism*, New York: Knopf, 2007.

43   Ú. O. Spring, *Gender and Disasters. Human, Gender and Environmental Security: A HUGE Challenge*, Bonn: UNU-EHS Intersection, 2008.

44   H. G. Brauch, *Environment and Human Security: Towards Freedom from Hazard Impacts*, Bonn: United Nations University Institute for Environment and Human Security Intersections No. 2, 2005.

45   See the numerous diverse contributions in H. G. Brauch, J. Grin, C. Mesjasz, P. Dunay, N. Chadha Behera, B. Chourou, Ú. O. Spring, P. H. Liotta and P. Kameri-Mbote (eds), *Globalisation and Environmental Challenges: Reconceptualising Security in the 21st Century*, Berlin: Springer-Verlag, 2008.

46   R. M. Leichenko and K. L. O'Brien, *Environmental Change and Globalization: Double Exposures*, Oxford: Oxford University Press, 2008.

47   International Commission on Intervention and State Sovereignty, *The Responsibility to Protect*, Ottawa: IDRC 2001.

48   R. Eckersley, 'Ecological Intervention: Prospects and Limits', *Ethics and International Affairs* 21 (3), 2007, 293–316.

49   W. Steffen, A. Sanderson, P. D. Tyson, J. Jäger, P. A. Matson, B. Moore III, F. Oldfield, K. Richardson, H. J. Schellnhuber, B. L. Turner and R. J. Wasson, *Global Change and the Earth System: A Planet under Pressure*, Berlin: Springer-Verlag, 2004.

50   W. Steffen, P. Crutzen and J. R. McNeill, 'The Anthropocene: Are Humans Now Overwhelming the Great Forces of Nature?', *Ambio* 36 (8), 2007, 614–21.

51   See T. Weir and Z. Virani, 'Three linked risks for development in the Pacific Islands: Climate change, disasters and conflict', *Climate and Development* 3 (3), 2011, 193–208.

52   W. Sachs and T. Santarius (eds), *Fair Future: Resource Conflicts, Security and Global Justice*, New York: Zed Books, 2007.

53   C. Webersik, *Climate Change and Security: A Gathering Storm of Global Challenges, Santa Barbara*, CA: Praeger, 2010.

54   D. Moran (ed.), *Climate Change and National Security: A Country-Level Analysis*, Washington: Georgetown University Press, 2011.

55   See www.oxfordresearchgroup.org.uk/ssp/sustainablesecurityorg (accessed August 2012).

56   P. Rogers, *Losing Control: Global Security in the Twenty First Century*, London: Pluto, 2010.

57   C. Webersik, 'Wars over Resources? Evidence from Somalia', *Environment* 50 (3), 2008, 46–58.

58   G. E. Smith, *In Search of Sustainable Security: Linking National Security, Human Security and Collective Security to Protect America and the World*, Washington, DC: Center for American Progress, 2008.

59   S. D. Beebe and Mary Kaldor, *The Ultimate Weapon is No Weapon: Human Security and the New Rules of War and Peace*, New York: Public Affairs, 2010.

60   P. Wapner, *Living through the End of Nature: The Future of American Environmentalism*, Cambridge, MA: MIT Press, 2010.

61   H. G. Brauch, S. Dalby and U. O Spring, 'Political Geoecology for the Anthropocene', in H. G Brauch, U. O Spring, P. Kameri-Mbote, C. Mesjasz, J. Grin, B. Chourou, P. Dunay and J. Birkmann (eds), *Coping with Global Environmental Change, Disasters and Security Threats, Challenges, Vulnerabilities and Risks*, Berlin: Springer-Verlag, 2011, pp. 1453–85.

62   S. Dalby, *Security and Environmental Change*, Cambridge: Polity, 2009.

63   K. O'Brien and R. Leichenko, 'Climate Change, Equity and Human Security', *Die Erde* 137, 2006, 165–79.

64   K. Conca and G. Dabelko (eds), *Environmental Peacemaking*, Washington, DC: Woodrow Wilson Center Press, 2002; M. Renner and Z. Chafe, *Beyond Disasters: Creating Opportunities for Peace*, Washington, DC: Worldwatch Institute, 2007; United Nations Environment Program, *From Conflict to Peacebuilding: The Role of Natural Resources and Environment*, Nairobi: United Nations Environment Program, 2009.

65   S. Dinar (ed.), *Beyond Resource Wars: Scarcity, Environmental Degradation, and International Cooperation*, Cambridge, MA: MIT Press, 2011.

7

ECOLOGICAL SECURITY

A conceptual framework

Dennis C. Pirages[1]

The early years of a new millennium offer an excellent opportunity for continuing a very important dialogue among scholars and policy-makers over an increasingly obsolete, but deeply entrenched, security paradigm. The security thinking that evolved over the last millennium focused almost exclusively on protecting states and citizens against foreign military threats. In the new millennium this security paradigm already has been modified somewhat to acknowledge the growing importance of new kinds of challenges to human well-being. Intense ethnic conflicts in many countries, terrorist attacks on the World Trade Center and the Pentagon in the United States, a series of natural disasters in many parts of the world, and more recently, rapidly rising energy and food prices followed by a global economic collapse, have all played a role in broadening thinking about the nature and causes of insecurity. Furthermore, an ongoing HIV/AIDS pandemic, a brief outbreak of a deadly SARS virus, and persisting fears that avian flu or other viruses will mutate into a form that could kill millions of people have focused attention on infectious disease as another important component of a nascent alternative security paradigm.

Since the 1972 Stockholm Conference on the Human Environment, increasing attention has been paid to alternative ways of thinking about and measuring human security. There has been an extensive debate among academics and policy-makers over the utility of expanding the security framework to include environmental threats to well-being.[2] More recently this debate has expanded to encompass threats from an array of infectious diseases.[3] The argument made here, however, pushes the discussion much further, hopefully making a convincing case for moving significantly beyond these rather timid and incremental changes in security perspectives. It is suggested that we start from scratch to empirically re-evaluate the nature of present and future threats to collective well-being and develop a research agenda and related methodologies to support this assessment. The suggested re-thinking begins with the seemingly reasonable assumption that increasing the security of human beings in an era of deepening globalization means moving beyond a narrow focus on things military and addressing a broad array of what now have come to be called non-conventional security challenges.

Traditional security policy quite understandably has evolved over time to have a very heavy focus on protecting people and property from predatory neighbours. "At its most fundamental level, the term security has meant the effort to protect a population and territory against organized force while advancing state interests through competitive behavior."[4] Given a very visible legacy of violent conflict among peoples and countries, it is understandable why security has been conceptualized and operationalized largely in military terms. Warfare is vivid, violent, and destructive. Foreign armies massed on borders conjure up visions of impending devastation, mayhem, and death. But other sources of insecurity, such as a wide variety of threats embedded in the physical environment (nature), which have been historically responsible for killing and injuring much larger numbers of people, have been much less researched and understood. Unlike more conventional security threats, remedies for these less visible but often more deadly challenges to human well-being, have not been readily apparent. Thus, the security paradigm that has dominated theory, research, and practice, historically has emphasized the application of military force to protect power and privilege while ignoring less well understood, but much more serious, ecological threats to human well-being.

Although the world remains a militarily dangerous place, promoting national interests and welfare in an era of deepening globalization increasingly involves economic, environmental, and ecological matters. Even the focus of conventional military efforts has shifted substantially from cross-border warfare to managing intra-national quarrels in ethnically divided and sometimes failing states. In keeping with this new thinking, the more inclusive assessment of threats to human well-being developed here rests on the notion that security is a very important public good, and that the primary purpose of security policy should be the management of a very broad array of potential threats to human well-being. A key assumption underlying this perspective is that in the world of the future, challenges to human well-being are more likely to come from destabilized relationships with nature than they are from hordes of aggressor troops swarming over national borders.

An ecological approach to defining and analysing security represents a much broader and arguably more relevant way of thinking about the most serious existing and future threats to human well-being. It is a logical successor to the human security perspective proffered by the United Nations Development Program in an important 1994 statement. The UN report argued that human security should be defined first as safety from such chronic threats to well-being as hunger, disease, and repression and second from harmful disruptions in patterns of daily life. It went on to identify four main characteristics of human security. The first is that human security is based on a concern for the welfare of all people everywhere. The second is that security is a multidimensional concept with all dimensions being related. The third is that human security is much more easily secured through early prevention rather than remedial action. And finally, the key referent is to the well-being of people rather than states.[5]

An ecological approach to conceptualizing security builds upon this human security perspective and continues with the observation that *Homo sapiens* is but one species, albeit an important one, among millions of species that are co-evolving within the ever-changing limits of a shared biosphere and its component ecosystems. The primary research questions suggested by this approach focus on how to best maintain the present and future well-being of all human beings embedded in ever-changing and increasingly integrated ecosystems. Perhaps if this chapter were being written by a chimpanzee, it might well focus directly on how best to protect the future well-being of chimps. But, while this chapter is clearly directed at enhancing the welfare of *Homo sapiens*, the ecological security approach also requires concern with the preservation of other species, as well as of the biosphere as a whole, since their persistence is ultimately crucial to the preservation of our own security.

The argument for thinking more broadly about sources of human insecurity developed in the rest of this chapter is logical and hopefully persuasive. As Mikhail Gorbachev put it in a 1988 statement to the UN General Assembly, "the relationship between man and the environment has become menacing. Problems of ecological security affect all, the rich and the poor."[6] Norman Myers has used the term "ultimate security" to refer to this multidimensional relationship between human beings and the physical environment. For him ultimate security includes not only protection from physical harm and injury, but access to water, food, shelter, health, employment, and other basic requisites that he argues are the due of every person on Earth.[7]

While these ultimate or ecological security challenges have always been a significant source of fear and concern in the past, the causes of such threats have not been readily understood and their origins thus have been considered to be mysterious, perhaps the work of God or gods. Therefore it has been assumed historically that they were well beyond the powers of human remedy. For example, the Black Death (bubonic plague) carried from Asia to Europe by rats and their companion fleas in the fourteenth century was surely a source of fear and insecurity to the threatened Europeans, particularly since nearly 40 per cent of the people exposed to the disease perished. Because the source of the disease was unknown, the only remedies that were thought to possibly work at that time were prayer and self-flagellation.[8]

In the contemporary world, however, the causes and consequences of these previously mysterious ecological threats to well-being are coming to be much better understood and it is now possible to do something about them. The task, therefore, is to transform the currently well-entrenched security paradigm, with its very heavy emphasis on managing conflict among humans, into a much broader one that focuses on these much more significant threats. This emphasis on enhancing ecological security not only raises a much larger and arguably more important agenda of survival issues, but over time could result in a more peaceful world because it also addresses many of the underlying causes of human conflict.

Both historical and contemporary data clearly support this argument for paradigm transformation. For example, the numbers of casualties from military combat over the centuries have paled before those from other sources of death and disability. It is estimated that all of the wars of the twentieth century took the lives of 111 million combatants and civilians, an average of about 1.1 million persons per year.[9] While this is a very significant and regrettable waste of mostly young lives, this number is not that much larger than the 0.75 million people estimated to have died annually from famine during that same century.[10] It is not easy to estimate precisely the average annual number of deaths from infectious diseases over the last century, mainly because of the large number of people who died unrecorded in remote parts of the world. But we do know that at the turn of this century more carefully gathered mortality statistics indicated that worldwide infectious diseases were taking the lives of nearly 15 million people annually.[11] This is fourteen times the average number killed annually as a result of military conflict during the twentieth century.

From a slightly different perspective, even as the world was caught up in the bloodshed of World War I, a much more deadly influenza virus was spreading around the world. Death toll estimates from this influenza pandemic vary, but it is thought that it took the lives of between 20 and 40 million people, many times the total lost due to military combat.[12] Furthermore, in recent years infectious and parasitic diseases have accounted for 26 per cent of all deaths worldwide, while warfare has accounted for only three-tenths of 1 per cent.[13]

Thus, a variety of challenges from nature, often aided and abetted by intemperate human behaviour, over time have been responsible for killing and disabling much larger numbers of people than has military conflict. Also, while these ecological sources of insecurity, taking the forms of plagues, pestilence, predation, famines, storms, floods, tsunamis, droughts, earthquakes, environmental collapse and so forth, historically have taken directly, and continue to take, very large numbers of lives, these challenges also have often been responsible for taking significant numbers of human lives indirectly since they often have been a significant cause of subsequent violence among people.[14] In the past, the dynamics linking such ecological problems with subsequent conflict or warfare have not been readily apparent. But as a result of ongoing research, they are now much better understood and there is little excuse for not addressing these causal linkages in a straightforward manner.

Conceptualizing ecological security

Homo sapiens historically has evolved within and identified with basic biological and social units that social scientists call societies (or ethnic groups) and biologists and ecologists call populations. The latter define populations as "dynamic systems of individuals . . . that are potentially capable of interbreeding with each other".[15] Even in an era of deepening globalization, the contemporary limits of thousands of older primordial human societies or populations, although fading, could still be mapped. But scientists need not go into the field armed with needles and test tubes in order to gather genetic material to ascertain such limits. Rather, they also can be detected by marked gaps in the efficiency of communication.[16]

These gaps among peoples have persisted over time and have been maintained by differences in language, values, beliefs, and other key aspects of culture. The remnants of such populations, societies, or ethnic groups that once were more clearly separated on the earth's surface are being increasingly pressured to assimilate into an emerging global system. And clashes among some of them represent a formidable obstacle to peace, particularly since the geographic boundaries of territories traditionally occupied by such groups often are not congruent with those of contemporary states. Witness the persisting problems associated with the Kurdish Diaspora in the Middle East or those associated with tribes sharing states in Africa.

Relationships among these human societies and between them and their sustaining ecosystems are continually in flux. As Simon Dalby has put it, "Arguably, the most important facet of the recent discussions relating ecology to security is that stability in systems is temporary and that long-term fluctuations are inherent in natural phenomena."[17] Keeping this observation in mind, ecological security can best be defined as *maintenance of dynamic equilibria in continually evolving relationships among human societies and between them and key components of the ecosystems in which they are embedded.*

Human societies have evolved over extended periods of time within ever-changing constraints of shared physical environments while interacting with other species, a wide variety of microorganisms, and other human societies. These interactions have been strongly influenced both by changes in human populations and related resource requirements as well as by changes in physical environments, in the activities of other species, and in the mix and mobility of potentially pathogenic microorganisms.

Building on this evolutionary perspective and keeping Dalby's observation in mind, the definition of ecological security can be refined further as the maintenance of dynamic equilibria between human societies and four key components of the ecosystems in which they are embedded:

(1) Between human societies and nature's resources and services
(2) Between human societies and pathogenic microorganisms
(3) Between human societies and populations of other animal species
(4) Among human societies.

Ecological security is optimized by maintaining stability or balances in these four critical relationships. These are dynamic equilibria because relationships among societies and between them and components of the ecosystems in which they are embedded are continuously in flux. Disequilibria result in ecological insecurity, which ultimately is manifest in human suffering and premature deaths. Levels of ecological security, thus defined, can be assessed (and measured) on the societal level, the level of the state, or in an era of deepening globalization, for humanity as a whole.

Ecological insecurity (disequilibria) can result either from changes in human activities or changes in nature. Societies have grown and declined over time, as have the sustaining capabilities of nature. Threats from pathogens, from populations of competitor species, and from other societies have similarly waxed and waned. From a human perspective, substantial disruptions in the balance between people and pathogens historically have been seen as plagues and pandemics. And disruptions in the balance between people and competitor species (pests) have been referred to as pestilence.

There are many ways that analysts can study and measure ecological insecurity. Insecurity can be measured in deaths due to infectious diseases, starvation, pestilence, or military conflict. The future research agenda involves identifying potential causes of instability in the four equilibria defining ecological security as well as devising remedies for them. While the four relevant equilibria ideally would be assessed at the more basic societal level, the availability of such information and data is now limited. Therefore, the qualitative and quantitative analysis of ecological security is most easily done at the more encompassing country level.

Because analysts, being human beings, have a primary concern with the well-being of people, the level of ecological security or insecurity at any time is indicated by changes in human life expectancy data, given that pollution, plagues, malnutrition, conflict, and other manifestations of disequilibrium (along with accidents) are the causes of almost all of the premature deaths of human beings (see Table 7.1). From the human point of view, long and increasing life expectancy is a general indicator of greater ecological security while short and declining life expectancy indicates the opposite. More refined data indicative of more specific aspects of ecological insecurity, that is, deaths and disabilities from malnutrition and starvation, infectious disease, military combat, and so on, are also available to enhance the study of ecological security.

Aggregate data can give a very general overview of worldwide patterns of ecological insecurity, but a deeper understanding of causes requires future interdisciplinary research focusing more closely on factors that disturb equilibria in any of the four dimensions defining ecological security. A better understanding of the causes of such disequilibria – rapid growth of human populations; changes in the deadliness of pathogens; increases in the size and distribution of populations of pests and predators; or even changes in natural systems themselves, such as climate change – can best be developed by interdisciplinary research teams drawn from the physical and social sciences.

Future security initiatives thus should be based on this research, developing a better understanding of factors that can destabilize these four key relationships. For example, the worldwide spread of an industrial way of life has increased dramatically demands for natural resources and put heavy strains on environmental services, initially in the industrializing countries, but currently on a global scale. Growing ecological insecurity recently has been manifest globally in rapidly fluctuating energy prices, dislocations in food markets, and in growing environmental disruptions, including global warming. Ecological security is strengthened to the extent that changes in population and per capita demands for non-renewable resources are sustainable; that is, they change in a manner that can be supported without damage to the environment over time.[18]

Future security policies can also be informed by anticipating challenges to existing equilibria through methodologies and research aimed at developing a better understanding of the origins of fluctuations in the other three dimensions of ecological security. It should

Table 7.1 Ecological security in selected countries

Highest Life Expectancy at Birth (years)		Lowest Life Expectancy at Birth (years)	
Japan	83	Afghanistan	44
Sweden	82	Zimbabwe	46
Switzerland	82	Guinea-Bissau	48
Israel	82	Lesotho	49
Australia	82	Swaziland	49
Spain	82	Zambia	49
France	82	Congo	49
Italy	81	Chad	50
Canada	81	Angola	50
Norway	81	Central African Republic	50

Source: Population Reference Bureau, *2008 World population data sheet.*

be obvious that security can be strengthened by eliminating sources of tension among societies, which, if unaddressed, can lead to conflict. This can involve dealing with less traditional sources of conflict such as income disparities, unequal resource distributions, and other kinds of demographic and economic dislocation before they create situations that result in conflict. Similarly, ecological security could be much enhanced through better understanding of the very crucial and ever-changing relationships between people and pathogens, and developing broader and wiser policies aimed at limiting the incidence and spread of infectious disease. Populations of *Homo sapiens* and a wide variety of pathogens have co-evolved locally over much of history, preserving, for the most part, local equilibria. But pathogens periodically have been able to get the upper hand, due either to changes in human behaviour or changes in pathogen mobility or virulence. In the contemporary era of deepening globalization, however, infectious diseases can spread rapidly from one part of the world to others along with rapidly moving people, thus shifting this co-evolutionary process from the local to the global level. Understanding factors leading to the emergence and spread of previously unknown infectious diseases, as well as developing better methods to anticipate and deal with increasing disease mobility, should be another crucial element of ecological security policy.[19]

Finally, in many parts of the world people still are in direct conflict with a variety of pests over crops and stored food supplies, and ecological security could be enhanced by learning to deal with them more effectively. From biblical times to the present, hordes of locusts, grasshoppers, aphids, and other pests have periodically demolished crops, thus creating food insecurity. In 2008, for example, hordes of rats attacked the rice crop in south eastern Bangladesh, putting 150 thousand people at risk of starvation.[20] Much could be done to minimize food losses by anticipating these attacks and taking quick action to reduce the impact of such pests on food supplies.

Devising strategies and methods to enhance future ecological security creates temptations to deploy indiscriminately the products of technological innovation in attempts to insulate people permanently from the assaults of pests, predators, and pathogens. Couldn't ecological security be optimized by launching pre-emptive, all-out, technology-based, scorched earth campaigns using the currently available arsenal of weapons to sterilize nature? Such a strategy would involve more powerful pesticides, herbicides, antibiotics, and other chemicals in an effort to wipe out all potential pests, predators, and pathogens. Since the onset of the Industrial Revolution, there has been a growing propensity for societies to pursue this technological option in dealing with nature, and sometimes even with other peoples.

Thankfully *Homo sapiens* is gradually coming to the humbling realization that the wisdom and knowledge required to carry out such a scorched earth campaign is lacking. Furthermore, this kind of strategy might well eventuate in greater insecurity because of unanticipated collateral damage to ourselves and the rest of the biosphere. For example, it is impossible to use this strategy to eliminate potentially dangerous microorganisms while simultaneously sparing those that are biologically essential to the preservation of ecosystems. And attempts to eliminate all potential pests and predators might well result in the inadvertent loss of "keystone" species that are essential to the functioning of the ecosystems that sustain all forms of life.[21]

Nature's challenges

Future fluctuations in ecological security will be caused by changes in the two sets of variables that have shaped and will continue to shape the four dynamic equilibria that define it. On one side of the equation are continuing changes in nature and the associated ecosystems that shape and sustain all forms of life. On the other side are shifts in the size, resource demands, composition, organization, mobility, and ingenuity of human societies.

Ecosystem change, whether originating solely in nature or abetted by human activities, will remain a key factor in shaping future ecological security. Many of the more violent and destructive changes in nature still lie beyond our ability to predict and remedy. Earthquakes, volcanic eruptions, and severe storms cannot be prevented. But methodologies exist to better anticipate them, and damage and deaths from them can be minimized through early warning and rapid response. But other potentially destructive changes in nature, such as global warming, are more the result of human activities and much can be done to mitigate their impact.

Over the last 730 thousand years, *Homo sapiens* (and close relatives) have endured eight ice ages and countless other traumatic periods of global and regional climate turmoil.[22] Little could have been done to prevent these changes and people painfully adapted to them whenever possible. Similarly, large-scale volcanic eruptions and earthquakes have played and will continue to play a major role in shaping the fortunes of civilizations.[23] Hurricanes, typhoons, and other storms cannot yet be prevented. But they can be anticipated and the destruction and loss of life caused by them can be minimized. Thus, Hurricane Katrina, which flooded New Orleans in the US in 2005, was a much-anticipated event. Many studies had been done on the city's susceptibility to a major hurricane. But taking steps to prepare for what seemed inevitable was not a high security priority and little was done to mitigate the impact of the storm.

Many of nature's challenges thus still remain beyond human control, but ingenuity, reflected in technological innovation, while often having unintended consequences, has been and will continue to be an important instrument in efforts to tame or even enhance nature. Technological innovation, for example, has over time been responsible for increasing nature's productivity in many ways. Over the last few decades repeated neo-Malthusian predictions of impending worldwide starvation due to the destabilizing impact of rapid population growth have been proved wrong. This is because agricultural productivity has been substantially increased through innovations that have permitted food production to keep up with growing food demands, at least to this point in time. Likewise, innovations ranging from better mousetraps to pesticides often have done an admirable job of keeping threats from predators and pests in check. And advances in medicine, pharmaceuticals, sanitation, and nutrition have helped to maintain some equilibrium in the relationship between people and pathogens over the last century. These advances frequently have offset changes in both human behaviour and pathogens that have opened up new vulnerabilities to disease. But technological innovation itself can be a destabilizing force and its impacts on the biosphere also must be carefully assessed.

Demographic change and ecological security

While progress has been made and much more surely will be made in research anticipating and dealing with a wide variety of security challenges originating in nature, the greatest progress in understanding and reducing ecological insecurity has been and will continue to be realized through research on the human side of the equation. Social scientists have learned a great deal about the causes and consequences of demographic change and much can now be done to minimize or avoid its harshest impacts. But changes in human populations still remain one of the biggest sources of ecological insecurity (see Table 7.2).

Any major demographic change is likely to have a significant impact on the four equilibria that define ecological security. Generally, the more rapid and substantial these changes are the greater the destabilization involved. Since the early twentieth century, rapid population growth has been considered to be the most obvious and destabilizing demographic force. But there are at least five other kinds of demographic shift and dislocation that can have an impact on security. "Youth bulges", which result from periods of rapid population growth, can lead to unemployment, social dislocation, political instability, and even violence. Differential population growth can be a source of conflict, both among ethnic groups that share states and among states themselves. Large-scale population movements, whether manifest in migration or rapid urbanization, bring another set of security challenges. Somewhat paradoxically, even declining fertility and associated population aging bring new kinds of biological and socioeconomic challenges. Finally, many industrialized countries now are facing an unanticipated twenty-first century demographic dilemma as they attempt to adjust to the socioeconomic and political challenges of significant population decline.

Over long stretches of history, the world's population grew very slowly and represented little threat to ecological security on a global scale. But since the onset of the Industrial Revolution, the number of people in the world has been growing very rapidly. There were two billion people on Earth in the year 1930. This number rapidly doubled to four billion by 1970. Growth slowed somewhat toward the end of the last century, and the world's population is now about 6.7 billion. The population continues to grow at 1.2 per cent annually, meaning that there are likely to be eight billion people on Earth by 2025.[24] But almost all of this population growth and related increased pressure on natural systems is now taking place in the Global South. The number of people in the more developed

Table 7.2 Ecological security: impacts of demographic change

Demographic Changes	Nature	Disease	Conflict	Species
Growth	X	X	X	X
Youth Bulges			X	
Movement	X	X	X	X
Differential Growth			X	
Aging		X	X	
Decline			X	

countries is expected to barely grow at all by the year 2050, while the number of people in the less developed countries (excluding China) is expected to grow by 59 per cent. Of a projected 2050 world population of 9.4 billion, only 1.3 billion will live in the presently developed countries.[25]

In this demographically divided world most of the wealthier countries have reached or passed through zero population growth. Europe's population as a whole is now stable. Russia's population is declining at three-tenths of a per cent annually, Germany's at two-tenths of a per cent, and Hungary's at four-tenths of a per cent. But on the other side of the demographic divide, many countries are still facing destabilizing rates of population growth. In Africa, the population growth rate remains stubbornly high at 2.4 per cent annually. The population of Africa will grow from its present population of 967 million to almost two billion by 2050.[26]

For much of the twentieth century rapid population growth was a primary demographic source of growing ecological insecurity, particularly in the Global South. Rapidly growing numbers of people there created a seemingly Malthusian situation as they overwhelmed food and water supplies and were responsible for significant environmental degradation. While the world's population as a whole has passed through its peak rate of growth, substantial areas of rapid growth and related environmental dislocation and food shortages persist in the Global South. And extremely rapid growth in demand for petroleum, metals, and agricultural products in China and India destabilized world commodity markets in 2007 and 2008. Population growth has levelled off in the industrialized countries, but growing affluence, an associated increase in resource consumption per capita, and growing greenhouse gas emissions are adding to global ecological insecurity.

Rapid population growth in the Global South continues to impact ecological security by creating serious dislocations in relationships between people and nature. This is often manifest in "natural disasters" as people are forced to live increasingly in marginal areas – coastal lowlands, areas prone to earthquakes, or land subject to periodic droughts.[27] Human pressures on nature's resources are growing apace. The amount of water available per person has declined substantially over the last four decades and one-third of the people in the world live in countries experiencing moderate to high water stress. Over one billion people don't have access to safe water supplies and 2.6 billion lack adequate sanitation.[28] Food is in limited supply in relation to growing needs and more than 800 million people do not get enough protein and calories in their diets. Worldwide, the lack of adequate nutrition accounts for 10 per cent of the global burden of disease.[29] Trees used for firewood are rapidly disappearing before axes wielded by growing numbers of people in the Global South and related deforestation is increasing soil erosion and flooding.

Population growth also remains instrumental in creating disequilibria between people and pathogens. Infectious diseases now take the lives of nearly 15 million people each year. Rapid population growth clearly leads to greater population density, thus making it easier for infectious diseases to spread. In addition, population growth in the Global South pressures people to move onto marginal territory, often land cleared at the edge of previously remote forests. Along with the millions of plant and animal species living in these forests are numerous potentially lethal pathogens. Before recent human incursions, these pathogens undoubtedly preyed mainly upon forest animals, only jumping to humans in rare cases.[30] New settlements and associated highways built to link these frontier areas to cities offer newly liberated pathogens, with which human immune systems have little experience, routes by which they can move to more densely populated areas. And, in an era of deepening globalization, they can move swiftly from there to the rest of the world. These dynamics have been at least partially responsible for the emergence of more than three dozen previously unknown diseases in the last four decades, including HIV/AIDS, SARS, and avian flu.

Research on the relationship between rapid population growth and human conflict indicates that it is not as straightforward as it might seem. Logically, population growth should be the cause of strife among ethnic groups and even states. But political, social, and economic forces serve to mute these relationships. Goldstone nicely summarizes key research findings linking demographic change and this traditional dimension of security. While rapid population growth brings degradation of forests, water resources, arable land, and other sources of misery, it is not necessarily a major or pervasive cause of international or domestic conflict. Rather, rapid population growth provides a context out of which conflict may or may not emerge. Whether or not violent conflict actively occurs is dependent upon whether the state has the capacity to channel and moderate elite conflicts.[31]

Rapid population growth also can result in youth bulges that are often a precipitant of social conflict. Youth bulges are defined as a cohort of young adults (15 to 29 years old) that comprises more than 40 per cent of a country's population. Youth bulges have previously proved to be a demographic bonus in some Asian countries where capital was available for significant economic growth. But they are often now liabilities in poorer Middle Eastern and African countries bogged down in the early phases of a demographic transition. Under these circumstances there usually are limited economic opportunities and restless youths can be a major source of civil conflict.[32]

Human population growth also is destabilizing the equilibria between *Homo sapiens* and other species. While pests and predators remain a threat to people in more remote parts of the world, it is now the growing numbers of people that threaten many other species with extinction, with unknown consequences to the biosphere. It is estimated that human activities now threaten the extinction of 18 per cent of all mammals, 11 per cent of birds, and 5 per cent of fish.[33]

Since the emergence of modern humans more than 100,000 years ago in Africa, this curious and industrious species has spent much time moving from one place to another. There are now two types of population movement that can have a significant impact on ecological security: urbanization and migration. Population growth and the worldwide spread of industrialization have spurred a large-scale movement of people from rural to urban areas. In 1965, 36 per cent of the world's population lived in cities. By 2007, more than 50 per cent of the people in the world were living in cities, with nearly 10 per cent living in "megacities" of 10 million people or more.[34] Projections indicate that by 2015 there will be 23 megacities in the world, 19 of which will be in the Global South.[35] Rapid urbanization creates significant ecological security problems because densely populated cities can become incubators for infectious diseases, other species get crowded out of their territory by urban expansion, and growing cities cast a large "ecological footprint" over the surrounding countryside where large quantities of food must be grown to feed urban dwellers.[36]

Large-scale migration among countries can also be a significant source of insecurity. In 2005, 191 million people, 3 per cent of the world's population, migrated from their countries of origin. Of these, 62 million moved from countries of the Global South to the industrialized countries and 61 million people moved from one country of the Global South to another.[37] International migration has a significant impact on two dimensions of ecological security. Migrants are not always welcome when they reach their destinations, conflict sometimes breaking out between migrants and those who see the newcomers as an economic or cultural threat. Also, migrants, both legal and illegal, can often be a source of infectious diseases in the countries to which they move.[38]

Differential population growth, both within and among states, is another type of demographic change that creates disequilibria that often result in conflict. Differential growth among neighbouring states can cause friction when more slowly growing states perceive themselves to be potential victims of rapidly growing neighbours. For example, the rapid population growth of neighbours is a persisting source of concern for Israel. Israel's population is growing at 1.6 per cent per year. By contrast, the population of the Palestinian Territory is growing at 3.3 per cent, that of Jordan at 2.4 per cent, and that of Egypt at 2.0 per cent.[39] But differing growth rates of religious and ethnic groups within countries can spark similar concerns. Again, using Israel as an example, the rapid growth of the population of ultra-orthodox Jews there, many of whom pay no taxes and are exempt from Israeli military service, threatens to overwhelm the non-orthodox Jews, potentially creating future internal and external security concerns.[40]

Finally, somewhat ironically, population aging and decline are expected to have destabilizing effects on the most impacted industrialized countries. The aging or greying of societies is a logical result of declining birth rates and longer life spans. Many industrial countries (and a small number of others) have reached or dropped through zero population growth. In fact, 45 countries are expected to experience declining populations or a population "implosion" between now and 2050.[41] This impending demographic transformation will have profound economic, political, and security impacts. For example, there will be far fewer people in the work forces of these countries to support a rapidly growing retired population, possibly leading to intense intergenerational conflict. Aging societies will be faced with mounting medical expenditures as older populations will experience more chronic diseases and will be more vulnerable to infectious diseases. On the other hand, the good news is that there will be less manpower to devote to traditional military pursuits.[42]

Conclusion

Building ecological security in this millennium requires confronting a novel agenda of research and methodology challenges. The first of these challenges is assessing the impact of deepening globalization on the four dimensions of ecological security. People are moving across borders more rapidly and in much larger numbers, disease organisms are often moving with them, "bioinvasion" is increasing dramatically in scale, and a more desperate global struggle for petroleum and other resources holds the potential for increased conflict. The second challenge is developing a better understanding of the environmental impact of the continuing spread of the Industrial Revolution to China and India. Industrializing these countries, with their enormous populations, exacerbates the ecological security challenges mentioned above and in other chapters of this book. These and other emerging challenges mean that there will be a much smaller margin for error in maintaining the integrity of the biosphere in the face of human activities in the future.

This chapter has suggested that an outmoded security paradigm, an example of what might be called social lag, is becoming a major obstacle to dealing with growing ecological security threats. A paradigm shift is clearly needed in order to deal effectively with more complex and potentially more destructive future challenges inherent in human interactions with nature. The onset of global warming, increasing threats of new pandemics, growing water shortages, an urgent need for new energy sources, and the threat of future more deadly natural disasters in more densely populated areas of the world argue convincingly for a major rethinking of security research and spending priorities. But while hundreds of billions of dollars are now being spent on questionable military adventures in the Middle East, a tiny fraction of this amount is being spent in preparation for these future ecological security challenges.

The framework for research laid out in this chapter suggests a much different and more logical way of conceptualizing security as well as a broader research agenda. Several of the chapters in this book fit quite well into this framework and the research discussed in them tells us much about the causes of some aspects of ecological insecurity. But much more needs to be learned about the linkages between likely future changes in societies, nature, and the four equilibria that define ecological security. This requires much closer cooperation between the physical and social sciences, something that is much discussed in theory, but often ignored in practice. In the end, however, all of this suggested research may be done in vain if it does not translate into a major transformation of an increasingly outmoded security paradigm and a related massive shift in security spending priorities. How this can be accomplished politically also remains a priority item on the social science research agenda.

Notes

1   I would like to acknowledge the assistance of Winta S. Gebremariam for her assistance with earlier drafts of this chapter.
2   L. Brown, *Redefining national security*, Washington, DC: Worldwatch Institute, 1977; D. Pirages, *Global ecopolitics: A new context for international relations*, North Scituate, MA: Duxbury Press, 1978; N. Myers, 'Environment and security', *Foreign Policy*, 74, Spring 1989, 23–41; M. Levy, 'Time for a third wave of environment and security scholarship', *Environmental Change and Security Project Report*, 1, Spring 1995, 44–47; N. P. Gleditsch (ed.), *Conflict and the environment*, Dordrecht: Kluwer Academic Publishers, 1997; G. D. Dabelko and P. J. Simmons, 'Environment and security: Core ideas and the US government initiatives', *The SAIS Review*, Winter/Spring 1997, 127–46; D. Deudney and R. Matthew, *Contested grounds: Security and conflict in the new environmental politics*, Albany, NY: SUNY Publishers, 1998; T. F. Homer-Dixon, *Environment, scarcity, and violence*, Princeton, NJ: Princeton University Press, 1999; J. Barnett, *The meaning of environmental security: Ecological politics and policy in the new security era*, London: Zed Books, 2001; S. Dalby, *Environmental security*,

Minneapolis: University of Minnesota Press, 2002; R. A. Matthew *et al.* (eds), *Global environment and change and human security*, Cambridge, MA: MIT Press, 2010.

3   A. T. Price-Smith, *The health of nations: Infectious disease, environmental change and their effects on national security and development*, Cambridge, MA: MIT Press, 2001; M. Moodie and W. J. Taylor, *Contagion and conflict: Health as a global security challenge*, Washington, DC: CSIS, 2000; National Intelligence Council, *The global infectious disease threat*, Washington, DC: National Intelligence Council, 2000; J. Ban, *Health, security and global leadership*, Washington, DC: Chemical and Biological Arms Control Institute, 2001; S. Elbe, *Security and global health*, Cambridge, UK: Polity, 2020.

4   G. D. Dabelko and D. D. Dabelko, 'Environmental security: Issues of conflict and redefinition', *Environmental Change and Security Project Report*, 1, Spring 1995, 3–13.

5   United Nations Development Program, *Human development report 1994: New dimensions of human security*, Oxford: Oxford University Press, 1994, pp. 23–24.

6   Cited in N. Myers, *Ultimate security: The environmental basis of political stability*, New York: W.W. Norton, 1992, p. 11.

7   Myers, *Ultimate security*, p. 31; See also P. Stoett, *Human and global security: An exploration of terms*, Toronto: University of Toronto Press, 1999.

8   W. H. McNeil, *Plagues and peoples*, Garden City, NY: Anchor Press/Doubleday, 1976, Chap. 4.

9   Anonymous, 'Millennium of wars', *Washington Post*, 13 March 1999, A-13.

10   S. Devereux, 'Famine in the twentieth century', *Institute of Development Studies Working Paper* 105, 2000, 6.

11   World Health Organization, *World health report 2000*, Geneva: World Health Organization, 2000, Annex.

12   L. Iezzoni, *Influenza 1918*, New York: TV Books, 1999, p. 204.

13   World Health Organization, *World health report 2004*, Annex Table 2.

14   See Homer-Dixon, *Environment, scarcity and violence*; M. T. Klare, *Resource wars: The new landscape of global conflict*, New York: Metropolitan Books, 2001; C. H. Kahl, *States, scarcity and civil strife in the developing world*, Princeton, NJ: Princeton University Press, 2006.

15   K. Watt, *Principles of environmental science*, New York: McGraw-Hill, 1973, p. 1.

16   K. Deutsch, *Nationalism and social communication*, Cambridge, MA: MIT Press, 1964, p. 100.

17   Dalby, *Environmental security*, p. 143.

18   See D. Pirages (ed.), *The sustainable society*, New York: Praeger, 1977; World Commission on Development and Environment, *Our common future*, Oxford: Oxford University Press, 1987; D. Pirages (ed.), *Building sustainable societies: A blueprint for a post-industrial world*, London: M.E. Sharpe, 1996.

19   D. Pirages, 'Nature, disease, and globalization: An evolutionary perspective', *International Studies Review*, Winter, 2007, 616–28; T. McMichael, *Human frontiers, environment, and disease*, Cambridge: Cambridge University Press, 2001.

20   Anonymous, 'Thousands receive food aid after rats devastate crops', *Wall Street Journal*, 14 July 2008, A-10.

21   P. R. Ehrlich and A. H. Ehrlich, 'The value of biodiversity', *Ambio*, 21 (3) May 1992, 219–26; M. E. Power *et al.*, 'Challenges in the quest for keystones', *Bioscience*, 46, September 1996, 609–20.

22   E. Ladurie, *Times of feast, times of famine: A history of climate change since the year 1000*, Garden City, NY: Doubleday, 1971.

23   J. Tainter, *The collapse of complex societies*, Cambridge: Cambridge University Press, 1988; J. Z. deBoer and D. T. Sanders, *Volcanoes in human history*, Princeton, NJ: Princeton University Press, 2004.

24   Population Reference Bureau, *2008 World population data sheet*, Washington, DC: Population Reference Bureau, 2007.

25   Ibid.

26   Ibid.

27   A. Wijkman and L. Timberlake, *Natural disasters: Acts of god or acts of man?*, London: Earthscan, 1984.

28   World Health Organization, *Ecosystems and human well-being*, Geneva: World Health Organization, 2000, p. 2.

29   Ibid.

30   A. Gibbons, 'Where are "new" diseases born?', *Science*, 261 (5122), 6 August 1993, 680–1.

31   J. A. Goldstone, 'Population and security: How demographic change can lead to violent conflict', *Journal of International Affairs*, 56 (1), Fall, 2002, 3–23.

32   R. P. Cincotta, R. Engelman and D. Anastasion, *The security demographic: Population and civil conflict after the cold war*, Washington, DC: Population Action International, 2003, Chap. 3; see also M. Gavin, 'Africa's restless youth', *Current History*, 106 (700), May 2007, 220–6.

33   United Nations Environment Program, *Global biodiversity assessment*, Cambridge: Cambridge University Press, 1995, p. 234.

34   E. Linden, 'The exploding cities of the developing world,' *Foreign Affairs*, January/February, 1996, 53; A. Aston, 'It's become a world of bright lights and big cities', *Businessweek*, 14 April 2008, 10.

35   E. Pianin, 'Around the globe cities have growing pains', *Washington Post*, 11 June 2001, A-9.

36   See M. Wackernagel and W. Rees, *Our ecological footprint*, Philadelphia, PA: New Society Publishers, 1996.

37   P. Martin and G. Zurcher, 'Managing migration: the global challenge', *Population Bulletin*, March 2008, 3.

38   M. Carballo and A. Nerukar, 'Migration, refugees, and health risks', *Emerging Infectious Diseases*, 7 (3), June 2001, 556–60.

39   Population Reference Bureau, *2008 World population data sheet*.

40   M. D. Toft, 'Differential demographic growth in multinational states: Israel's two front war', *Journal of International Affairs*, 56 (1), Fall 2002, 71–94.

41   Population Reference Bureau, *2008 World population data sheet* ; P. G. Peterson, 'The shape of things to come: Global aging in the twenty-first century', *Journal of International Affairs*, 56 (1), Fall 2002, 189–99; N. Eberstadt, 'The population implosion', *Foreign Policy*, 123, March/April 2001, 42–53.

42   M. L. Haas, 'A geriatric peace? The future of U.S. power in a world of aging populations', *International Security* 32 (1), 2007, 112–47.

8

GENDER AND ENVIRONMENTAL SECURITY

Nicole Detraz

In March 2012, the Intergovernmental Panel on Climate Change (IPCC) issued a special report on *Managing the Risks of Extreme Events and Disasters to Advance Climate Change Adaptation*.[1] This report outlined the range and impacts of natural disasters that are associated with climate change. It integrated several examples of securitized language when discussing climate change impacts, including several mentions of food security; discussions of "environmental resources that support human welfare and security";[2] the impacts of natural disasters on "psychological well-being and sense of security";[3] "lack of security" impacting vulnerability to climate change;[4] and concerns about "local water conflicts."[5] In fact, there is an entire section of the report on climate change and violent conflict. This report is not unique in linking the environment to security, but instead is one among a flurry of publications that links environmental change with elements of security. Climate change in particular has inspired a host of securitized language by various actors,[6] however security–environment connections span to multiple environmental issue areas.

The myriad publications that define, assess, and predict the linkages between security and the environment have been termed "environmental security studies" (ESS). The starting point for this chapter is that ESS scholarship reflects a multitude of discourses, or "specific ensembles of ideas, concepts and categorization that are produced, reproduced and transformed in a particular set of practices and through which meaning is given to physical and social realities."[7] These discourses give clues about the shape and scope of the topics that we are confronting. They shape how we understand security–environment connections by defining the terms of debate. These discourses are not static, but instead are fluid by-products of continual renegotiations of the world around us based on new information, and reassessments of existing discussions and areas of salience. While these discourses contain a diverse array of narratives that reflect the complexity of security–environment connections, there is a significant gap in each – gender. Despite the important shifts in security and environment scholarship over the past few years, there is still a noticeable silence on the ways that these topics intersect with gender. When gender appears in discussions of security and environment linkages, it has tended to be brief and at times essentialist in nature. For example, the IPCC report mentioned above includes numerous mentions of "gender" and "women"; however, the discussions highlight only the vulnerability of women in nearly every case.

The challenge for the ESS community is to incorporate gender into security and environment discourses in ways that reflect the multiplicity of experiences of both men and women around the globe. This chapter proceeds from the perspective that environmental damage has important insecurity potential for multiple actors at multiple levels of analysis. This is consistent with calls within security studies to evaluate security with the overall goal of revealing insecurities so that they may be addressed and removed. This kind of endeavor is emancipatory. This is particularly important in the area of environmental security due to the fact that vulnerability to environmental damage typically intersects with multiple sites of marginalization which are challenging to overcome. To this end, I explore the existing silences on gender within existing security and environment discourses while also suggesting ways that gender can inform this important area of scholarship. Revealing gender in ESS requires rethinking terms and concepts, problematizing power relations, and asking new questions.

Security, the environment, and feminism

For gender to become a central component of security–environment discourses, it is essential that we understand the complex ways that existing understandings of both "security" and "environment" are gendered. Feminist perspectives are essential to efforts to rethink central components of scholarship in this area. It is important to stress that there is not a single feminist perspective, but rather feminism is best thought of as an umbrella over a diverse group of positions. What unites feminisms is a concern about gender, or the socially constructed ideas about what men and women *ought* to be. The notion of gender as a social construct implies that ideas about masculinity and femininity vary across time and place. There are multiple masculinities and femininities that exist within even a single society. As a whole, feminism contains perspectives that are concerned with gender equality and gender emancipation. There will be disagreement among feminists about what these goals look like and the best means to achieve them; however, there is a widespread argument that men and women often find themselves in different positions within society because they are associated with their gender. This is typically a result of that which is associated with masculinity being more highly valued than that which is associated with femininity. This unequal relationship is remarkably similar across parts of the globe.[8]

Feminist scholars use gender analysis to assess some of these trends across issue areas. Gender analysis involves examining gender-based divisions in society and differential control of/access to resources. This is different than an approach seeking to bring women into an analysis, which can isolate women from the broader socio-cultural context in which behavioral norms are embedded. Therefore, scholars must not only explore the particular position of women and men within the context of world politics, but also investigate the objects of study and the specific language used in the present discourses for examples of gendered implications.

Feminist perspectives have gained prominence within International Relations and other disciplines for the past several decades. It is often argued that the end of the Cold War opened up space for multiple perspectives to enter into attempts to explain and understand the world around us.[9] Feminism is characterized as a critical theory that engages in problematization in multiple ways. This entails challenging established research questions, definitions, and levels of analysis. Feminist scholars have studied a wide range of topics, including two that directly relate to ESS – the environment and security.

Gender and the environment

Several voices in both environmental scholarship and policymaking have called attention to the important connections between gender and the environment. Some feminist scholars highlight the specific associations between the relative position of people in society and the ways that they experience and/or contribute to environmental change. These feminist authors often claim that the systems of domination that contribute to the marginalization of women are often the same systems of domination that contribute to treating the non-human world as inferior. According to Val Plumwood:

> [a]n ecological form of feminism must be willing to mount a more thorough challenge to the dominant models of culture and humanity which define them against or in opposition to the non-human world, treating the truly human as excluding characteristics associated with the feminine, the animal and nature.[10]

Additionally, feminist environmental scholars specifically seek to understand the unique experience of women *and* men in the face of environmental damage. Most acknowledge that both women and men are often negatively impacted by worsening environmental conditions; however, these impacts are typically gendered. Rather than assume that environmental change impacts everyone similarly, or even that it impacts the marginalized in the same ways, feminist environmental scholars conclude that our relationships to nature are gendered – and that this often serves to make women experience environmental change more acutely than men in the same society. Studies along these lines focus on issues like women having to travel farther from home to collect water or fuel wood, women's unique experiences as environmental refugees, or women suffering food insecurity in greater numbers than men.[11] These kinds of studies set out to explore how gender norms shape lived experiences.

At the same time, environmental policymaking often reflects gendered understandings of environmental issues. Much environmental scholarship and policymaking treats environmental damage as a gender-neutral phenomenon, which masks the complexity of human–nature connections as well as opportunities for effective and just environmental policies. For example, research on water policy demonstrates that women are often adversely affected by the prevailing tendency towards privatization of water sources.[12] Feminist scholarship claims that water is essential for the survival of all life, and therefore the gender dynamics of water governance need to be explored by both scholars and policymakers. Without exposing the relevance and presence of gender in these kinds of discussions, important debates may continue without the inclusion of a key element.

Gender and security

Feminist scholars have also produced a varied and interesting body of scholarship on gender and security. Feminist security studies focuses on understanding the myriad ways that gender intersects with security, broadly defined. This literature asks questions about the persistent use of gendered language to describe and explain security topics as well as explores the experiences of men and women in conflict, peacekeeping, terrorism, and so on. Most feminist security scholarship fits into the "broadening" and "deepening" of security studies.[13] This means that it opens security studies up by expanding on ideas of threats and vulnerabilities while also looking at security at multiple levels.

As this volume demonstrates, there is no single definition of security accepted by scholars. Feminist security scholars often argue that expanding security scholarship beyond militarized notions of state security is necessary and beneficial. For example, Ann Tickner argues that:

> security must be analyzed in terms of how contemporary insecurities are being created and by a sensitivity to the way in which people are responding to insecurities by reworking their understanding of how their own predicament fits into broader structures of violence and oppression.[14]

Similarly, Lene Hansen and Louise Olsson argue that the goal of feminist security studies is twofold: "to critique the field of security studies for its inherent male biases and to trace how particular political practices *produce* collective conceptualizations that constrict or enable what can be recognized as legitimate problems of the individual."[15] Feminist security studies concentrate on the ways world politics can contribute to the insecurity of individuals, especially individuals who are marginalized and disempowered.[16]

While it is true that feminist security studies routinely calls attention to marginalization, it is important to note that this should not be equated to the essentialization of experiences. A simplistic analysis that automatically views women as victims in times of insecurity is counterproductive. This is just as true for characterizing all women as vulnerable to environmental change as it is for assuming that all women experience violent conflict similarly. Feminist scholarship strives to identify particular gendered patterns of experience tied to ideas of masculinity and femininity while at the same time being reflexive about differences and being mindful of the need to trace the processes that contribute to lived experiences of insecurity. Security scholarship that automatically associates men with war and women with peace is a false representation of complex stories.[17] This reflexive scholarship on both security and gender can make important contributions to the attempts by scholars, policymakers, and the media to understand the connections between security and the environment.

Feminist reconceptualizations of security and environment discourses

There have been important shifts in security and environment discourses since the 1990s.[18] The field has grown in both size and scope to include a range of important questions about how environmental damage may be a threat to security. The very fact that ESS is a research area that has both shifted over time and includes a relatively wide variety of voices suggests that further shifts are possible. In the introduction to this volume, Floyd and Matthew explain that they see "diversity as a strength and necessary basis for intellectual and policy innovation" in the area of environmental security studies. I share this view of the necessity of multiple ways of thinking about security–environment connections. Part of this aspiration for necessary diversity is incorporating gender into ESS scholarship. As the above discussions of feminist (re)conceptualizations of environment and security illustrate, making gender a central part of ESS would require asking new questions and problematizing key concepts. This volume addresses several of the most important debates that are and have been central to ESS, and each of these have important links to gender. These debates include whether conflict is driven by scarcity or abundance; the debate about whether ESS should focus on state security or human security; and debates about whether securitization is a useful strategy for discussing environmental problems.

This final point is particularly important when evaluating where gender fits in to ESS. Elsewhere I have argued that some feminists will be uncomfortable with using securitized discourses to think about environmental change.[19] Some will view the particular insecurities that militarization bring for society in general as reason enough to steer clear of the concept. This is similar to the arguments made by several non-feminist authors who criticize this connection for its potential to militarize the environment and further expand the realm of issues that are seen as the purview of the state.[20] However, I believe that presenting a counter-discourse to traditional security studies can be performed in a way that highlights the gendered assumptions of mainstream perceptions and calls attention to the specific issues that both men and women face in the current era of environmental politics. If one of the original goals of environmental security discourses was to continue the "broadening" and "deepening" of security studies, then presenting a gendered discourse on security and the environment can contribute to a similar expansion of environmental security thinking. The following sections reflect gender analysis of three central discourses that link security and the environment – *environmental conflict*, *environmental security as human security*, and *ecological security*. Each of these discourses reveals unique connections between broadly conceptualized ideas of security and environment. Each are made up of multiple narratives which focus on particular elements of the overall security–environment linkages, and each intersects with gender in specific ways.

Gendering environmental conflict

The environmental conflict discourse is one of the earliest ways that scholars made connections between security and the environment. Narratives within this discourse include concern about the potential for populations to engage in violent conflict over access to natural resources, and the ramifications of these resource conflicts for states. Conceptualized in this way, the environmental conflict discourse encompasses narratives that focus on conflicts due to both scarcity and abundance of resources. This discourse is the most closely linked to traditional notions of security of the three discussed in this chapter. For this reason, it is likely to require some important shifts in order for gender to be significantly incorporated. There are several concepts/topics that pervade the environmental conflict discourse that become problematized through gender lenses – these include the focus on state security, conceptualization of resource conflict, and assumptions about the causes of environmental scarcity.

In the first place, the narratives associated with the environmental conflict discourse are closely tied to notions of state security. Even when scholars discuss the potential for resource conflict at sub-state levels, there is a tendency to relate this phenomenon to the security and stability of the state. This state-centrism makes it difficult to integrate concerns about both threats and vulnerabilities to people and ecosystems into security–environment connections. This concern with threats and vulnerabilities is consistent with feminist goals of gender equality/emancipation and largely lacks a primary focus with the security of the state. This is not to suggest that feminist scholars are unconcerned with the fate of states, but rather that their primary goal is not the maintenance of state security. In fact, some feminist scholarship actually calls attention to the various ways that states can contribute to the insecurity of their populations.[21] For these reasons feminist scholars tend to look at multiple levels when assessing insecurity.

A second area of feminist concern about the environmental conflict discourse is prevailing conceptualizations of resource conflict. Most ESS scholarship discusses resource conflict as if it were a gender-neutral phenomenon. It is possible to look to feminist scholarship to give clues about ways in which gender analysis can be integrated into these debates. There is a widespread tendency within feminist security studies to analyze what happens during wars and conflicts as well as paying attention to their causes and endings. Feminist security scholars reveal that men and women typically experience violent conflict differently. This is often due to their positions in society which relate to prevailing assumptions about masculinity and femininity. For example, men have been more likely to serve in militaries and violent insurgent groups in higher numbers than women. This means that they have been more likely to directly experience fighting, while women have often experienced the indirect results of conflict, including being counted among civilian casualties or the wartime raped.[22] This trend holds for both traditional ideas of interstate conflict, as well as intrastate conflicts. It is important that environmental conflict discourses similarly incorporate attention to the gendered differences of experiencing resource conflict, either between states or, more likely, between groups within states.

At the same time, using discourses infused with conflict narratives could result in raising environmental concerns into the realm of "high politics" without careful reflection on the ramifications of this move. As mentioned above, increased militarization is heavily critiqued by many within feminist security studies.[23] The defense establishment got a post-Cold War boost with the rise in environmental conflict discourses.[24] Despite the fact that environmental conflict discourses have been conceptualized as a way to "green the military" or raise the profile of environmental degradation on par with issues like terrorism or nuclear proliferation, the military is an institution that is infused with militarized masculinities while also having a huge environmental footprint.[25] Additionally, it is an institution that has traditionally been in the business of identifying threats rather than being concerned with vulnerabilities. If security–environment connections are being used to concentrate power and authority into the hands of states and their militaries, the human security aspects of environmental change may be ignored.

Finally, feminists will point out that many of the proposed causes of resources conflict are themselves gendered. Several environmental conflict narratives identify particular phenomena that are linked to resource scarcity and therefore resource conflict. These include population growth, globalization, unequal resource distribution, and human migration.[26] It is essential to reveal the gendered nature of each of these processes if we are to truly understand the complexity of potential resource conflict. For example, concern with environmental migration, and climate migration in particular, has a long history. Many security scholars have specifically incorporated migration into the category of "high politics," particularly after the attacks on 9/11.[27] This means that migration has been "securitized" by several high-profile actors both inside and outside of ESS. What has been missing from most existing environmental migration studies is attention to the important ways that decisions to migrate, experiences during migration processes, and experiences in a new destination are gendered.[28] It is essential that the gendered nature of environmental migration, and other phenomena linked to resource conflict, are understood if these narratives are to continue within the environmental conflict discourse.

There are some significant challenges to making gender a central part of the environmental conflict discourse as it is currently formed. Part of this is explained by the close ties between the environmental conflict discourse and traditional ideas of security, including narratives of state security and conflict. This discourse has been influential in popularizing the idea of security–environment connections; however, its rather restricted narratives present some obstacles to the integration of gender. The main contribution that gender analysis makes to ideas of environmental conflict is encouraging reflection on its key narratives.

Gendering environmental security

The environmental security as human security discourse (hereafter environmental security discourse) is primarily focused on the negative implications of environmental change for humans. While the environmental conflict discourse is closely associated with state security, the environmental security discourse is related to ideas of human security. This means that while it is still anthropocentric (i.e., focused on humans), the primary referent object of security is people rather than states. This discourse includes a concern that environmental damage, often but not exclusively caused by human behavior, undermines the health and well-being of populations around the globe. An example of a scholar using an environmental security discourse is the following quote by Simon Dalby:

> By focusing on people's vulnerabilities and what makes them insecure in the first place, attention is drawn to their context and where they are situated in the flow of artificial and natural energies and materials. Such considerations raise a host of issues for scholars of security policies.[29]

Here Dalby highlights the challenge that human security presents to traditional security scholarship. He also calls attention to both the idea of vulnerability as being linked to insecurity as well as the links between vulnerability and power relations.

ESS scholars who use an environmental security discourse have been some of the most reflexive on issues of vulnerability and power relations. Reflexive scholarship involves assessing power dynamics and questioning existing assumptions. For example, Jon Barnett claims that environmental security is "the process of peacefully reducing human vulnerability to human-induced environmental degradation by addressing the root causes of environmental degradation and human insecurity."[30] These concerns are important, as is a discussion of the ways in which "human vulnerability" is gendered, as is "human-induced environmental degradation." Gendering the environmental security discourse will require assessing the processes of vulnerability and processes of environmental degradation through gender lenses.

Of all of the security and environment discourses, environmental security is the one that has the most potential for gender incorporation. It is likely for this reason that when gender is seen in ESS work it is typically linked to human security approaches to security–environment connections. In particular, scholars associated with the Global Environmental Change and Human Security (GECHS) initiative have explicitly made a connection between gender, security, and the environment.[31] These authors have highlighted the important contributions that gender makes to understanding the complexity of environmental security. In particular, Úrsula Oswald Spring has advocated thinking about these issues in terms of "human, gender and environmental security" (HUGE).[32] This concept reflects on the ways that understanding gender helps us to re-evaluate priorities. Importantly, this means getting away from a focus on maintaining existing social orders which often contribute to insecurity, and instead focusing on human vulnerabilities.

It is important to engage in this analysis in ways that avoid automatically viewing women as victims in the face of environmental change. This victim narrative has been recurring in painting women's place within the environment. Beyond the victim narrative, some claim that women can play an important role in environmental protection, some that women are often to blame for environmental degradation because of their roles as fuel wood gatherers, and so on. This means that women have been repeatedly cast in the roles of "agents, victims and saviours in relation to environmental change."[33] It is therefore important to be reflexive about the connections between gender and the environment, and strive for a more nuanced understanding of the ways that women and men both contribute to and address environmental damage. This caution is echoed by many feminists who argue against simplistic binary notions of nurturing or life-giving women and destructive men.[34] The automatic connection of women with environmental protection paints a simplistic, and inaccurate, picture of environmental issues. The story of environmental change and environmental protection is a very complex one that is deeply and intimately connected to socially constructed ideas of "nature" – much the same way that the story of gender is tied to socially constructed ideas of masculinity and femininity.

A second important area for gender analysis of the environmental security discourse is the evaluation of the way that proposed "causes" of environmental damage are gendered. Scholars who use an environmental security discourse have put forth some important critiques of the way that "threats to environmental security" have been conceptualized in past ESS scholarship.[35] Narratives about population are essential to these debates. Writing from a political ecology perspective, Betsy Hartmann argues that:

> Subsumed into the analytic frame of population pressure, women, through their fertility, become the breeders of environmental destruction, poverty, and violence. They are the invisible heart of environmental scarcity, made visible only when policies to ease "population growth-induced scarcity," such as "family planning and literacy campaigns" are put forward.[36]

She and other scholars are critical of using environmental concerns to legitimate an increase in attempts to control women's bodies. In many instances, it is the fertility of women from the global South that is regarded as a threat to environmental security. While being careful not to essentialize, it is important to point out that reflexive ESS scholarship requires untangling the complex relationships between poverty, vulnerability to environmental damage, and environmental policymaking.

While some scholars who use the environmental security discourse have articulated critiques to narratives that have direct gendered impacts, like those on population, it is important that these critiques become a central component of the discourse as a whole. This is one way that the environmental security discourse intersects with feminist perspectives. Additionally, bringing gender into the discourse is consistent with the existing narratives of human security, vulnerability, and power relations.

Gendering ecological security

The final discourse discussed in this chapter, ecological security, is the only one that is ecocentric in nature. This means that the primary referent object of security in this discourse is the environment itself. The central concern is the negative impacts of human-induced environmental damage for the stability and survival of ecosystems. Actors have used this discourse both to call attention to the threats facing the planet which can be traced to human behavior, and to specifically critique traditional conceptualizations and practices of security for their environmental devastation. These critiques result in important areas of overlap between this discourse and both feminist security studies and ecofeminist perspectives.

Deep green ecology's principle of biocentric equality, that all species are considered equal, is an important narrative within the ecological security discourse. There is disagreement about whether ecological security should focus on the security of humans as well as, or over the non-human world, or whether focusing on humans is counter to the whole discursive effort. These debates have interesting ties to feminist environmental perspectives. In the first place, the ecocentric nature of the ecological security discourse is consistent with ecofeminist critiques of assuming a separation between humans and their environment. Ecofeminism represents a widely discussed lens to view the combination of gender issues and the environment.[37] The term ecofeminism traces back to 1974 when French feminist Françoise d'Eaubonne used the word *ecoféminisme* to refer to the movement by women necessary to save the planet. The 1970s and 1980s saw the tendency for scholars and activists to use the term "ecofeminist" to refer to their struggle to link feminism and ecology. Scholars associated with ecofeminism, particularly Val Plumwood, have called attention to the links between anthropocentrism (i.e., human centeredness) and androcentrism (i.e., male centeredness).[38] From this perspective, the structures and discourses that allow for the continued domination of humans over the environment echo the structures and discourses that allow for the continued domination of males over females. Feminist environmentalist voices could potentially be useful allies in the goal of highlighting the close relationships (at times positive and at times negative) between humans and ecosystems.

On the other hand, narratives within the ecological security discourse that stress the security of the environment with no specific attention to its human inhabitants would lack the focus on gender emancipation, which is central to feminist scholarship. Ecological security narratives that reject privileging humans over the non-human world will reject the idea that human security should be a guiding principle when engaging in environmental policymaking.

A second interesting tie between the ecological security discourse and feminist perspectives is the tendency within both to problematize concepts like "security" and "scarcity." Scholars who use an ecological security discourse at times critique existing conceptualizations of security for being incompatible with environmental protection. It has long been acknowledged that military activities in both wartime and peacetime can have devastating impacts on the environment.[39] If security is limited to narrow ideas of state security with no reflection on the insecurities that may arise from policies designed to achieve the stability of the state, then there are very real limitations to that notion of security. Likewise, if conceptualizations of scarcity are restricted to satisfying the needs of human communities with no attention to the needs of ecosystems as a whole, then we end up with a rather incomplete concept which fails to take into consideration the power relations involved in determining allocation and distribution of resources.

For these reasons, there are some potential connections between feminist perspectives and the ecological security discourse, as long as narratives allow for a concern about the security and well-being of both humans and the non-human world. Ecological security narratives that critique militarized ideas of security and simplistic assumptions about scarcity have important counterparts within feminist scholarship. If ecocentrism is read as privileging whole ecosystems, including those with human inhabitants, then there is space for the integration of feminist concerns.

A gendered environmental security discourse

As the above sections illustrate, the connections between gender, security, and the environment are both numerous and important. There are significant areas of both overlap and contention which exist between feminist perspectives and each security–environment discourse. This section puts forward some general trends that can potentially inform a gendered environmental security discourse.

I specifically advocate for a gendered *environmental security* discourse because this is the existing perspective that has already made some progress in incorporating feminist goals and concerns, as well as having the most space available for an expansion of the integration of gender. Early ESS work had close associations with "mainstream" security scholarship, so it should therefore not be surprising that gender was not a very visible part of the discourses used to understand these links. Feminist security scholars have long been critical of traditional security studies for its lack of gender awareness and the insistence of some of its practitioners that their work is gender neutral. The environmental conflict discourse has narratives that closely echo traditional security studies, and likewise proceeded with little to no gender analysis. As ESS has progressed and questions have been raised regarding the primary referent of security, the implications of state-centric discourses, and the ways that security–environment links intersect with justice, space has opened up for feminist perspectives on these important issues. For these reasons, when gender has entered in to ESS, it has tended to be located in scholarship using the environmental security discourse.

Examining security–environment connections through gender lenses involves asking how the social constructions of masculinity and femininity impact how we relate to nature, the perceived "appropriate" roles for men and women in addressing environmental damage, and the unique experiences that men and women may face during times of environmental insecurity. I argue that the following components are necessary for a gendered environmental security discourse: 1) multilevel analysis of security and the environment, 2) broad and critical conceptualizations of key terms, and 3) reflexive solutions to environmental insecurity.

One of the reasons that gender has appeared frequently in the environmental security discourse over the years is the fact that the narrative of human security is so central to the discourse as a whole. Shifting the discourse away from the level of the state to include multiple levels is key not only to expanding our understanding of security–environment connections, but also to reflecting on how these connections are gendered. Thinking about security at multiple levels is again consistent with a broadened and deepened understanding of security. In particular, human security reorients debates about security in ways that call attention to both threats and vulnerabilities.[40] This perspective is guided by the idea that security is a contested term and considering humans to be the primary referent object of security offers an essential redefinition of security that better captures the complexity which pervades everyday threats to human health and well-being. This does not mean that the state should be ignored in debates about environmental insecurity. On the contrary, it is important to reflect on the multitude of actors and actions that contribute to both environmental damage/insecurity and solutions to environmental insecurity. Part of this reflexive scholarship involves contemplating how these actors and actions are gendered, as well as revealing the gendered ways that environmental insecurity is experienced.[41]

Problematization and concept reformation are two of the significant contributions that feminist perspectives offer academic scholarship. These tools are essential for rethinking our understandings of key concepts in ESS. Feminist work on gender and security, and gender and the environment aid in developing critical concepts in this area. Terms like "security," "environment," and "scarcity" are widely discussed by scholars using a variety of security–environment discourses; however, it is essential that these concepts are approached critically. For example, the environmental conflict discourse largely conceptualizes security as maintaining state security in the face of resource conflict. The ecological security discourse, on the other hand, conceptualizes security as stability and well-being of ecosystems. A gendered environmental security discourse will include expansive and fluid ideas of security which reflect the varied sources of threats and vulnerabilities which determine insecurity, as well as the complex relationships between groups in society, and between humans and ecosystems. This suggests that expanding notions of security also involves widening our interpretations of insecurity. Developing a gendered environmental security discourse forces a reassessment of the ways scholars and policymakers have conceptualized the occurrences of environmental insecurity. Each of the existing security and environment discourses has a tendency to treat these experiences as gender neutral. This understanding of environmental insecurity masks the important differences in the ways people witness insecurity in their daily lives.

Likewise, each security–environment discourse conceptualizes "environment" and "scarcity" in unique ways. A gendered environmental security discourse draws on conceptualizations of environment that include human and non-human nature as well as considering the places where people live. The environment is not a distant, external entity, nor is it a storehouse of resources that exist to be consumed by humans. Conceptualizing the environment requires acknowledging the close relationships that exist between humans and ecosystems. This means that scarcity must also be thought of as a notion that intersects with power dynamics and the politics of distribution.

Finally, a gendered environmental security discourse is one in which reflexive solutions to environmental insecurity are entertained and advocated. Emancipation is a driving goal for much feminist scholarship. If a gendered environmental security discourse is similarly guided by the goal of emancipation, then environmental policymaking will be guided by a desire to halt environmental damage, but also to remove the constraints to choice that accompany insecurity. This notion of removing obstacles and constraints stems from Ken Booth's ideas of emancipation as a concept with close links to security.[42] Security, in this framework, goes beyond simply ensuring a lack of violent conflict. Instead it involves empowering people to freely make choices.

Emancipatory environmental policymaking requires increasing stakeholders in environmental decision making, and evaluating gendered processes of policymaking. This means broadening our interpretations of who is regarded as having a say in the policy process. Both men and women are routinely impacted negatively by environmental damage. Because of this, there has been a call to view both men and women as stakeholders in environmental decision making.[43] This inclusive approach to the policy process has not always been implemented in practice, however. Women are often underrepresented, or unrepresented, in environmental decision making. Emancipatory environmental policymaking requires paying attention to the knowledge and experiences of both men and women. Part of this requires valuing both local knowledge and scientific knowledge on environmental issues.

Conclusion

This chapter has explored the current place of gender in discourses that link security and the environment. It has argued that gender is currently only a very small part of the debates about security–environment connections. There are areas of overlap and contention between feminist perspectives and each security–environment discourse as they currently exist. At present, the environmental security as human security discourse is the most likely choice for increasing the integration of gender into security–environment debates. A gendered environmental security discourse builds on existing narratives, but specifically uses multilevel analysis of security and the environment, broad and critical conceptualizations of key terms, and advocates for reflexive solutions to issues of environmental insecurity.

Incorporating gender into security and environment discourses offers a number of advantages to the field of ESS. First, making gender a foundational element of security and environment discourses, and the environmental security discourse in particular, aids in the association of environmental security with environmental justice. If a key goal is to understand the human security aspects of security–environment connections, then gender offers an important piece to the story. As the study of global environmental politics has progressed, there has been increased attention paid to connections between environmental damage and race, class, and gender.[44] It is often claimed that men and women have unique relationships with the environment due to socially constructed gender roles. It is essential that ESS proceed with a full understanding of the ways in which socially constructed roles in society influence environmental insecurity.

Second, the critical character of gender analysis helps to reveal the policy implications of using one security and environment discourse over another. Discourses shape how we understand problems and impact the range of solutions likely to be proposed for how to solve them. Using gender lenses to understand the potential for environmental conflict necessitates reflecting on the gendered impacts of war and conflict. On the other hand, gendering environmental security involves understanding how sources of human insecurity associated with environmental change intersect with gender. Both of these concerns have a central place within ESS; however, they result in very different sets of policies. Gender analysis gives us a language of problematization to reflect on the intended and unintended consequences of environmental policymaking that is guided by any of the security–environment discourses.

For these reasons, future scholarship and policymaking on security and environment connections should incorporate gender. It should do this in ways that avoid essentialization, and instead critically examine the sources of threats, vulnerabilities, and insecurity to understand how each intersects with gender. The environmental security discourse has already made some steps in this direction with fruitful results. Gender has not been integrated into the environmental conflict and ecological security discourses in the same way; however, utilizing gender lenses allows us to critically reflect on key aspects of these debates as well. Gender should become a central element as ESS continues to expand and produce interesting and policy-relevant scholarship.

Notes

1 IPCC, *Managing the Risks of Extreme Events and Disasters to Advance Climate Change Adaptation*, Cambridge: Cambridge University Press, 2012.
2 Ibid., p. 32.
3 Ibid., p. 42.
4 Ibid., p. 48.
5 Ibid., p. 251.

6   N. Detraz, 'Threats or Vulnerabilities? Assessing the Link Between Climate Change and Security', *Global Environmental Politics* 11 (3), 2011, 104–20; N. Detraz and M. Betsill, 'Climate Change and Environmental Security: For Whom the Discourse Shifts', *International Studies Perspectives* 10, 2009, 304–21.

7   M. Hajer, *The Politics of Environmental Discourse: Ecological Modernization and the Policy Process*, London: Oxford University Press, 1995, p. 45.

8   V. Peterson and A. Runyan, *Global Gender Issues in the New Millennium*, Boulder, CO: Westview Press, 2010; J. Tickner, *Gendering world politics: issues and approaches in the post-Cold War era*, New York: Columbia University Press, 2001.

9   B. Buzan and L. Hansen, *The Evolution of International Security Studies*, New York: Cambridge University Press, 2009.

10 V. Plumwood, 'Feminism', in A. Dobson and R. Eckersley (eds), *Political Theory and the Ecological Challenge*, Cambridge: Cambridge University Press, 2006, p. 54.

11 V. Scharff (ed.), *Seeing Nature through Gender*, Lawrence: University of Kansas Press, 2003.

12 T. Wallace and A. Coles, 'Water, Gender and Development: An Introduction', in A. Coles and T. Wallace (eds), *Gender, Water and Development*, New York: Berg, 2005, pp. 1–20.

13 K. Krause and M. Williams, 'Broadening the Agenda of Security Studies: Politics and Methods', *Mershon International Studies Review* 40, 1996, 229–54.

14 Tickner, *Gendering world politics*, p. 47.

15 L. Hansen and L. Olsson, 'Guest Editors' Introduction', *Security Dialogue* 35, 2004, 406.

16 C. Enloe, *Maneuvers: The International Politics of Militarizing Women's Lives*, Berkeley: University of California Press, 2000; C. Enloe, *Globalization and Militarism: Feminists Make the Link*, New York: Rowman & Littlefield, 2007; C. Enloe, *Nimo's War, Emma's War: Making Feminist Sense of the Iraq War*, Berkeley: University of California Press, 2010; B. Reardon and A. Hans (eds), *The Gender Imperative: Human Security vs State Security*, New Delhi, India: Routledge, 2010.

17 Tickner, *Gendering world politics*; A. Wibben, *Feminist Security Studies: A Narrative Approach*, New York: Routledge, 2011.

18 J. Barnett, *The Meaning of Environmental Security: Ecological Politics and Policy in the New Security Era*, New York: Zed Books, 2001.

19 N. Detraz, 'Environmental Security and Gender: Necessary Shifts in an Evolving Debate', *Security Studies* 18, 2009, 345–69.

20 K. Conca, 'In the Name of Sustainability: Peace Studies and Environmental Discourse', in J. Käkönen (ed.), *Green Security or Militarized Environment*, Dartmouth: Aldershot, 1994, 7–24; N. Gleditsch, 'Armed Conflict and the Environment: A Critique of the Literature', *Journal of Peace Research* 35, 1998, 381–400; M. Levy, 'Is the Environment a National Security Issue?', *International Security* 20, 1995, 35–62.

21 L. Sjoberg and J. Peet, 'A(nother) Dark Side of the Protection Racket: Targeting Women in Wars', *International Feminist Journal of Politics* 13, 2011, 163–82.

22 B. Sutton and J. Novkov, 'Rethinking Security, Confronting Inequality: An Introduction', in B. Sutton, S. Morgen and J. Novkov (eds), *Security Disarmed: Critical Perspectives on Gender, Race, and Militarization*, New Brunswick, NJ: Rutgers University Press, 2008, pp. 3–29.

23 Enloe, *Maneuvers*; Enloe, *Nimo's War, Emma's War*.

24 J. Barnett, 'The prize of peace (is eternal vigilance): a cautionary editorial essay on climate geopolitics', *Climatic Change* 92, 2009, 1–6; C. Raleigh and H. Urdal, 'Climate change, environmental degradation and armed conflict', *Political Geography* 26, 2007, 674–94; R. Floyd, *Security and the Environment: Securitisation Theory and US Environmental Security Policy*, Cambridge: Cambridge University Press, 2010, pp. 116–20.

25   S. Whitworth, 'Militarized masculinity and Post-Traumatic Stress Disorder', in J. Parpart and M. Zalewski (eds), *Rethinking the man question: sex, gender and violence in international relations*, New York: Zed Books, 2008, pp. 109–26.

26   Barnett, *The Meaning of Environmental Security*.

27   F. Adamson, 'Crossing Borders: International Migration and National Security', *International Security* 31, 2006, 165–99.

28   N. Piper (ed.), *New Perspectives on Gender and Migration: Livelihood, Rights and Entitlement*, New York: Routledge, 2008.

29   S. Dalby, *Security and Environmental Change*, Malden, MA: Polity Press, 2009, p. 107.

30   Barnett, *The Meaning of Environmental Security*, p. 129.

31   Ú. O. Spring, *Human, Gender and Environmental Security: A HUGE Challenge*, Bonn, Germany: UNU Institute for Environment and Human Security, 2008; H. Goldsworthy, 'Women, Global Environmental Change, and Human Security', in R. Matthew *et al.* (eds), *Global Environmental Change and Human Security*, Cambridge, MA: MIT Press, 2010, 215–36.

32   Spring, *Human, Gender and Environmental Security*.

33   M. Awumbila and J. Momsen, 'Gender and the environment: Women's time use as a measure of environmental change', *Global Environmental Change* 5 (4), 1995, 337–46.

34   C. Sandilands, *The Good-Natured Feminist: Ecofeminism and the Quest for Democracy*, Minneapolis: University of Minnesota Press, 1999.

35   Barnett, *The Meaning of Environmental Security*; S. Dalby, *Environmental Security*, Minneapolis: University of Minnesota Press, 2002; Dalby, *Security and Environmental Change*.

36   B. Hartmann, 'Will the Circle Be Unbroken? A Critique of the Project on Environment, Population, and Security', in N. Peluso and M. Watts (eds), *Violent Environments*, Ithaca, NY: Cornell University Press, 2001, p. 60.

37   K. Warren, *Ecofeminist Philosophy: A Western Perspective on What it is and Why it Matters*, Boulder, CO: Rowman & Littlefield, 2000.

38   V. Plumwood, 'Androcentrism and Anthropocentrism: Parallels and Politics', in K. Warren (ed.), *Ecofeminism: Women, Culture, Nature*, Bloomington: Indiana University Press, 1997, pp. 327–55.

39   M. Paterson, *Understanding Global Environmental Politics: Domination, Accumulation, Resistance*, New York: Palgrave, 2001; J. Seager, 'Patriarchal Vandalism: Militaries and the Environment', in J. Silliman and Y. King (eds), *Dangerous Intersections: Feminist Perspectives on Population, Environment, and Development*, Cambridge, MA: South End Press, 1999, pp. 163–88.

40   UNDP, *'New Dimensions of Human Security: Human Development Report 1994'*, United Nations, 1994. Online: http://hdr.undp.org/en/reports/global/hdr1994/chapters/ (accessed August 2012).

41   N. Detraz, *International Security and Gender*, Malden, MA: Polity Press, 2012.

42   K. Booth, *Critical Security Studies and World Politics*, Boulder, CO: Lynne Rienner Publishers, 2005.

43   C. Nellemann, R. Verma and L. Hislop (eds), 'Women at the Frontline of Climate Change: Gender Risks and Hopes', 2011. Online: www.unep.org/pdf/rra_gender_screen.pdf (accessed August 2012).

44   P. Newell, 'Race, Class and the Global Politics of Environmental Inequality', *Global Environmental Politics* 5, 2005, 70–94.

9

UNDERSTANDING WATER SECURITY

Patrick MacQuarrie and Aaron T. Wolf

Introduction

As human populations and economies grow, the amount of freshwater in the world remains roughly the same as it has always been. The total quantity of water in the world is immense, but most is either saltwater (97.5 percent) or locked in ice caps (1.75 percent). The hydrological system pumps over 44,000 cubic kilometers of water onto lands each year, putting the amount available per person at over 6,500 cubic meters per year. However, the amount economically available for human use is only about 13,500 cubic kilometers (0.007 percent of the total on Earth), reducing the amount available to just over 2,300 cubic meters per person per year – a 37 percent drop since 1970.[1] This increasing *water scarcity* is made more complex because almost half of the globe's land surface lies within international watersheds – a landscape that contributes to the world's over 276 transboundary waterways. The political boundaries are further complicated by both water quantity and water quality degradation, and rapidly growing rural and urban populations in developing countries. The world is now struggling to respond to the hard facts:

- More than one billion people lack access to safe water supplies.
- Almost three billion do not have access to adequate sanitation.
- Three to five million people die each year from water-related diseases or inadequate sanitation.
- 20 percent of the world's irrigated lands are salt laden, degrading crop productivity.[2]

These stressors on water resources development lead to intense political pressures, often referred to as water stress, a term coined by Malin Falkenmark,[3] or water poverty as suggested by Feitelson and Chenoweth.[4] The combination of increasing populations and fixed water supply results in decreasing water availability per capita. Peter Gleick predicts that by 2025, over thirty countries will be unable to provide at least 1,000 cubic meters per person per year, a figure regarded as the minimum necessary for an adequate quality of life in a moderately developed country. Nineteen countries will be unable to provide even 500 cubic meters per person per year.[5]

Water ignores political boundaries, evades institutional classification, and eludes legal generalizations. Water demands are increasing, groundwater levels are dropping, surface-water supplies are increasingly contaminated, and delivery and treatment infrastructure is aging. Collectively, these issues provide compelling arguments for considering the *security implications* of water resources management. A huge and growing literature speaks to the human and ecological disasters attendant on the global water crisis.[6] In conjunction with these crises, come the political stresses that result as the people who have built their lives and livelihoods on a reliable source of freshwater are seeing the shortage of this vital resource impinge on all aspects of the tenuous relations that have developed over the years. Considering all of this, it is not surprising that water, or rather its scarcity, is increasingly hailed as a variable in violent conflict. This chapter analyzes the likelihood of so-called water wars. It is argued that while water will almost certainly play a role in future conflict, the incentives for cooperation ultimately outweigh those of conflict.

The geo-politics of water

The twentieth century has seen an unprecedented movement towards decolonization and self-determination. As a result of these geopolitical rearrangements, many rivers, lakes, and groundwater aquifers, once managed from central governments and authorities, are now increasingly shared by two or more states. The Register of International River Basins defines a river basin as "the area that contributes hydrologically (including both surface- and groundwater) to a first order stream, which in turn, is defined by its outlet to the ocean or to a terminal (closed) lake or inland sea."[7] A basin is international if any perennial tributary crosses the political boundaries of two or more states. Similarly, the 1997 UN Convention on Non-Navigational Uses of International Watercourses defines a watercourse as "a system of surface and underground waters constituting by virtue of their physical relationship a unitary whole and flowing into a common terminus."[8] Within each international basin, demands from environmental, domestic, and economic users increase annually, while the amount of freshwater in the world remains roughly the same as it has been throughout history. Given the scope of the problems and the resources available to address them, avoiding water conflict is vital. Conflict is expensive, disruptive, and interferes with efforts to relieve human suffering, reduce environmental degradation, and achieve economic growth. Developing the capacity to monitor, predict, and pre-empt trans-boundary water conflict is key to promoting human and ecological security in international river basins, regardless of the scale at which they occur.

A closer look at the world's international basins gives a greater sense of the magnitude of the issue. First, the problem is growing. There were 214 international basins listed in a 1978 United Nations study,[9] the last time any official body attempted to delineate them, and there are over 276 today.[10] The growth is largely the result of the internationalization of national basins through political changes, such as the breakup of the Soviet Union and the Balkan states, as well as access to today's better mapping sources and technology. One way to visualize the dilemmas posed by international water resources is to look at the number of countries that share each international basin. Table 9.1 shows that nineteen basins are shared by five or more riparian countries. The Danube basin has eighteen riparian countries; five others are shared by between nine and eleven countries; and the remaining thirteen basins have between five and eight riparian countries.

Table 9.1 Number of countries sharing a basin

Number of Countries	International Basins
3	Asi (Orontes), Awash, Cavally, Cestos, Chiloango, Dneiper, Dniester, Ebro, Essequibo, Gambia, Garonne, Gash, Geba, Har Us Nur, Hari (Harirud), Helmand, Hondo, Ili (Kunes He), Icomati, Irrawaddy, Juba-Shibeli, Kemi, Lake Prespa, Lake Titicaca-Poopo System, Lempa, Maputo, Maritsa, Maroni, Moa, Neretva, Ntem, Ob, Oueme, Pasvik, Red (Song Hong), Rhone, Ruvuma, Salween, Sanaga, Schelde, Seine, St. John, Sulak, Talas, Torne (Tornealven), Tumen, Umbeluzi, Volga, and Zapaleri
4	Amur, Daugava, Drin, Elbe, Indus, Komoe, Lake Turkana, Limpopo, Lotagipi, Swamp, Narva, Oder (Odra), Ogooue, Okavango, Orange, Po, Pu-Lun-T'o, Senegal, Struma, and Vardar
5	La Plata, Neman, and Vistula (Wista)
6	Aral Sea, Ganges-Brahmaputra-Meghna, Jordan, Kura-Araks, Mekong, Tarim, Tigris-Euphrates (Shatt al Arab), and Volta
8	Amazon and Lake Chad
9	Rhine and Zambezi
11	Congo, Niger, and Nile
18[a]	Danube

[a] Increased from 17 to 18 with the addition of Kosovo.

Source: Updated from A. T. Wolf, J. A. Natharius, J. J. Danielson, B. S. Ward and J. K. Pender, "International River Basins of the World," *International Journal of Water Resources Development* 15, 1999, 387–427.

The increase in shared water basins coupled with the decrease of freshwater availability makes it necessary for states to develop strategies that manage this resource. One such strategy is to treat water as a security issue. Several respected and widely read reports, such as Kofi Annan's *In Larger Freedom* (2005), the UN High-level Panel on Threats, Challenges, and Change's *A More Secure World* (2004), the UNDP's *Human Development Report* (2006) focusing on power, poverty, and the global water crisis, and Worldwatch Institute's *State of the World – Redefining Global Security* (2005), have linked water to security.[11] To underscore the importance water plays in securing our future, one of the six clusters of threats listed in *A More Secure World* is environmental degradation; with the authors linking the degradation of the environment to the quality, quantity, and availability of water. Movement towards making water a national security issue, however, is not straightforward. Experiences on the upper Nile, Tigris-Euphrates, the Indus, and the Colorado in the United States illustrate that once states move water onto their security agenda many marginalized groups are left out. Hence, instead of placing water in the hands of the state's security apparatus, some are arguing for managing water on a local level, while interacting with regional, national, and international actors. Be that as it may, it is first of all necessary to look at the reality of water scarcity.

The reality of water scarcity

Whether one is an optimist or a pessimist, access to available or "easy" water is becoming increasingly problematic everywhere. In the introduction to this chapter we outlined the *global* water situation in the starkest of terms. When calculated across the entire world system, however, the figures indicate relative water abundance, on average at over 2,300 cubic meters per person per year. If a reasonable amount of water is over 1,000 cubic meters per person per year, then what is all the fuss about? Why are there so many studies, papers, and reports written about the world's water supply? It seems it is not the total amount of water available that is the issue, but rather the way it is used and where it is located. Falkenmark's scale describes *water stress* as the situation when a country's water availability falls to less than 1,700 cubic meters per person per year. It describes *water scarcity* as the situation when less than 1,000 cubic meters per person per year are available, whilst *absolute scarcity* is described as the situation when less than 500 cubic meters per person per year are available.[12] There is some debate about the accuracy of Falkenmark's scale, specifically because it falls short of addressing a) spatial variability within a country, b) water demand, and c) the quality of water actually utilized and not wasted (water efficiency). Despite these limitations, Falkenmark's scale remains the predominant indicator for understanding and predicting water stress, which is why we chose to use it here.

Figure 9.1 shows Falkenmark's projected water stress in trans-boundary basins in Africa up to 2025. An estimated sixteen basins in Africa, or 25 percent of all international basins on the continent, will have absolute water scarcity (less than 500 cubic meters per person) by 2025, compared with eight basins from Asia, or 12 percent of the total international basins on the continent.

Figure 9.2 suggests that changes in water stress are driven mainly by population growth. Thirty percent of African basins will experience over 50 percent more water stress by 2025 compared with over 40 percent of basins in Asia, a significant development, indeed. Of those African basins, seven already experience absolute water scarcity and given that they are spanning over three climate zones, they will get worse. Another eleven basins will go from relative water scarcity to absolute water scarcity.[13] Asia's basins are in somewhat better shape, yet even here twenty-four basins will experience an over 50 percent increase in water stress by 2015, putting both Asia and Africa on alert for water conflict across various sectors in the near future.[14]

Water needs are driving withdrawals of water in almost every sector, but predominantly in agriculture in underdeveloped and developing countries, and in industry and domestic sectors in developed countries. Gleick calculates a daily water need of 20 cubic meters (50 liters) per person per day for basic needs such as drinking water, washing, personal hygiene, and food preparation (not food production).[15] When food production is factored into requirements, the figures explode to almost 1,000 cubic meters (2,700 liters) per person per year – 98 percent of basic needs including food production. Water use to produce food also varies widely based on diet. It takes one cubic meter of water to produce one kilogram of grain compared with 13.5 cubic meters to produce one kilogram of meat,[16] with the difference between meat and vegetarian diet ranging from 5,400 to 2,500 cubic meters per person per year, respectively.[17] Also important is that domestic water is largely returned and recycled whereas over 50 percent of water used in agriculture is lost to evapotranspiration. Consequently, countries with water availability less than 1,500 cubic meters per person per year tend to rely more on cereal imports than those with greater water availability. It is easy to see where difficulties arise as developing countries begin

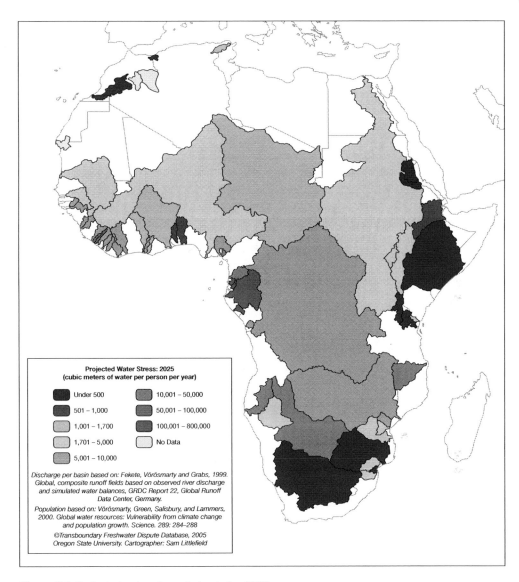

Figure 9.1 Projected water stress in basin for 2025.

to adopt more water intensive diets – if people in the Middle East dined like Americans, they would require over 2,000 cubic meters per person per year just for food. Table 9.2 shows that water supplies in the Middle East and North Africa are well below this figure – posing a spatial dilemma for water resource management.

Table 9.2 also gives an indication of how *natural* internal renewable water resources are distributed and usage rates by region. The first noticeable distinction is that Latin America and Asia have enormous amounts of freshwater resources – over half of the world's renewable freshwater – due to the massive river basins on both continents. Water availability per capita, however, is a different story. Asia, while containing over a quarter of the world's internal renewable freshwater, has one of the lowest per capita freshwater

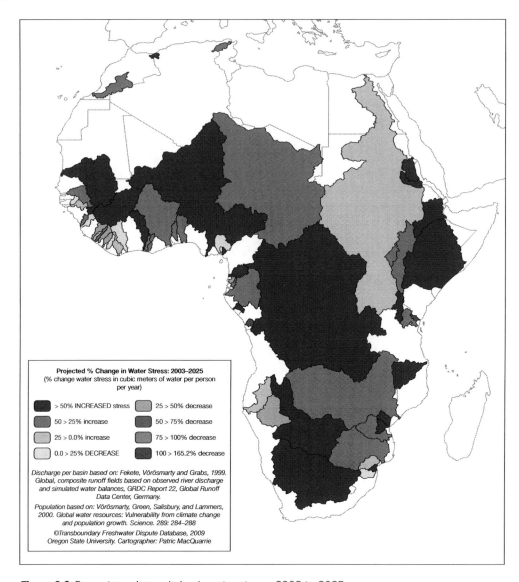

Figure 9.2 Percentage change in basin water stress: 2003 to 2025.

supplies at 3,000 cubic meters per person per year due to the continent's large population. While this is still well above the 1,000 limit for personal well-being, this point illustrates the dilemma facing the world. The severe regional implications of water distribution are no better illustrated than in the Middle East and North Africa, or MENA region.

Table 9.2 further illustrates that the differences in internally available freshwater between MENA and Asia are tremendous.[18] The MENA region only possesses slightly greater than 1 percent of the world's total internal freshwater supply but has a rapidly growing population estimated at over 400 million, or roughly six percent of the world's population.[19] It is thus easy to see why the Middle East has to import water while still experiencing chronic water scarcity. Furthermore, many of the countries have had spurious relations

Table 9.2 Internal renewable water resources and withdrawals by region and sector

Continent or Region	IRWR (km³)	% of world IRWR	Per caput in year 2003 (m³)	Freshwater withdrawal by sector (% of total)			With-drawal as % of IRWR
				Domestic	Industrial	Agriculture	
Africa	3,936	9.0	4,600	10	4	86	5.5
Asia	11,594	26.6	3,000	7	11	81	20.5
Latin America	13,477	30.9	26,700	19	10	71	1.9
Caribbean	93	0.2	2,400	23	9	68	14.4
North America	6,253	14.3	19,300	13	48	39	8.4
Oceania	1,703	3.9	54,800	18	10	72	1.5
Europe	6,603	15.1	9,100	15	53	32	6.3
World	43,659	100	6,900	10	20	70	8.8
Arid ME	*22*	*0.05*	*320*	—	—	—	—
ME Total[b]	*380*	*0.90*	*1,930*	—	—	—	—
NA Total[c]	*85*	*0.19*	*523*	—	—	—	—
MENA Total[a]	*465[d]*	*1.1*	*1,200[d]*	—	—	—	—

[a] MENA data taken from T. Allan, *The Middle East Water Question: Hydro-politics and the Global Economy,* I.B. Tauris, 2001, p. 48.

[b] ME Total figures are distorted due to high flows from Euphrates and Tigris for Turkey and Iraq.

[c] Note MENA countries rely significantly on external (trans-boundary) water sources not included here.

[d] FAO puts IRWR at 487.7 and per caput IRWR at 1,897 based on 2003 population estimates, FAO, "Review of Water Resources by Country (Water Reports)," *Food and Agriculture Organization of the United Nations,* 2003, 58.

Source: FAO (2006), *AQUASTAT database.* Online: www.fao.org/ag/aquastat.

throughout history. Israel, for example, which has experienced conflict in the region, has a reported per capita water availability of less than 200 cubic meters per person per year. The Palestinians reportedly two to three times less than that (80 cubic meters per person per year) in the West Bank and still less in Gaza (less than 15 cubic meters per person per year).[20] The spatial variability of water is not only found in the Middle East; parts of the American Southwest are experiencing record droughts, as are Northern China, Australia, Africa, and South America. These regions are all predicting increased water withdrawals in the future due to both population growth and increased irrigation needs.

Water security: threats and challenges

Studies on the effects of climate change on water security have led to some dramatic results. Two primary themes have come out of the literature and model simulations: 1) dry areas are becoming drier and wet areas wetter, and 2) there will be increased variation in water flows (precipitation), most likely linked to more frequent and extreme weather events.[21] The consequences of these two effects will be increased risk to rural communities and agriculture, resulting in a reduction of farm yields, increased poverty and malnutrition, increased risk and vulnerability due to more frequent and higher intensity floods and droughts, and longer-term reduction of flows to snow-melt fed watersheds, resulting in lower dry/summer time flows and rising sea levels affecting low-lying countries.[22] A real concern in these predictions is the number of people that are or will become vulnerable under the changing environmental conditions. Thus in almost all cases, the risk is greatest for the most vulnerable groups, rural communities, and the poor. The World Bank predicts that rainfall variability will push over 12 million people into absolute poverty. Fischer et al. predict that climate change could increase global malnutrition by up to 25 percent by 2080.[23] The IPCC scenarios themselves predict a set of outcomes that suggest water availability from rainfall events will decline by 30 percent or more in the developing world. This includes Southern Africa, the long strip across Senegal and Mauritania, and much of Brazil (up to 55 percent reduction in yields), Venezuela, and Colombia.[24]

In addition to the increased vulnerability due to climate change, resource extraction is already having a significant effect on ground waters across the globe. In India alone, 16 million farmers depend on groundwater as a reliable source of irrigation and drinking water. The effects of over pumping are mounting, driving up costs of extraction, rendering lands unusable due to salinization, polluting aquifers, and in some cases, forcing securitization of the resource.[25] Groundwater levels are declining by over one meter a year in places like Mexico, North China, the Midwest of the U.S., and northwest India, while in other locations they are sinking more rapidly, such as Indonesia, the Middle East, coastal regions of the Mediterranean – water tables in northern China are over 50 meters lower than they were forty years ago.[26] An important point is that ground waters are being increasingly developed not only in regions where surface waters are allocated (notably in Europe and North America), but also in areas where technology make ground water extraction more economically viable, such as India, Pakistan, and western Asia.

In addition to increasing extraction, water quality has a direct bearing on quantity, or the amount of water available for use by ecological systems and by humans. All of India's fourteen major river systems are badly polluted, as are many Southeast Asian rivers. Rivers flowing through urban landscapes also have significant problems, for example, many are suffering from the disposal of untreated sewage and industrial waste. Although water pollution is not exclusively a problem in the undeveloped and developing world, unlike the developed world, these states do not possess the necessary infrastructure to clean up polluted waters. Combined with reduced surface water flows, increased population pressures, and poor policy, water quality issues have the potential to threaten economic growth, human development, and ecological sustainability.

Water wars

Water is a unique and vital resource for which there is no substitute. It ignores political boundaries and it has multiple and conflicting demands on its use – problems compounded in the international realm by the fact that the international law concerned with governing water relations is poorly developed, contradictory, and unenforceable. It is no wonder then that water is perpetually suspected as the resource that will bring combatants to the battlefield in the twenty-first century. What is the likelihood that "the wars of the next century will be about water," as some have predicted?[27]

In order to cut through the prevailing anecdotal approach to the history of water conflicts, researchers at Oregon State University undertook a three-year research project, which attempted to compile a dataset of *every* reported interaction between two or more states, whether conflictive or cooperative, that involved water as a scarce and/or consumable resource or as a quantity to be managed, that is, where water was the *driver* of the events[28] over the past fifty years.[29] The study documented a total of 1,831 interactions, and found the following. First, despite the potential for dispute in international basins, the record of acute conflict over international water resources is historically overwhelmed by the record of cooperation. The last fifty years have seen only thirty-seven acute disputes (those involving violence). In fact, the only "water war" on record occurred over 4,500 years ago between the city-states of Lagash and Umma in the Tigris-Euphrates basin.[30] Over 67 percent of the total number of water related events between states of any magnitude are likewise weighted towards cooperation, implying that violence over water is neither strategically rational, nor hydrographically effective or economically viable.[31] Second, despite the occasionally fiery rhetoric of politicians—perhaps aimed more often at their own constituencies than at the enemy—most actions taken over water are mild. Of all the events, some 43 percent fell between mild verbal support and mild verbal hostility. If the next level on either side, such as official verbal support and hostility – is added, the share of verbal events reaches 62 percent of the total. Thus almost two-thirds of all events were only verbal and more than two-thirds of those had no official sanction.[32]

Third, most issues, including water quantity, quality, economic development, hydro-power, and joint management, lend themselves to cooperation and not conflict. In contrast, almost 90 percent of the conflict-laden events related to quantity and infrastructure. Furthermore, almost all extensive military acts (the most extreme cases of conflict) fell within these two categories.

Fourth, despite the lack of violence, water can act as an irritant, as well as a unifier. As an irritant, water can make good relations bad and bad relations worse. Despite the complexity, however, international waters can act as a unifier in basins with relatively strong institutions. Thus historical records clearly suggest that international water disputes do get resolved, even among enemies, and even as conflicts erupt over other issues. Some of the world's most vociferous enemies have negotiated water agreements or are in the process of doing so, and the institutions they have created often prove to be resilient, even when relations are strained. The Mekong Committee, for example, established by the governments of Cambodia, Laos, Thailand, and Vietnam as an intergovernmental agency in 1957, exchanged data and information on water resources development throughout the Vietnam War. Israel and Jordan have held secret "picnic table" talks on managing the Jordan River since the unsuccessful Johnston negotiations of 1953–55, even though they were technically at war from Israel's independence in 1948 until the 1994 treaty. The Indus River Commission survived two major wars between India and Pakistan. All ten Nile Basin riparian countries are currently involved in senior government-level negotiations to develop the basin cooperatively, despite "water war" rhetoric between upstream and downstream states.

All of this shows that there is little reason to anticipate violence at the international level. This is further supported if we consider what the goal of a water war would be. For an effective water war, the aggressor would have to be both downstream and the regional undemocratic hegemon – an upstream riparian would have no cause to launch an attack and a weaker state would be foolhardy to do so. The downstream power would then have to decide whether to launch an attack, if the project were a dam, destroying it would result in a wall of water rushing back on down-stream territory; were it a quality-related project, either industrial or waste treatment, destroying it would probably result in even worse quality than before. All of this effort would be expended for a resource that costs at most about a US dollar per cubic meter to create from seawater, costs of which are dropping every year. There are "only" 276 international watersheds, in only a handful of which the above scenario is even feasible (the Nile, Plata, and Mekong), and many of those either have existing treaties or ongoing negotiations towards a treaty. Finding a site for a "water war" turns out to be as difficult as accepting the rationale for launching one. We have seen worst case scenarios which will likely become more common. In the Jordan basin, for example, where there is tremendous hostility both across and within borders (to this day, all riparians will not even sit in the same room) complicated by the fact that the basin "ran out" of water (i.e. demand reached supply) in 1968, and yet the last shot fired across international boundaries over water was in 1970. In other words, even in this arid and hostile setting, where population growth has been exponential, dialog and collaboration, and not violence, have been the determining feature of water relations.

Water as a dependent variable in conflict

If then there is little violence between states over their shared waters, what is the problem? The problem is that we know with certainty that water can cause or exacerbate tensions between states. One of the ways in which this happens is when there is a time lag between when states first start to impinge on each other's water planning and when agreements are finally, arduously, reached. A general pattern has emerged for international basins over time. Riparians of an international basin implement water development project unilaterally – first on water within their own territory, in attempts to avoid the political intricacies of the shared resource. At some point, one of the riparians (generally the regional hegemon) will implement a project that impacts at least one of its neighbors. In the absence of relations or institutions conducive to conflict resolution, the project can become a flashpoint, heightening tensions and regional instability, and requiring years, sometimes decades, to resolve. The Indus treaty, for instance, took ten years of negotiations, the Ganges treaty thirty years, and the Jordan treaty forty years, and all the while, water quality and quantity degraded so that the health of dependent populations and ecosystems were damaged or destroyed.

The timing of water flow is another important issue as the operation of dams is also contested. For example, upstream users might release water from reservoirs in the winter for hydropower production, while downstream users might need it for irrigation in the summer. In addition, water quantity and water flow patterns are crucial to maintaining freshwater ecosystems that depend on seasonal flooding. As awareness of environmental issues and the economic value of ecosystems increases, particularly in developing countries, claims made in the name of the natural environment's water requirements are growing. For example, in the Okavango Basin, Botswana's claims for water to sustain the Okavango Delta and its lucrative ecotourism industry have contributed to a dispute with upstream Namibia, which wants to use the water passing through the Caprivi Strip on its way to the delta for irrigation.

Water quality problems include excessive levels of salt, nutrients, or suspended solids. Salt intrusion can be caused by groundwater overuse or insufficient freshwater flows into estuaries. For example, dams in the South African part of the Incomati River basin reduced freshwater flows into the Incomati estuary in Mozambique and led to increased salt levels. This altered the estuary's ecosystem and led to the disappearance of salt-intolerant flora and fauna important for people's livelihoods. Excessive amounts of nutrients or suspended solids can result from unsustainable agricultural practices, eventually leading to erosion. Nutrients and suspended solids pose a threat to freshwater ecosystems and their use by downstream riparians, as they can cause eutrophication and siltation; both of which can lead to loss of fishing grounds or arable land. Suspended solids can also cause the siltation of reservoirs and harbours: for example, Rotterdam's harbour had to be dredged frequently to remove contaminated sludge deposited by the Rhine River. The cost was enormous, and consequently led to conflict over compensation and responsibility among the river's users. Although negotiations led to a peaceful solution in this case, without such a framework for dispute resolution, siltation problems can lead to upstream/downstream disputes such as those in the Lempa River basin in Central America.[33]

As water quality degrades, or quantity diminishes, over time, the effect on the stability of a region can be unsettling. For example, for thirty years the Gaza Strip was under Israeli occupation. During that time water quality deteriorated steadily, saltwater intrusion degraded local wells, and water-related diseases took a rising toll on the people living there. In 1987, the *intifada*, or Palestinian uprising, broke out in the Gaza Strip, and quickly spread throughout the West Bank. Was water quality the cause? Whilst it would be too simplistic to claim direct causality, it almost certainly was an irritant exacerbating an already tenuous situation.

Two-thirds of the world's water use is for agriculture so, when access to irrigation water is threatened, one result can be movement of huge populations of out-of-work, disgruntled men from the country-side to the cities, an invariable recipe for political instability. In pioneering work, Sandra Postel identified those countries that rely heavily on irrigation, and whose agricultural water supplies are threatened either by a decline in quality or quantity. The list coincides precisely with regions of the world community's current security concerns, where instability can have profound effects: India, China, Pakistan, Iran, Uzbekistan, Bangladesh, Iraq, and Egypt.[34]

Water management in many countries is also characterised by overlapping and competing responsibilities among government bodies. Disaggregated decision making often produces divergent management approaches that serve contradictory objectives and lead to competing claims from different sectors. Such claims are most likely to contribute to disputes in countries where there is no formal system of water-use permits, or where enforcement and monitoring are inadequate. Controversy also often arises when management decisions are formulated without sufficient participation by local communities and water users, thus failing to take into account local rights and practices. Protests are especially likely when the public suspects that water allocations are diverting public resources for private gain or when water use rights are assigned in a secretive and possibly corrupt manner, as demonstrated by the violent confrontations in 2000 following the privatization of Cochabamba, Bolivia's water utility.[35]

Intra-state water conflicts

The second set of security issues occur at the sub-national level. Indeed, if there is a history of water-related violence, and there is, it is a history of incidents at the sub-national level, generally between tribes, water-use sectors, or states' provinces. Recent research at Oregon State University suggests that, as the scale drops, the likelihood and intensity of violence rises.[36] There are many examples of internal water conflicts, ranging from interstate violence and death along the Cauvery River in India, to Californian farmers blowing up a pipeline meant for Los Angeles, to inter-tribal bloodshed between Maasai herdsmen and Kikuyu farmers in Kenya. The inland, desert state of Arizona even commissioned a navy (made up of one ferryboat) and sent its state militia to stop a dam and diversion on the Colorado River in 1934. Likewise, in 2001, farmers, ranchers, Native Americans, and irrigators brought their guns to protest water allocations in the Klamath Basin in Oregon State.

Another contentious issue at this level of analysis is water quality, which is also closely linked to water quantity. Decreasing water quality can render it inappropriate for some uses, thereby aggravating its scarcity. In turn, decreasing water quantity concentrates pollution, while excessive water quantity, such as flooding, can lead to contamination by sewage. Low water quality can pose serious threats to human health. It is estimated that between 2.2 and 5 million people die each year from water-related diseases or inadequate sanitation. Water quality degradation is often a source of dispute between those who cause degradation and the groups affected by it. As pollution increasingly impacts upon livelihoods and the environment, water quality issues can lead to public protests.

One of the main causes of declining water quality is pollution, for example through industrial and domestic wastewater or agricultural pesticides. In Tajikistan, for example, where environmental stress has been linked to civil war (1992–97), high levels of water pollution have been identified as one of the key environmental issues threatening human development and human security. Another example would be that of water pollution in the Palar Basin of the Indian state of Tamil Nadu, where the tanning industry has made water within the basin unfit for irrigation and consumption. The pollution contributed to an acute drinking water crisis, which led to protests by the local community and activist organizations, as well as to disputes and court cases between tanners and farmers.[37]

Given that incidences of intra-state violence over water have increased in recent years and considering further that water is becoming ever scarcer it is reasonable to expect a moderate rise in intra-state water conflicts in coming years. Disenfranchised peoples within states without the infrastructure or economy to mitigate water scarcity or degradation may well find their needs driving conflict with competing populations.

Cooperation

Despite the expected rise in intra-state water conflicts there are reasons for optimism. Experiences from inter-state water problems show that water is and can be a rich source for cooperation between states. In the last two decades there has been significant movement in the protection of water systems. Significant progress has been made at the international level to protect human rights with respect to water. The argument over treating water as a human right is not a new one. The 1977 Del Mar Plata Conference declared the right of people to have access to drinking water to meet their basic needs, the 1986 Declaration on Right to Development declared that the state should provide that protection, and the 1989 Right of the Child Convention linked human health to access to water (for the child). In the 1997 Non-navigation International Watercourse Treaty, vital human needs with respect to water were clarified. *Agenda 21* at the World Summit on Sustainable Development (WSSD)[38] in Johannesburg in 2002 explicitly added "the provision of clean drinking water and adequate sanitation is necessary to protect human health and the environment" to the already drafted eight Millennium Goals. The latter did not only make protecting water explicit, it linked the protection of human well-being (here understood as the condition where basic needs are being met) to water. [39]

While water was emphasized as an essential requirement for development in 1986, the importance of water for ecosystems first appeared in 1992 at the Earth Summit. It was affirmed that water resource priorities must be to meet basic needs and safeguard ecosystems. The question was – how? The answer is not obvious – however, the international agenda has been to utilize Integrated Water Resource Management, or IWRM, as the vehicle for protecting ecosystems. First introduced in 1977, the concept has been clarified in the last ten years to include ecosystems as a competing use. Water budgets are slowly beginning to include ecosystems on their balance sheets, yet this is only half the battle. There is a monumental amount of infrastructure working against ecosystems. With almost one out of three fish at risk of extinction, two-thirds of freshwater mussels at risk, and over 50 percent of crayfish disappearing, ecosystems are collapsing.[40] An example of this is the Aral Sea in Central Asia where Soviet planners diverted the Syr Dar'ya and Amu Dar'yr Rivers for irrigation for cotton, causing the world's fourth largest inland body of water to nearly disappear.

Without doubt, the two most crucial factors determining the nature of the cooperation on water any one state will propose are a) the position of a state within a given system (e.g. is it an upstream/downstream state, is it a regional hegemon, etc.), and b) its existing relationship with other states in any given basin. Naturally, a regional power which also has an upstream riparian position is in a greater situation to implement projects, which may be contentious to downstream riparians. In contrast, development plans of an upstream riparian may be held in check by a downstream power. The perception of unresolved non-water related issues with one's neighbor can complicate sharing benefits over water between states.

Moreover, different levels of development or socio-economic status within a watershed can exacerbate the hydro-political setting. As a country develops, personal and industrial water demands tend to rise, as does demand for previously marginal agricultural areas. While this can be somewhat balanced by more access to water-saving technology, a developing country often will be the first to develop an international resource to meet its growing needs. Other factors that may influence a regional hegemon's behavior with respect to water development, and particularly the handling of transboundary issues, is its geopolitical standing. An example of this would be Turkey's aspiration to join the European Union. Turkey has had to tread carefully as it develops its water resources in Southern Anatolia so as not to impede its nomination into the regional body.

New frameworks are emerging to describe and understand hydropolitics and international behavior over water resources. Two of the more popular ideas are so-called "hydro-hegemony theory" and "hydro-political security complex theory."[41] Hydro-hegemony theory holds that hegemony at the river basin level is achieved through water resource control strategies such as resource capture, integration, and containment. These strategies are executed through an array of tactics including coercion, pressure, treaties, knowledge construction, and others that are enabled by the exploitation of existing power asymmetries, usually within a weak institutional context. Regions where there are relatively abrasive relations, significant power differentials between states, combined with general water scarcity, such as the Middle East, make good candidates for this theory. Egypt's dominance on the Nile, for example, can be explained through power tactics, as can Israel's diversion of the Jordan.

A hydro-political security complex involves states that are geographically part owners and technical users of rivers, and as a consequence, consider the rivers to be a significant national security issue. This theory is a useful tool that has enabled some analysts to develop a deeper understanding of the political dynamics in various international river basins, such as Southern Africa, where water scarcity is a prominent concern. Applied to the Orange Basin, for instance, this theory explains how and why the nine Southern African provinces are increasingly relying on inter-basin transfers for their water supply, which is significantly linked to their gross geographic product, although some are just now coming to this realization.[42] This theory can also be used to explain in part events in Southeast Asia, where China's dam building regime on the Lancang, or upper Mekong, has precipitated a dam building regime in the Lower Mekong Basin. Thus, driven by a regional demand for hydropower and irrigation, Thailand and Vietnam, followed by Laos and Cambodia, linked together their economic, energy, and environmental concerns, into a regional security complex.[43]

Conclusion

Debates over *water security* inevitably start with the question of whether or not there is water scarcity in the first place. This chapter shows that a combination of increased population growth, unsustainable agricultural policies, and climate change have rendered water scarcity more common. Water scarcity, in turn, brings with it the issue of violent conflict over water. While many have hailed water wars the type of conflict indicative of our time, however, the evidence does not support this claim. International water wars are rare with states seemingly not prepared to go to war solely to improve their water security. Indeed, states are much more likely to cooperate over water scarcity issues. Nevertheless, evidence does suggest that water scarcity increasingly plays an aggravating role in inter-state conflict and that intra-state conflict over water is not uncommon.

Aside from leading or contributing to conflict, water scarcity has serious repercussions for human well-being and the sustainability of ecosystems. The way we use and manage water must certainly evolve if we are to overcome the scarcity of water and its effects on the planet, the environment and the human condition. This puts the emphasis back on society to manage water more effectively and sustainably. To do this it is important that we understand the nature of how we use water and how and at what scales conflict arises.

Water scarcity is very much an issue whereby not one policy fits all, but where a nuanced approach is necessary to meet water resources requirements. The chapter shows that perhaps the most promising entity for the advancement of water management and security lies at the regional level of analysis. The challenge is that most conflicts in shared basins also present opportunities to create greater cooperation implemented through regional basin institutions.

Ultimately the challenge of water security is rooted in water ethics.[44] We have been complacent and have historically separated our environment from human systems. Bridging this gap will be the most important task of the future – how to restore essential ecosystems while also meeting the most basic human needs. How we manage competing needs, both social and ecological, both economic and political, and how we protect water as a vital resource will be the key to our water secure and sustainable future.

Notes

1   United Nations, 'Water in the 21st Century: Comprehensive Assessment of the Freshwater Resources of the World', *WMO/Stockholm Environmental Institute*, 1997. Gleick puts this number at 0.01 percent, P. Gleick, *The World's Water 2000–2001*, Washington, DC: Island Press, 2000, p. 21.

2   P. H. Gleick, 'Basic water requirements for human activities: meeting basic needs', *Water International* 21, 1996, 83–92.

3   M. Falkenmark, 'Fresh waters as a factor in strategic policy and action', in A. H. Westing (ed.), *Global Resources and International Conflict: Environmental Factors in Strategic Policy and Action*, New York: Oxford University Press, 1986, pp. 85–113.

4   E. Feitelson and J. Chenoweth, 'Water poverty: towards a meaningful indicator', *Water Policy* 4, 2002, 263–81.

5   P. H. Gleick (ed.), *Water in Crisis: A Guide to the World's Fresh Water Systems*, New York: Oxford University Press, 1993.

6   See, for example, P. H. Gleick, *The World's Water: The Biennial Report on Freshwater Resources*, Washington DC: Island Press, 1998–2009; S. Postel, *Last Oasis: Facing Water Scarcity*, New York: Norton, 1992; UNEP Program/Oregon State University, *Atlas of International Freshwater Agreements*, Nairobi: UNEP Press, 2002; A. Carius, M. Feil and D. Taenzler, 'Addressing environmental risks in Central Asia: risks, policies, capacities', *UNEP Program*, 2004; UNEP/Woodrow Wilson Center, 'Understanding Environment, Conflict, and Cooperation', *UNEP-DEWA*, 2004; A. T. Wolf (ed.), *Hydropolitical Vulnerability and Resilience along International Waters*, Nairobi: UN Environment Program, 2006–07.

7   Register of International River Basins, 'Transboundary Freshwater Dispute Database', 2002. Online: www.transboundarywaters.orst.edu/database/interriverbasinreg.html (accessed August 2012).

8   United Nations, 'UN Convention on Non-Navigational Uses of International Water-courses', 1997. Online: http://untreaty.un.org/ilc/texts/instruments/english/conventions/8_3_1997.pdf (accessed August 2012).

9   United Nations, *Register of International Rivers*, New York: Pergamon, 1978.

10   P. MacQuarrie and A. Wolf, 'Trends in International River Basins', *TFDD Database*, 2009 (forthcoming paper).

11   K. Annan, *In Larger Freedom: Towards development, security and human rights for all (Report of the Secretary General A/59/2005)*, New York: United Nations, 2005; United

Nations, *A More Secure World: Our shared responsibility*, New York: United Nations, 2004; UNDP, 'Human Development Report 2006: Beyond Scarcity: Power, poverty and the global water crisis', *United Nations Development Program/Palgrave Macmillan*, 2006; Worldwatch Institute (ed.), *State of the World 2005: Redefining Global Security*, Washington, DC: Worldwatch Institute, 2005.

12   M. Falkenmalk, J. Lundquist and C. Widstrand, 'Macro-scale water scarcity requires micro-scale approaches: Aspects of vulnerability in semi-arid development', *Natural Resources Forum* 13, 1989, 258–67.

13   The Dra, Gash, Juba-Shibeli, Lake Natron, Limpopo, Orange, and Umbeluzi basins.

14   Asia's basins that have less than 500 cubic meters per person and span over three climate zones are the Jordan, Atrak, Indus, Pu Lun T'o, and Tarim.

15   P. H. Gleick, 'The human right to water', *Water Policy* 1, 1998, 487–503, 496.

16   F. R. Rijsberman, 'Water scarcity: fact or fiction?', *Agricultural Water Management* 80, 2006, 4.

17   F. R. Rijsberman, 'Water scarcity: fact or fiction?', 4.

18   Under the AQUASTAT groupings, the Middle East (ME) is grouped under Asia and North Africa (NA) is naturally with Africa. Due to the variability of climate zones in the MENA region, these grouping are not the most efficient when analyzing climate change impacts.

19   Population of the MENA region estimated at 432 million by UN Population Division, 'World Population Prospects: The 2006 Revision', 2007. Online. Available at www.un.org/esa/population/publications/wpp2006/WPP2006_Highlights_rev.pdf (accessed August 2012).

20   Palestine Water Authority, 'Water Resources Statistical Records in Palestine', *Economic and Social Commission for Western Asia, United Nations* June 2007, 10–12. AQUASTAT Israel and Palestinian Territory country reports give Total Actual Renewable Water Resources (TARWR) per inhabitant of 265 cubic meters per person per year for Israel, 333 for the West Bank, and 51 for the Gaza Strip.

21   See, for instance, R. K. Dixon, J. Smith and S. Guill, 'Life on the Edge: Vulnerability and Adaptation of African Ecosystems to Global Climate Change', *Mitigation and Adaptation Strategies for Global Change* 8, 2003, 93–113; G. Fischer, M. Shah, F. N. Tubiello and H. van Velthuizen, 'Socio-economic and Climate Change Impacts on Agriculture: An Integrated Assessment, 1990–2080', *Philosophical Transactions of the Royal Society B, Biological Sciences* 360, 2005, 2067–83; Stern Review on the Economics of Climate Change, 'What is the Economics of Climate Change? (Discussion Paper)', 2006. Online: www.hm-treasury.gov.uk/media/213/42/What_is_the_Economics _of_Climate_Change.pdf (accessed August 2012).

22   N. W. Arnell, 'Climate Change and Global Water Resources: SRES Emissions and Socio-economic Scenarios', *Global Environmental Change* 14, 2004, 31–52; N. W. Arnell and C. Liu, 'Hydrology and Water Resources', in J. J. McCarthy, O. F. Canziani, N. A. Leary, D. J. Dokken and K. S. White (eds), *Climate Change 2001: Impacts, Adaptation and Vulnerability*, Cambridge, UK: Cambridge University Press for the Intergovernmental Panel on Climate Change, 2001; UNDP, 'Human Development Report 2006'.

23   G. Fischer, M. Shah and H. van Velthuizen, 'Climate Change and Agricultural Vulnerability (Report prepared for the World Summit on Sustainable Development, Johannesburg)', *International Institute for Applied Systems Analysis*, 2002.

24   IPCC (Intergovernmental Panel on Climate Change) and R. T. Watson and the Core Writing Team (eds), *Climate Change 2001: Synthesis Report. A Contribution of Working Groups I, II, and III to the Third Assessment Report of the Intergovernmental Panel on Climate Change*, Cambridge, UK and New York: Cambridge University Press, 2001, p. 49.

25 For general groundwater issues see reports by M. Moench, J. Burke and Y. Moench, 'Rethinking the Approach to Groundwater and Food Security', *Food and Agriculture Organization of the United Nations*, 2003; for Pakistan, see World Bank, 'Pakistan's Water Economy: Running Dry (Report 34081-PK)', *Agriculture and Rural Development Unit*, 2005; for India, see B. Vira, R. Iyer and R. Cassen, 'Water', in R. Cassen, L. Visaria and T. Dyson (eds), *Twenty-first Century India: Population, Economy, Human Development, and the Environment*, Oxford: Oxford University Press, 2004; for Indonesia, see G. Kurnia, T. W. Avianto and B. R. Bruns, 'Farmers, Factories and the Dynamics of Water Allocation in West Java', in B. R. Bruns, C. Ringler and R. S. Meinzen-Dick (eds), *Negotiating Water Rights*, London: Intermediate Technology Publications, 2000.

26 D. Shen and R. Liang, 'State of China's Water (Research Paper)', *Third World Centre for Water Management with the Nippon Foundation*, 2003, p. 52.

27 World Bank vice-president I. Serageldin, *New York Times*, 10 August 1995. His statement is probably most often quoted.

28 Excluded are events where water is incidental to the dispute, such as those concerning fishing rights, access to ports, transportation, or river boundaries. Also excluded are events where water is not the driver, such as those where water is a tool, target, or victim of armed conflict.

29 A. T. Wolf, S. B. Yoffe and M. Giordano, 'International waters: identifying basins at risk', *Water Policy* 5, 2003, 29–60.

30 A. T. Wolf, 'Conflict and cooperation along international waterways', *Water Policy* 1, 1998, 251–65.

31 Wolf *et al.*, 'International waters', 39. There were 507 conflict-related events versus 1,228 cooperative events, 96 were considered neutral.

32 Ibid.

33 A. Lopez, 'Environmental conflicts and regional cooperation in the Lempa River basin: the role of Central America's Plan Trifinio (EDSP Work, Paper 2)', *Adelphi Research*, 2004.

34 S. L. Postel and A. T. Wolf, 'Dehydrating Conflict', *Foreign Policy* September/October 2001, 60–67.

35 Ibid.

36 See M. Girodano, M. Girodano and A. T. Wolf, 'The geography of water conflict and cooperation: internal pressures and international manifestations', *The Geographical Journal* 168, 2002, 293–312.

37 A. Carius, G. D. Dabelko and A. T. Wolf, 'Water, Conflict, and Cooperation (Policy Briefing Paper for United Nations & Global Security Initiative)', *United Nations Foundation*, 2004.

38 United Nations, 'World Summit on Sustainable Development', 2002. Online: www.un.org/events/wssd/ (accessed August 2012).

39 See the evolution of water as a human right in international law by P. H. Gleick, 'The human right to water', *Water Policy* 1, 1998, 487–503; S. C. McCaffrey, 'A human right to water: Domestic and international implications', *Georgetown International Environmental Law Review* 5, 1992, 1–24; A. K. Biswas, 'Water as a Human Right in the MENA Region: Challenges and Opportunities', *Water Resources Development* 23, 2007, 209–25; E. B. Bluemel, 'The Implications of Formulating a Human Right to Water', *Ecology Law Quarterly* 31, 2005, 957–1006.

40 S. Postel, 'Entering an Era of Water Scarcity: The Challenges Ahead', *Ecological Applications* 10, 2000, 943; S. Postel and B. Richter, *Rivers for Life: Managing Water for People and Nature*, Washington, DC: Island Press, 2003.

41   See, for example, M. Zeitoun and J. Warner, 'Hydro-hegemony – a framework for analysis of trans-boundary water conflicts', *Water Policy* 8, 2006, 435–60; M. Lowi, *Water and Power – The Politics of a Scarce Resource in the Jordan River Basin*, Cambridge, UK: Cambridge University Press, 1993; A. P. Elhance, *Hydropolitics in the Third World, Conflict and Cooperation in International River Basins*, Washington, DC: US Institute of Peace, 1999; S. Dinar, 'Negotiations and International Relations: A Framework for Hydropolitics', *International Negotiation* 5, 2000, 375–407; M. Schulz, 'Turkey, Syria and Iraq: a hydropolitical security complex', in L. Ohlsson (ed.), *Hydropolitics: conflicts over water as a development constraint*, London: Zed Books, 1995, pp. 91–122; for Southern Africa, see A. R. Turton, 'Hydropolitics and Security Complex Theory: An African Perspective (Paper presented at the 4th Pan-European International Relations Conference)', *University of Kent*, 2001; for the Mekong, see K. Bakker, 'The politics of hydropower: developing the Mekong', *Political Geography* 18, 1999, 209–32; E. Goh, 'China in the Mekong River Basin: the regional security implications of resource development on the Lancang Jiang (IDSS working paper)', *Institute of Defence and Strategic Studies, Nanyang Technological University* 69, 2004.

42   Turton, 'Hydropolitics and Security Complex Theory', p. 19.

43   Goh, 'China in the Mekong River Basin', 213–21.

44   For more along this thread, see J. Kolars, 'The Spatial Attributes of Water Negotiation: The Need for a River Ethic and River Advocacy in the Middle East', in H. A. Amery and A. T. Wolf (eds), *Water in the Middle East: A Geography of Peace*, Austin: University of Texas Press, 2000, pp. 245–50; S. Postel, 'A Water Ethic', in A. T. Wolf (ed.), *Conflict Prevention and Resolution in Water Systems (The Management of Water Resources Series)*, Northampton, MA: Edward Elgar Publishing, 2002; M. R. Lowi and B. R. Shaw (eds), *Environment and Security: Discourses and Practises*, Basingstoke: Macmillan, 2000, pp. 33–101.

10

CONSERVATION, SCIENCE AND PEACE-BUILDING IN SOUTHEASTERN EUROPE

Saleem H. Ali and Mary C. Watzin

Introduction: defining ecological versus political space

Much of the early discourse around environmental security attempted to elevate ecological issues to salience by focusing on their potential as the fundamental source of conflict between countries. Cases that might otherwise have been considered by policymakers as rooted in issues of territorial contentions, based on ethnic tensions or interests of economic or political hegemony over a region, were instead presented as the result of resource scarcity.[1] While this had a short-term impact of bringing environmental issues to prominence in conservative defence circles, the long-term policy impact was relatively limited. A focused analysis of each case quickly found enough intervening variables that any causal connection to resource scarcity was widely contested. Two parallel schools of thought developed and became increasingly polarized in their perception about the environment-conflict linkage.[2] While more liberal politicians continued to champion the importance of the environment as a security imperative, many political realists began to dismiss the connection. In some ways this debate mirrored the polarization between conservative and liberal elements within the conventional political spectrum.

Some of the more analytically driven research institutions such as the Peace Research Institute in Oslo (PRIO) began to present empirical evidence that countered earlier assumptions about the connection between environmentally driven resource scarcity and conflict. In an editorial following the awarding of the Nobel Peace Prize to Wangaari Maathai in 2004, two prominent PRIO researchers stated:

> Environmental destruction and scarcity of renewable resources can present a danger to life and livelihood in many third-world countries. But these hazards are not primarily linked to a danger of war. Exaggerating the security aspects of environmental decay hardly helps our efforts to overcome the negative effects of resource scarcity.[3]

Amidst all these conversations, an underlying issue was missed – the ecological constraints that ultimately limit all political constituents could still serve to foster cooperation even if the conflict in question had not been caused by environmental factors.

Several questions emerged. Is it possible to create recognition of inherent ecological space out of previously defined political space so as to transform a conflict? Can the cooperation on ecological factors between players itself transcend that limit and allow for a more lasting peace? What might be the defining variables for such a transformation? In this chapter we hypothesize that such a transformation is possible with a confluence of scientific decision making and carefully focused mediation.

An important area to study such interactions is southeastern Europe. Here nascent governance regimes are rapidly evolving in newly democratized or seceded states, but the region is still challenged by a legacy of prior tensions and conflicts. There is tremendous interest from external agents to consider environmental factors, and new countries aspiring towards linkages with regional governance structures such as the European Union are considering anew both their natural resource policies and their interactions with the global environmental community. Within this region, the transboundary politics surrounding Lake Ohrid and Lake Prespa in the Balkan peninsula were studied to examine our hypotheses about how natural resource scarcity and environmental impairment interact. While we recognize the limits of applying insights from this case across all geographic regions, the structural factors that emerge here in terms of the role of scientists and the emergence of regional governance mechanisms have broad applicability.[4]

Post-Cold War cooperation and the EU imperative

The lifting of the Iron Curtain, following the demise of the Soviet Union in 1991, was a momentous time for environmental planners as there had been serious concerns about the interest and ability of communist enterprises to manage and control environmental pollution. Around the same time, the European Union strengthened and the Maastricht Treaty was signed in 1992. Given the strong sense of suspicion and the long history of conflict along ethno-linguistic lines in this region, the treaty was premised on the concept of "subsidiarity" which was reformulated in the Treaty of Nice in 2003 and reads as follows:

> In areas which do not fall within its exclusive competence, the Community shall take action, in accordance with the principle of subsidiarity, only if and in so far as the objectives of the proposed action cannot be sufficiently achieved by the Member States and can therefore, by reason of the scale or effects of the proposed action, be better achieved by the Community.

The recently promulgated European Constitution (2007) in Article 9 further strengthens this concept as:

> *Under the principle of subsidiarity, in areas which do not fall within its exclusive competence the Union shall act only if and insofar as the objectives of the intended action cannot be sufficiently achieved by the Member States, either at central level or at regional and local level, but can rather, by reason of the scale or effects of the proposed action, be better achieved at Union level.*

This principle would initially suggest that environmental issues would be confined to the local level as implied by the aphorism of "thinking globally but acting locally". However, the inherent transboundary aspects of ecological systems soon surfaced as paramount among European policymakers and environmental issues began to quickly trump subsidiarity. At the same time, one would have envisaged the post-communist states to be more receptive to the notion of environmental governance, having endured enormous pollution during the communist era. Yet as Robert Darst has noted in his study of East–West environmental relations, there was a paradoxical resistance to embracing environmental concerns. Instead what he observed was an "instrumental manipulation of external environmental concerns" by the post-communist states because they realized that many of their woes were not dependent upon external cooperation for solution since they were largely "self-inflicted", and not downstream impacts from the West.[5] Furthermore, in his detailed study of the European Union's participation in the Convention on Long-range Transboundary Air Pollution, Stacy VanDeveer found that environmental assessments generally do not "enhance public debate" or influence policy "except in the event that they contain cost estimates for additional regulations that attract policy maker attention".[6] The analysis also revealed that there was a considerable difference within Europe between the core western European states and the "peripheral" states of eastern Europe where the efficacy of environmental assessments in shaping regional policy and potential cooperation is possible only when they are linked to non-environmental policy goals. In the context of many developing nations or indeed former communist bloc countries, this entails the potential for membership in an elite international institution such as the European Union.

While the goal of instrumental cooperation through environmental knowledge-sharing might be more challenging at the state level, it has already gained considerable traction at the civil society level. The need to establish infrastructure across the EU prompted several civil society organizations that previously had no relations with each other to form alliances. Hein-Anton van der Heijden has termed this phenomenon "multi-level environmentalism" that provides an opportunity for constructive confrontation between various stakeholders in environmental conflicts.[7] In the short term this may be perceived as a threat to cooperation, but in the long run such efforts provide ways of developing collective social capital around environmental causes that can move from the grassroots to the policymaking level.

Theoretical aspects of science and security

Environmental policy inevitably relies on science as a touchstone for authenticity since ecological issues have an inherent scientific premise. Yet the objectivity of science is still questioned repeatedly by social theorists and has led to the emergence of an entire field of inquiry called "science and technology studies" (STS) which emerged from the writings of biologist Ludwik Fleck[8] and others. Scholars such as Fleck contended that scientific "fact" emerged out of a complex social process which was not always objective. Each decision made by a scientist during empirical inquiry has certain subjective attributes. While this subjectivity is circumscribed by conventional scientific methods, the incremental impact of repeated decisions by scientists can greatly erode objectivity. This has also been revealed in work by French sociologist Bruno Latour in his studies of endocrinology research. Later, Latour developed "action-network" theory to show how such subjectivity is manifest in human institutions such as scientific laboratories more generally.[9] Other protégés of Fleck such as Thomas Kuhn expanded this "social constructionism" to also consider the trajectory of scientific progress, which could be punctuated rather than gradual. Thus social concerns such as physical security could stimulate scientific inquiry in spurts of enterprise but also make it more biased in a particular direction. The rapidity of research in nuclear technology during the Cold War is emblematic of this trend. Science was thus politicized and became an instrument of preexisting security mindset rather than framing security discourse itself.

During the last few decades, the process of scientific inquiry has been democratized at multiple levels. The availability of data through rapid communication methods and internet technologies has empowered individual citizens to consider scientific impacts independently. Often such citizens can also play the role of "citizen" or "street scientists" in diagnosing environmental impacts as noted by the work of environmental planner Jason Coburn.[10] This process of democratization of science has the potential to negate some of the effects of centrally planned and controlled scientific enterprise.

Nevertheless, many states are still contending with tremendous challenges around the cooptation of science by security interests. Concerns raised by social theorists such as Michael Foucault about the use of environmental knowledge, in particular, towards such ends by government agencies has been noted in theories of "ecogovernmentality".[11] Political ecologists such as Arun Agrawal have further developed theories of "environmentality" in the context of forest policy.[12] What our analysis in the case study of Lake Ohrid and Lake Prespa shows is that scientists themselves may be knowingly complicit in this process of ostensible manipulation. Thus even with the structural safeguards in scientific inquiry that critics of social construction such as Ian Hacking have propounded,[13] the potential for the politicization of science in matters of spatial security remains high.

Comparative case analysis: Lake Ohrid and Lake Prespa watersheds

The Lake Ohrid–Lake Prespa ecosystem is located in the southern part of the Balkan peninsula, and includes territory governed by the Former Yugoslavia Republic of Macedonia, Albania, and Greece (Figure 10.1). Lake Ohrid is an ancient lake, formed by tectonic forces 2–3 million years ago: it is shared by two countries, Macedonia and Albania. Lake Prespa is younger, but was also formed by tectonic forces, and is shared by Macedonia, Albania, and Greece. The two lakes are connected hydrologically; about half the water in Lake Ohrid comes from springs that originate in Lake Prespa and flow through underground karst channels into Lake Ohrid.[14]

Although the species vary, both lakes share a rich biodiversity and unique cultural heritage. Lake Ohrid is one of only a few "ancient" lakes, isolated by surrounding hills

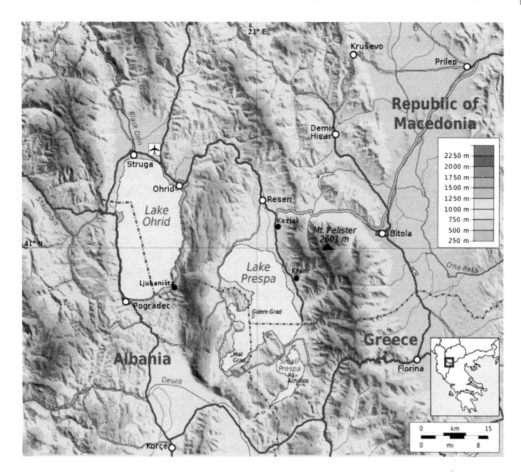

Figure 10.1 Ohrid and Prespa Lakes topographic map.

Source: Future Perfect at Sunrise (2009), *Wikimedia Commons*. Online: http://commons.wikimedia.org/wiki/File:Ohrid_and_
Prespa_lakes_topographic_map.svg.

and mountains, and containing a high level of endemism and many relict species.[15] Lake
Prespa, though younger, contains a number of rare and threatened water birds, including
the largest remaining nesting population of Dalmatian pelicans. People have lived around
both lakes for thousands of years; in mediaeval times, the town of Ohrid was the cultural
centre of much of the region and the Macedonian side of Lake Ohrid has been designated
a UNESCO Cultural Heritage site. There are also several designated RAMSAR sites in
the Macedonian, Albanian, and Greek portions of the Lake Ohrid–Lake Prespa watershed.

Political tensions have likewise also existed in this region for centuries, as Romans,
Ottomans, and Austrians traversed and occupied the land. After World War II, communist
rule isolated Macedonia, Albania, and Greece. There was no communication between
Macedonia and Albania during this era, and thus no coordinated management on the lakes.
Although the communist regimes ended in both Macedonia and Albania in the early 1990s,
national, ethnic and religious rivalries, weak governments, and organized criminal
networks continue to threaten regional security.

Lake Ohrid

In 1994, the World Bank, in cooperation with the Republics of Albania and Macedonia, began preparation for a Global Environment Facility (GEF) grant to fund the incremental costs of a Lake Ohrid Conservation Project (LOCP). In November 1996, Albania and Macedonia concluded a Memorandum of Understanding (MOU) concerning the Lake Ohrid Conservation Project. The MOU established a joint Lake Ohrid Management Board (LOMB) that was "responsible for the preparation of the regulations related to its activities" and authorized to approve projects "based on the previously prepared Feasibility Study". The Parties agreed to "coordinate and adopt laws and regulations necessary for the protection of Lake Ohrid with regard to pollution prevention, water use and fisheries management, etc."; to follow appropriate international pollution prevention regulations and standards; to develop a long-term plan to establish separate monitoring facilities; and to strengthen and develop protection institutions. When the grant was awarded, each party also agreed to carry out the activities needed to implement the LOCP, but they were given no real authority to do so. The LOCP began in late 1998 with a combined budget of about $4.3 million available from the GEF.

From the beginning, the LOCP served as a vehicle to bring officials from the governments of Albania and Macedonia together to stimulate the development of citizen environmental groups in both countries and to reconnect peoples that were isolated by post-World War II divisions and events. However, the initial LOMB had limited authority and the representatives that both countries sent to the LOMB changed regularly. Infrequent meetings made it difficult for these representatives to understand the complexity of the issues surrounding the lakes, or to make the difficult decisions that needed to be made. This led to an interest in creating a more structured management board, one that included high-level representatives of all major stakeholders on the lakes, and which was empowered with specific authorities.

In June 2004, a new transboundary treaty, the "Agreement for the Protection and Sustainable Development of Lake Ohrid and its Watershed", was signed by the Prime Ministers of Macedonia and Albania. The treaty was ratified by both countries in 2005 and is currently being implemented. The transboundary agreement called for the creation of an international "Lake Ohrid Watershed Committee" that would coordinate and direct management activities on the lake and in the watershed. The joint bodies created by the LOCP and the former Lake Ohrid Management Board, including the Lake Ohrid Monitoring Task Force, the Watershed Management Committee, the Organization of Fishery Management, and the Prespa Park Coordinating Committee (discussed in the following section) all continue their responsibilities under the Committee. The work of the Committee is being implemented by a Secretariat, which coordinates activities and sets the agendas for the Watershed Committee.

Lake Prespa and the Prespa Park

As the LOCP worked with a primary focus on Lake Ohrid, a coalition of Park interests was also growing around Lake Prespa. In an international effort, in February 2000, the

Prime Ministers of Albania, Macedonia, and Greece issued a Declaration announcing the creation of "Prespa Park" as the first transboundary protected area in southeastern Europe. The joint declaration stated that Lake Prespa and its surrounding watershed were unique for their "ecological wealth and biodiversity". The Declaration promised "enhanced cooperation" and "joint actions" to:

> a) maintain and protect the unique ecological values of the "Prespa Park," b) prevent or reverse the causes of its habitat degradation, c) explore appropriate management methods for the sustainable use of the Prespa Lake water, and d) to spare no efforts so that the "Prespa Park" becomes a model of its kind as well as an additional reference to the peaceful collaboration among our countries.

In the following year, the Ministers of the Environment of the three countries established the Prespa Park Coordination Committee. While the three governments made no financial commitments to the Committee, it received some funding from the Greek Government and from international donors, including the Swiss and German Governments. In 2001, the 40-year old Galicica National Park in the Macedonian part of the watershed, and the new Prespa Park in Albania signed a partnership agreement to share information and experiences and cooperate in joint management of their common ecosystem. In 2005, the GEF, through UNDP, initiated its project "Integrated Ecosystem Management in the Prespa Lakes Basin of Albania, FYR-Macedonia, and Greece" to catalyse the adoption and implementation ecosystem management interventions in the region. The full implementation of this project is currently under way.

The role of scientists versus community organizers in conflict resolution

The role of science in environmental conflicts has been studied in considerable detail within the United States policy context because of a presumption in many regulatory agencies that it can objectively be used as an "arbitrator" in decision making.[16] However, since environmental research is still often divorced from the larger discourse of security, the salience of ecological research as a means of fostering cooperation is still dependent on a primary recognition by authorities that environmental factors are of consequence in economic or political terms. Furthermore, there can be disagreements within the scientific community that have to be reconciled often through community consensus around a preponderance of evidence as well as risk perception. Both the LOCP and the Prespa Park network reached out to local citizen organizations and scientists. The citizens groups that have become involved in each effort differ and Greece has only been interested in becoming involved in the Prespa effort, where it felt it had a direct ecological and jurisdictional stake.[17] As the Prespa Park network developed, the participants sought independent identity and their own international profile, separate from the Lake Ohrid Conservation Project. Although most participants in both the LOCP and the Prespa Park network acknowledged and accepted the hydrologic data showing a connection between the two lakes, they did not believe that this was a compelling reason to treat the two basins as a single ecosystem. Instead these individuals felt that the cultural and political differences between the two regions and the equal cultural importance of the Prespa basin argued for separate management, with incremental steps towards coordination. Conversely, those more closely associated with the LOCP believed that the hydrologic connection clearly argued that the two basins should be managed as one ecosystem.[18]

When the LOCP began, a joint monitoring programme was considered essential to provide a scientific basis for guiding the work of other project components, and both jurisdictions worked to establish appropriate programmes. However, in order to provide management support to the projects and their governmental implementors, it was essential that the lead scientists collecting the data communicate regularly with policymakers. This communication was challenging in the LOCP.[19] Because of the legacy of the communist era, when the LOCP began there was no regular communication between the scientists in Macedonia and Albania. Although the scientists in both countries enthusiastically endorsed the LOCP, the lack of existing personal relationships and the legacy of isolation and mistrust made the early dialogue difficult. As the monitoring programme was implemented, each country kept its data to itself and neither was especially willing or adept at sharing their findings with the other or with the broader community of stakeholders and policymakers.[20] In part, it seemed that data were viewed as commodities that should be purchased, not as information that might be freely shared to support sound public policy.[21] Even among the academic community, data were not always freely shared; for example, a 2006 hydrologic analysis of the groundwater connections between Lake Prespa and Lake Ohrid used only Macedonian data, noting that data from Greece and Albania were "not available".[22]

To facilitate data sharing and to foster joint interpretation of data, considerable time and attention were directed to preparing a "State of the Environment" report in 2002, including hiring an outside expert to facilitate communication between the Macedonian and Albanian teams and to focus the analysis on key management concerns. This effort culminated in the publication of "Lake Ohrid and its Watershed: A State of the Environment Report" in October 2002.[23] This report represented the first time that Albanian and Macedonian data were used in a common assessment of the ecological conditions in the basin. Forty-nine Albanian and Macedonian scientists and other specialists contributed to this report and through its preparation, got to know each other and learned to work together more effectively. This kind of collaboration is essential for the ecosystem-level assessments needed for comprehensive management of a large watershed and lake system. In 2004, the monitoring programmes in both countries were restructured and incorporated into the state monitoring efforts of both.

Since the release of the State of the Environment Report, regular conferences have brought together scientists from both countries and other interested professionals, and collaborations are growing. Beginning in 2004, the BALWOIS Project (Water Observation and Information System for Balkan Countries), sponsored by the Ministry of Environment and Physical Planning of the Republic of Macedonia and the French Embassy in Skopje, has hosted four international conferences that included scientists from the region and around the world. The Fifth International Scientific Conference on Water, Climate and Environment will occur during the summer of 2012. Later that same summer, UNESCO will host its Seventh Conference on Sustainable Development of Energy, Water and Environmental Systems in Ohrid, Macedonia.

Although scientific communication is increasing, challenges do remain. For example, the scientific data suggest that over the long term, the higher levels of eutrophication in the Lake Prespa watershed might threaten the health of Lake Ohrid.[24] Currently, about 65 per cent of the phosphorus that flows out of the Lake Prespa into the karst ground water channels is retained by the aquifer, however, potential increases in the phosphorus load to Lake Ohrid from either agricultural development or water losses to irrigation, which increase phosphorus concentrations in Lake Prespa, could change this situation.[25] Therefore, coordinated management is an ecological imperative if both countries want to maintain their unique biodiversity and the ecosystem services the lakes provide.

As personal interactions increase across the border and trust continues to be built between the participating scientists, greater progress may be possible on some of the more difficult and complex management issues. It is also likely that the technical community will be more comfortable in designating leaders who might participate in the public debate as individuals begin to know one another. This may help ensure that the issues are increasingly framed by the science that is available. As scientists from both countries are freer to participate in international scientific conferences and other venues, confidence in regional analyses and expertise will also be legitimized. If the experts participate more freely and openly in the public discourse, showing how their data are useful, public and governmental support for data collection efforts and science-based decision making might also grow.

The LOCP invested considerable effort into public outreach and education and these efforts were quite effective in increasing awareness about the environmental problems on the lake. In the critical first years of the project, the work concentrated on increasing the number and capability of citizen groups in the basin. Workshops were held to build the capacity of the NGOs, focusing on organization skills, meeting facilitation skills, public outreach and involvement, and other topics. "Green Centres" were established in both Macedonia and Albania. These Centres served as clearinghouses to connect the NGOs to each other and to provide the critical information they need to mobilize public interest and public action. As a result of this effort, the number of NGOs on the Albanian side increased from 13 to 19, and on the Macedonian side, from 19 to 42 during the implementation of the LOCP.[26] With the financial support of the LOCP, the NGOs in both Macedonia and Albania carried out a variety of activities including summer eco-camps, education in the schools, clean-ups along the shoreline of Lake Ohrid, reforestation on tributary streams in the watershed, the production and distribution of public education materials, hosting round table discussions and workshops, and marking hiking trails in Galicica National Park in Macedonia. In Albania, 51 grants were made.[27] Individual environmentalists and community organizers continue to speak out, but organized activities continue to be limited because funding to support them is difficult to obtain.

During the LOCP, the watershed management committees in each country filled the void in leadership left by the scientific community and worked to develop watershed action plans that outlined some of the environmental changes that were needed. These action plans were combined into a Joint Watershed Action Plan in 2003 and endorsed by the

LOMB. The Action Plan stressed working in partnership, using an ecosystem-based, watershed approach that integrated environmental and economic goals, pollution prevention, a consensus-based, collaborative approach to management, and flexibility. The many small on-the-ground efforts that were part of the LOCP brought people together for a common purpose, one that transcended ethnic differences or other debates that were part of the governmental transitions.

Although enormous progress was made during the period of GEF investment in the Lake Ohrid watershed, local funding was not sufficient to support the full implementation of the Watershed Action Plan. Immediately after 2005, the LOCP was significantly reduced in scope and focus was shifted to the Lake Prespa basin, where GEF funds are still available. Unfortunately, the continuing economic challenges in southeastern Europe still threaten the full implementation of an ecosystem-based management approach in the region.

The third party imperative in environmental cooperation

In the Lake Ohrid basin, it is unlikely that steps would have been taken towards ecosystem management without the intervention of the GEF and the funding it has provided. These resources provided the necessary motivation to allow the formation of informal management structures. When these funds are expended, it is also unlikely that the countries themselves will have the necessary resources to fully sustain the programmes that are created, despite the best intentions.

Borders can be strong barriers that isolate communities from one another. Because local communities often have strong connections to their surroundings, there is a natural interest in environmental protection and public discourse about environmental issues can sometimes transcend other ethnic and political divisions. The Regional Environmental Centre, a non-partisan, non-advocacy, not-for-profit international organization with a mission to assist in solving environmental problems in central and eastern Europe, writes:

> Shared natural entities, when deeply embedded in the local culture, help unite communities divided by country borders. They provide a neutral topic for trans-boundary exchanges, which, when led by an independent facilitator, can serve as a bridge in politically, ethnically and economically difficult cross-border situations.[28]

While the average annual per capita income in the Greek part of the Lake Ohrid and Lake Prespa watersheds is higher than that of Macedonia or Albania ($10,000, compared to $2000), it is low for Greece as a whole,[29] and the people who live in the Prespa basin share a similar rural lifestyle. The investments of the GEF have attracted national and international interest and assisted in efforts towards sustainable development in this otherwise isolated, and internationally ignored region. Without these investments, it is unlikely that the same progress would have been made.

Building on the work of Dean Pruitt on negotiation and mediation theory, Steven Burg has noted that in the case of post-communist eastern Europe, there has been relative success of what might be termed "weak mediating institutions" whose main tool is "the realm of communication and formulation rather than manipulation . . . they include the capacity to transmit and interpret messages, to bring realism to the parties' conceptions of each other, to reframe the issues, and to make suggestions for settlement". Using the example of the Project on Ethnic Relations, Pruitt presents five key principles for finding success in such cases: i) create credible, neutral forums for dialogue early; ii) maintain momentum; iii) work within political realities; iv) encourage indigenous solutions from within existing processes; and v) act with the backing of powerful states.[30] In the case of the LOCP, these principles were largely met during the period of GEF investment. Other donors joined the GEF; for example, the LOCP project was also implemented in Macedonia and Albania under the "Mavrovo Process" (named after the resort where the negotiations occurred). At the initiative of the Macedonian and Albanian governments and with the help of the Swiss and German governments, complementary investments in water and sewage infrastructure, environmental monitoring and impact assessment, and other related areas have shown considerable success.[31]

However, the larger global political situation also created challenges. The collapse of the former USSR left Eastern Europe in a paradoxical predicament: many obstacles for environmental cooperation such as the militaristic focus of the previous years were removed, while at the same time the economic collapse and fragmentation of the former satellite states prevented new environmental initiatives from taking flight. Western donors attempted to address this perplexing situation through "transnational subsidization" of various aid programmes; however, a clear purpose was lacking and, as noted by Robert Darst, this policy also had many negative security implications such as "moral hazard (complacent behaviour when insulated from threat), polluter life extension, and donor vulnerability to environmental blackmail".[32]

Eco-regional policy formulation

As the influence of the former USSR continues to wane, the interest and progress among the Balkan states in joining the European Union offers opportunities to revive the momentum of nascent environmental movements. River basin or watershed management and public participation are central to implementation of the European Union Water Framework Directive, something both Albania and Macedonia must address as they approach ascendancy. Albania as well as Macedonia and all the countries of the former Yugoslavia have begun the major institutional and legal adjustments necessary to move towards meeting the environmental and social requirements of the EU Water Framework Directive. Transparency and capacity building among stakeholder groups and within communities are central parts of these changes, which foster connection and reduce tensions and conflicts.

Moving from the initial phase of new legal structures to the much more challenging phase of fully implemented national environmental policies, regulations, and monitoring and enforcement structures is an enormous additional step, one that even the countries of northern Europe have not fully achieved despite their much more powerful economies and governments.[33] The countries of the Balkan peninsula are among the poorest in Europe and their economies and governments are still democratizing and transitioning towards a market economy. Strategic choices to address poverty and unemployment as the highest national priorities can mean that environmental concerns remain lower on the agenda.

In Macedonia and Albania, the environmental laws that have been passed are still largely frameworks which will require more specific and detailed follow-up legislation before real changes will become apparent. Currently, responsibilities for water management are fragmented among different institutions and management agencies in both Macedonia and Albania, and therefore effective communication and coordination between different ministries and countries is difficult. The process of decentralization presents additional challenges in that local governments have been delegated significant responsibilities for water supply, water protection, and environmental management, but the technical assistance and funding to address these challenges has not followed. Most of these communities have very limited capacity to act on their new responsibilities however well-intentioned they may be. Where communities are organized around historic ethnic ties that reinforce cultural differences, tensions between groups can lead to environmental challenges as well.

In the last decade, much of the environmental focus of Macedonia, Albania, and Greece has centred on the designation of protected areas, rather than on active management of their valuable natural resources like water or endangered fish and wildlife. While cooperation is necessary to create a park or other protected area, there is perhaps less inherent tension in such an action than in delegating water rights, negotiating water quality agreements, or establishing and harmonizing fishing harvest quota. The establishment of a park can create positive energy that, over time, might be extended to other more difficult issues.

In Macedonia, parks and other protected areas must still fund the majority of their management activities on their own, receiving little support from the central government. In Albania, a protected status precludes local managers from managing their own finances. In Macedonia, Albania, and Greece, there are no mechanisms to give local authorities or user groups more of a stake in the benefits of conservation,[34] so local governments have little motivation to address more difficult issues or reach across borders to create transboundary partnerships. For the transboundary Prespa Park, the Prespa Park Coordination Committee functions as a trilateral semi-institutional structure for collaboration, and clearly the goodwill that has developed across all three national boundaries has facilitated its efforts to explore an ecosystem approach to management.

While there are still substantial political disagreements between Macedonia, Greece, and Albania, a desire to protect the unique natural treasures in the Lake Ohrid and Lake Prespa watersheds has brought these nations and their people together. Over time, it may also help to transcend ethnic, cultural, and political divisions that exist both within and between the countries. If continued decentralization and economic growth provide for community ownership and responsibility, the goodwill that has been engendered may ultimately provide for better conservation and mutually beneficial sustainable development strategies. The potential for environmental cooperation as an instrumental means of conflict resolution is thus gaining greater empirical support and deserves greater attention by researchers and practitioners alike.

Conclusion

The experiences of lacustrine cooperation in southeastern Europe suggest that conservation science is increasingly being recognized as a means of fostering cooperation. However, it is yet to play a deterministic role and overriding political grievances can still trump any lasting cooperation unless enshrined within some clear legal process. The peace-building that occurs through such cooperative processes is often just as much in terms of people-to-people interactions as it is on the scientific basis. Indeed, it is also possible for scientists to securitize the discourse and prevent the full potential of peace-building. The role of international eco-regional management institutions is thus still highly relevant and efforts should be made to strengthen mechanisms which allow such institutional growth.

Notes

1   Thomas Homer-Dixon, *Environment, Scarcity and Violence*, Princeton, NJ: Princeton University Press, 1997.
2   See this volume, chapters one and two.
3   N. P. Gleditsch and H. Urdal, 'Roots of conflict: Don't blame environmental decay for the next war', *International Herald Tribune*, 22 November 2004.
4   S.H. Ali (ed.), *Peace Parks: Conservation and Conflict Resolution*, Cambridge, MA: MIT Press, 2007.
5   R. Darst, *Smokestack Diplomacy: Cooperation and Conflict in East-West Environmental Politics*, Cambridge, MA: MIT Press, 2001.
6   S. VanDeveer, 'European politics and a scientific face: framing, asymmetric participation and capacity in LRTAP', in A. E. Farrell and J. Jager (eds), *Assessments of regional and global environmental risks: designing processes for the effective use of science in decision-making,* Washington, DC: Resources for the Future Press, 2006.
7   H-A. van der Heijden, 'Multi-level environmentalism and the European Union: The case of trans-European transport networks', *International Journal of Urban and Regional Research* 30 (1), 2006, 23–37.
8   Fleck's most important work in this regard is *The Genesis and Development of a Scientific Fact*, which was published in German as *Entstehung und Entwicklung einer wissenschaftlichen Tatsache. Einführung in die Lehre vom Denkstil und Denkkollektiv*, Schwabe und Co., Verlagsbuchhandlung, Basel. The first English translation of the book (edited by T.J. Trenn and R.K. Merton, foreword by Thomas Kuhn) was published in Chicago: University of Chicago Press, 1979.
9   B. Latour and S. Woolgar, *Laboratory Life: the Social Construction of Scientific Facts*, Los Angeles, CA: Sage, 1979. Also see B. Latour, *Reassembling the social: an introduction to Actor-network theory*, New York: Oxford University Press, 2005.
10 J. Coburn, *Street Science: Community Knowledge and Environmental Health Justice*, Cambridge, MA: MIT Press, 2005.
11 M. Foucault, 'Governmentality', in G. Burchell, C. Gordon, and P. Miller (eds), *The Foucault Effect*, London: Harvester Wheatsheaf, pp. 87–104. See also T. W. Luke, 'Environmentality as Green Governmentality', in E. Darier (ed.), *Discourses of the Environment*, Malden, MA: Blackwell Publishers, 1999, pp, 121–51.
12 A. Agrawal, *Environmentality: Technologies of Government and the Making of Subjects*, Durham, NC: Duke University Press, 2005.

13 See I. Hacking, *The Social construction of What?*, Cambridge, MA: Harvard University Press, 1999.
14 M. Watzin, V. Puka, and T. B. Naumoski (eds), *Lake Ohrid and its Watershed, State of the Environment Report*, Tirana, Albania and Ohrid, Macedonia: Lake Ohrid Conservation Project, 2002; A. Matzinger, M. Jordanoski, E. Veljanoska-Sarafiloska, M. Sturm, B. Mueller and A. Wuest, 'Is Lake Prespa jeopardizing the ecosystem of ancient Lake Ohrid?', *Hydrobiologia* 553, 2006, 89–109; C. Popovska and O. Bonacci, 'Basic data on the hydrology of Lakes Ohrid and Prespa', *Hydrological Processes* 21 (5), 2006,. 658–64.
15 S. Stankovic, 'The Balkan Lake Ohrid and its Living World', *Monographiae Biologiae* 19, 1960, 357.
16 C. Ozawa, *Recasting Science: Consensual Procedures in public Policy Making*, Boulder, CO: Westview Press, 1991.
17 A. Antypas and O. Avramoski, 'Polycentric environmental governance: towards stability and sustainable development', *Environmental Policy and Law* 34, 2004, 87–93.
18 Ibid.
19 M. Watzin, 'The role of law, science and public process: practical lessons from Lake Champlain (USA and Canada) and Lake Ohrid (Macedonia and Albania)', *Pacific McGeorge Global Business and Development Law Journal* 19 (1), 2006.
20 Ibid.
21 Global Environment Facility, *Assessment of the Management of Shared Lake Basins in Southeastern Europe*, Athens, Greece: GEF IW:LEARN Activity D2, 2006.
22 Popovska and Bonacci, 'Basic data on the hydrology of Lake Ohrid', pp. 658–64.
23 Watzin et al., Lake Ohrid and its watershed.
24 Matzinger *et al.*, 'Is Lake Prespa jeopardizing the ecosystem of ancient Lake Ohrid?'
25 Ibid.
26 M. C. Watzin, O. Avramoski, S. Kycyku, T. Naumoski, D. Panovski, V. Puka and L. Selfo, *Lake Basin Management Initiative Experience and Lessons Learned Brief*, Lake Ohrid, Macedonia and Albania, Regional Workshop for Europe, Central Asia, and the Americas. Final Report, 2003.
27 Ibid.
28 Regional Environmental Centre, *Trans-boundary cooperation through the management of shared natural resources*, Szentendre, Hungary: self-published, 2007.
29 UNDP, *Integrated Ecosystem Management in the Prespa Lake Basin of Albania, FYR-Macedonia, and Greece*, Governments of Albania, FYR-Macedonia, United Nations Development Programme, Global Environment Facility, 2004.
30 S. L. Burg, 'NGOs and ethnic conflict: Lessons from the work of the project on ethnic relations in the Balkans', *Negotiation Journal* 23 (1), 2007, 7–33.
31 For an enterprise analysis of cooperation in this region see M. Dmiitriv *et al.*, 'Cross-border cooperation in southeastern Europe – The enterprises' point of view', *Eastern European Economics* 41 (6), 2003, 5–25.
32 Darst, *Smoke Stack Diplomacy*, p. 199.
33 R. Perkins and E. Neumayer, 'Implementing Multilateral Environmental Agreements: An Analysis of EU Directives', *Global Environmental Politics* 7 (3), 2007, 13–41.
34 Global Environment Facility, Assessment of the Management of Shared Lake Basins in Southeastern Europe, 2006.

11

POPULATION AND NATIONAL SECURITY

Jennifer Sciubba, Carolyn Lamere,
and Geoffrey D. Dabelko

Introduction

Population is an integral part of environmental security studies. As part of the environment, humans have a powerful ability to shape nature and in turn be shaped by it. Ironically, it was man's ability to remove himself from nature that led to greater understanding of co-dependence. In December 1972, astronaut Harrison Schmitt aboard Apollo 17 took what has now become one of the most recognizable images in human history. This photo, nicknamed the 'Blue Marble,' showed the Earth's fully illuminated face for the first time. This view of the planet, and subsequent missions from further away that show the Earth as nothing but a 'pale blue dot,' jolted many into realizing that rather than a planet of limitless resources, Earth was like a spaceship hurtling through space on a long-term mission with no hope of resupply.[1] So was born the concept of Spaceship Earth—the notion of ecological limits based in part on both consumption and population growth.

Tensions between humankind and Earth's ability to provide resources are only part of the story of population, environment, and security. The issue of population fits squarely into traditional, meaning state-centric, aspects of security. This chapter will describe the effects of population on national security, which is to say on 'the ability of states to maintain their independent identity and their functional integrity.'[2] It is important to understand that there are numerous population trends at work in every society—differential growth of ethnic groups, changing size of the population, an aged or youthful age structure, for example. Throughout this chapter, we comment on size, structure, distribution, composition, changes in any of these, and interactions between population and other variables. The possibilities may seem daunting, but all population trends are driven by changes in only three variables: fertility, mortality, and migration.

Threats to national security can arise in all five sectors of security; that is in the military, political, environmental, societal, and economic sectors of security. Security sectors are analytical lenses through which the security field can be ordered.[3] Our choice to focus on national security is driven purely by analytical reasons and influenced by the preclusions of space. To be clear, demographic factors are also an issue of concern for other referent objects of security. It is obvious that human numbers are likely to severely compromise biosphere or ecological security, while it is also the case that human security can be adversely affected by population factors. Some of these are discussed in the context of national security, for example, when we discuss the connection between violent conflict and population, as any kind of conflict also jeopardizes human security. Many recent defense strategies, including the United States', Russia's, and the European Union's, recognize the value of a comprehensive assessment, including aspects of human security.[4] Though many theorists and policymakers, especially those from the realist tradition, have resisted a broadening of the definition of national security from its traditional military/power focus, there is increasing recognition among policymakers that a multitude of threats emanating from a variety of sources affects national security. For the U.S., the terrorist attacks on September 11, 2001, were one of the first such wakeup calls. The state had to react to a non-state actor and, in subsequent years, through involvement in Iraq and Afghanistan, scholars have often argued that ties between poverty and insecurity in those communities may have created conditions that motivated terrorists. It is also the case that militaries are being called upon to provide order when individuals are in danger from structural failures, such as US and Canadian military responses to the natural disaster in Haiti in January 2010. In other words, the military is increasingly providing human security. The interesting connection among these kinds of security is that if demography affects national security in a way that weakens it (e.g. when there are not enough military age people, or too many fiscal constraints), then the military, as the arm of the state, will be unable to continue its 'human security' missions, such as delivering aid and building schools.

The following sections will demonstrate the various ways in which fertility, mortality, and migration function as components, indicators, and multipliers of a state's national security. We review examples of the population security challenges arising in all five sectors of security. We conclude by summarizing the argument and by restating the importance of population for security research.

Military security

Leaders have always considered the importance of population to the security of their empire, kingdom, city-state, and nation. In this sense, including population in an assessment of military security is not broadening the definition at all, but rather reflects a very traditional conception of the relationship between people and the security of the state. The category of military security includes conventional state concerns about protection from external threats, avoiding or prevailing in interstate war, and the ability of a state to project power and pursue its international objectives. Some forms of civil conflict also fall into this category, though there is overlap here with the environmental sector of security.

Population and (military) capability

One of the oldest conceptions of population and security concerns its influence on the power a state can project in the international system. There is a long history of statesmen praising the contribution of a large population to power and it wasn't until the eighteenth century, culminating in the Reverend Thomas Malthus' *An Essay on the Principle of*

Population in 1798, that people started to think that explosive growth might not always be positive.[5] Even within the International Relations discipline, the population–power relationship once held a prominent place. Early International Relations texts brought demography into the political science fold by focusing on ways that population trends, as a means to power, might contribute to yet another world war. Many scholars had seen connections between population trends and both wars of that early century, and were especially disturbed by Hitler and the Nazi Party's obsession with eugenics and population growth as a means to international power. The scholarship of this time period shares a couple of characteristics that make it relevant to any contemporary analysis of demographic security.[6]

First, these authors argue that population can be given equal treatment alongside economic and political variables. One of the most influential formulations was power transition theory, first posed by A.F.K. Organski. At its most basic level, power transition theory argues that three variables comprise state power: population size, productivity, and political capacity. When combined with preferences, changes in these variables affect the hierarchy of states in the international system. Changes in the hierarchy—or power transitions—then lead to conflict. Population has a direct effect on power because people are the soldiers and laborers vital to any functioning state. As is likely obvious, the logical implication is that the bigger the population, the more power a state will have. Yet, Organski wasn't so linear in his supposition. He argued that though size is a key element of power because it provides the potential resource pool a state could mobilize, size alone does not guarantee great power status. An inept government that does not put its resources to good use has no advantage. According to power transition theory, 'In order to be truly powerful the population also must be productive . . . But those advantages cannot be realized without political capacity, defined as the ability of governments to extract resources to advance national goals.'[7] So, while population is an important determinant of power in the long term, according to power transition theory it is not the sole determinant of a state's position in the international system; economic productivity and political capacity matter as well.

The population–power literature points out that population is actually a complex variable. Not only size, but also composition, rate of change, and distribution of the population matter. In Harold and Margaret Sprout's excellent 1945 book, *Foundations of National Power*, they discuss how demography, like economic development, is a means to national power, but they consider how population can be a burden and describe how 'quality'—productivity, education, and so on—matters as much as 'quantity.'[8] One of the major arguments of their book is that any population trend can be a boon or a burden, meaning that a list of the most populous countries only tells part of the demographic story. Population can contribute positively to a state's power and security, or can be a force for insecurity and conflict. Demography is important, but other political variables are as well.

Another lesson from the population–power literature is a caution against determinism. Many of these scholars theorize how a range of intermediate variables translate population into political power and influence. They emphasize that though demography is important, it should be considered alongside other important trends and factors. Just because a population makes demands on a state does not guarantee the state will respond to those demands.

A couple of decades later, the population 'crisis' literature of the 1960s focused less on ways that domestic population trends can contribute to a state's power and more on ways that population growth in less developed countries could be a threat to advanced industrial democracies, which were starting to see declines in fertility after the post-World War II baby boom. Most of the scholars of this decade focus on the dangers of growth, particularly the population–resource balance; this resurgence of Malthusianism was most notably encompassed in Paul Ehrlich's *The Population Bomb*.[9]

While more traditional views of security like those prescribed by realism are still at the forefront, it is no longer possible to ignore the impact the natural environment can have on the security of a state, a country's government, and the inhabitants thereof. The recent increase in publications related to environmental issues reflects the ubiquity of the environment and its impact on both domestic politics and international relations.

How do these lessons apply to contemporary demographic trends and national security considerations? The intermediate variables many of these authors point to remain important for translating population into military power and global influence. One example is political institutions, including the party system, the role of interest groups, and the role of supranational bodies. Let's say we want to assess the impact of one population variable, population aging, on a state's ability to allocate money towards the military as the number of dependents requiring pensions and other entitlements grows. One of the worries about population aging is that large cohorts of elderly citizens, particularly older voters, will hijack domestic politics. A zero-sum game between the aged and the military could then result. An aging state with politically powerful elderly could see that the national budget allocates more money towards entitlements and less towards defense, thus weakening national security.[10] Though this sounds logical, for a truly comprehensive and accurate assessment we must consider the domestic institutional context.

Europe, home to some of the world's oldest states, is one of the major areas of concern about trade-offs between military and entitlement spending. European states are major powers in the world system, allies to the United States, and frequently contribute to humanitarian missions. Numbers alone—proportions of elderly and current entitlement projections—lend credence to the zero-sum projections.[11] In any country, government programs must compete for slices of the budgetary pie. In an aging state, the pie itself does not necessarily get bigger because GDP growth is lower in aging states, and if no policies change then the portion allotted to entitlements grows, leaving less for the military. However, the way that European states utilize their domestic institutional processes could play a role in determining how much money is available and how it is allocated.

Policies that increase efficiency and productivity and try to bring under-utilized segments of the population, such as women, more into the labor force, could help states avoid a shrinking pie. European states that limit the power of interest groups may not see the same politics of aging as the United States, where the retired persons' interest group AARP is very influential, and thus avoid a bulging slice of the pie that goes towards entitlements. On the other hand, preferences and the ability to mobilize also matter. Mark Haas, writing on aging and security, cites evidence that aging European states have limited ability to cut pensions despite the increasing burden on the budget.[12] Throughout 2011 and 2012, Europe continued to struggle with these issues. Finally, the role of the European Union matters as well. The Lisbon Treaty mandates major changes in states' retirement ages, labor policies towards older workers, and entitlements.[13] So, even as reform may be difficult domestically because of powerful or numerous constituencies, membership in the EU forces states to comply with tough measures. If the EU is permanently weakened by financial crisis, then the Lisbon Treaty will become unimportant and states will probably turn less attention to their militaries and more towards growing their domestic economies. When we step outside of Europe and assess the likely impact of China's aging on their ability to project military might we must still factor in the role of political institutions as an intermediate variable. Regime type—ranging from authoritarian to fully democratic—may matter as well. Democratic leaders must factor in the demands of various interest groups and the power of voters, whereas less democratic governments can ignore some of these demands. Thus, unlike Europe, China may not face stress from population aging if its elderly are marginalized in the political process.

In addition to budget constraints, other military security concerns with aging populations include the state's ability to staff a military as proportions of that age group shrink and even the ability to maintain the state itself as the overall population shrinks. Russia's leaders appear to obsess about their demographic dilemmas and the effects these will have on military security. With extremely high mortality, especially for males, plus low fertility, political economist Nicholas Eberstadt has argued that Russians are committing 'ethnic self-cleansing.'[14] The potential implications frighten Russian leaders, since as Eberstadt says, 'history offers no examples of a society that has demonstrated sustained material advance in the face of long-term population decline.' Vladimir Putin has said that Russia's demographic crisis is the number one problem facing the country and their 2009 Security Strategy gives the odd suggestion that the national security forces should 'create conditions for . . . stimulating fertility.'[15] Despite these deficiencies, Russia has continued to behave assertively in international politics as they oppose eastward expansion of NATO and doggedly pursue control of the Arctic, as just two examples. Russia illustrates that demography is only one factor of military security; a state's political goals are equally important.

What we learn from this review of the population and power literature is that an assessment of population's effect on military security must be much broader than size alone. Size matters, but it is only part of the equation that calculates a state's power, and political response is the primary factor in determining whether any population trend is a benefit to or a drain on a state.

Civil conflict and war

One of the trends most associated with conflict is a youthful or youth bulge age structure. Henrik Urdal, one of the foremost researchers on youth and conflict, found that 'when youth make up more than 35 percent of the adult population, which they do in many developing countries, the risk of armed conflict is 150 percent higher than in countries with an age structure similar to most developed countries.'[16] Six out of nine new outbreaks of civil conflict between 2000 and 2006 occurred in countries with very young or youthful age structures.[17] Why are societies with large proportions of young adults so volatile? Richard Easterlin's theory of relative cohort size 'posits that, other things constant, the economic and social fortunes of a cohort (those born in a given year) tend to vary inversely with its relative size, approximated by the crude birth rate in the period surrounding the cohort's birth.'[18] Richard Cincotta points our attention to large numbers of men in search of limited resources and affirmation—jobs, education, guidance, respect, identity, and women.[19] A large youth cohort lowers the cost of recruitment, since potential soldiers are abundant and the opportunity cost of joining a rebel group is low for a young person when there are no other opportunities available in the society.[20] The size of the cohort matters as well, as those who are born into a large youth cohort face an even lower opportunity cost than those born into a smaller youth cohort.

Pakistan provides an example of a country with a young and growing population and decreasing political or regime security. High rates of population growth there and a young age structure—35 percent of the population under the age of 15 and a total fertility rate of 3.6 children per woman—provide a context ripe for recruitment.[21] Certainly, only a fraction of unemployed youth will choose militancy, but in one of the world's largest countries, a tiny fraction of youth is still a considerable number. Over the last several years, Pakistan has had many governance challenges and is home to a growing Taliban insurgency. Throughout 2009, newspapers were flooded with stories of Taliban takeovers of strategic areas near the capital, Islamabad.[22] This civil war not only undermines the Pakistani leadership's ability to govern, it also connects to the international conflict underway throughout the region.

Revolution and rebellion

Youth are often stereotyped as rebellious, and famous times in history, such as America's 1960s civil rights, feminist, and anti-war protests or China's 1989 student protests in Tiananmen Square, are frequently cited examples that seem to prove the rule. Those concerned with security may look to demography to anticipate countries likely to experience uprisings that will bring new leaders to power. For example, Guinea's coup in December 2008 took place in a demographically dangerous context. Guinea's median age is 18 years, total fertility rate is 5.5, and almost 20 percent of the population is aged 15–24 years.[23] This demographic profile, plus a history of instability, should have made the coup no surprise, and indeed for most watchers, it was not. As researchers at Population Action International have found, countries with age structures in which 60 percent or more of the population is younger than age 30 are the most likely to face outbreaks of civil conflict.[24]

Clearly, demography is an important variable in intra-state conflict. Thomas Homer-Dixon, one of the leading voices on environmental scarcity and conflict, found that 'Environmental and population changes that increase the risks of violence have been shown to matter almost solely for *internal* or domestic conflicts: ethnic struggles, regional rebellions, or local group conflicts.'[25] In the following section, we turn our attention to the literature that covers such struggles.

Environmental security

Wars and violent conflict

While population trends like large youth bulges are closely associated with conflict, the tie between the environment and conflict is less clear. Scholars have sought to tie environmental degradation to conflict, though with only limited success. While Homer-Dixon has claimed that environmental factors will soon be the leading cause for conflict, other scholars have struggled to find the evidence that supports such a claim. Diehl and Gleditsch find population to be a more convincing factor driving conflict; they cite Wenche Hauge and Tanja Ellingsen's 2001 study, which found environmental scarcity only marginally increases conflict when removed from other variables. Hague and Ellingsen point to high population density, income inequality, and poverty as the main driving factors.[26]

While traditional theories, like those outlined by Malthus and neo-Malthusians, discuss the link between scarcity and conflict, others, especially those from economics, argue that an abundance of resources is actually more likely to lead to conflict. The 'honey pot hypothesis,' for example, discusses the propensity for armed struggle to capture valuable resources like mines, while the 'resource curse hypothesis' deals with the economic imbalance of focusing unduly on few exports. Colin Kahl has emphasized that these types of conflicts are more prevalent in mineral states—the struggles over African diamond mines come to mind. But the presence or absence of resources is not sufficient to predict conflict; other factors, like the actions and sympathies of elites, the presence or absence of grievances between groups in the state, and the strength or weakness of the government also impact the likelihood of conflict within a state.[27]

While population factors do affect national security, some scholars shy away from heavily emphasizing just one aspect of the conflict. For example, Betsy Hartmann resists this focus on demography as an essential player in international politics. For Hartmann, population and the environment play a role that is secondary to economic and political issues.[28] She argues that while demography is a convenient scapegoat for donor countries, an undue focus on population control comes at the expense of more holistic health and development programs. Additionally, she suggests that this kind of attitude is dehumanizing for the impoverished, and the securitization of this issue can even present them as the implicit enemies who create global environmental insecurity.[29] In some respects, Hartmann is correct: focusing on any one aspect of a state risks missing the bigger picture. But scholars should not seek to rectify this kind of inattention by forgoing all study of population and the environment, as both can greatly contribute to one's understanding of international politics in general and the study of a particular country.

Environmental security as human security

Population can also adversely affect environmental security as human security. And human security is relevant for national security, as without the security of individuals and the environment they live in, the state itself is insecure. Increasingly, human security matters for planners of national security, so that the state remains the key provider of human security. However, human security can also be provided by other actors, such as the United Nations. The ecological security approach as proposed by Dennis Pirages and Teresa DeGeest makes many of these connections clear. They call for a paradigmatic shift to recognize diffuse threats. They say that this paradigm:

> begins with the not unreasonable observation that promoting human security in an era of globalization means moving beyond the rather moribund post-Cold War agenda that was originally shaped by fear of communism, and addressing a much broader array of nonconventional security challenges. It also suggests that a more robust assessment of the causes of and potential cures for widespread human misery and premature human deaths should initially inform the discussion.[30]

According to them, a complete picture of the security environment must take these diffuse threats into account. Great power politics was increasingly unable to explain international relations after the Cold War and thus, as they argue, the end of the Cold War opened intellectual space to consider a range of variables—including population—in a different light. According to another pair of demographic scholars, during the Cold War, 'Issues like internal conflicts, intrastate and interstate migratory flows, and environmental problems were more or less subsumed under the security policies and interests of nation-states.'[31] One groundbreaking article, Jessica Tuchman Mathews' 'Redefining Security' in *Foreign Affairs* in 1989, captures this shift. Since that time, many scholars have tried to look at the varying links between different demographic trends and different kinds of conflict.[32]

As Pirages says in a much earlier book, 'Shifts in dominant social paradigms [like the agricultural and industrial revolutions] result from social anomalies or dilemmas, conditions for which the prevailing DSP cannot provide an explanation.'[33] The rising number of civil conflicts in the 1990s also drove the evolution of human security within academia. What many saw as the failure of the prevailing US security paradigm, which focused on interactions between states, to account for, or indeed 'fix,' many security problems is an example, and opens space for population as a factor in national security. States around the world 'face a debilitating combination of rising competition for resources, severe environmental breakdown, the resurgence of infectious diseases, poverty and growing wealth disparities, demographic pressures, and joblessness and livelihood insecurity.'[34] Pirages and DeGeest say that such threats have traditionally 'been seen as matters better addressed through prayer than defense spending.'[35] Michael Renner calls these 'problems without passports.'[36]

Many of these scholars recognize that population is part of a complex nexus that can be related to conflict.[37] Population and environmental concerns can be a cause of poverty, a challenge for governance, and therefore a distal cause of civil conflict. Renner argues that such problems 'cannot be resolved by raising military expenditures or dispatching troops. Nor can they be contained by sealing borders or maintaining the status quo in a highly unequal world.'[38] In their book *Too Poor For Peace?* Lael Brainard, et al. show how such challenges can balloon into global security problems. They say the danger is when:

Poor, fragile states can explode into violence or implode into collapse, imperiling their citizens, regional neighbors, and the wider world as livelihoods are crushed, investors flee, and ungoverned territories become a spawning ground for global threats like terrorism, trafficking, environmental devastation, and disease.[39]

Youthful and growing populations, rapid urbanization, or large migration flows often tax the ability of the state to provide for its inhabitants. Such demographic challenges occur in states that already have weak governance. One example where underlying population–environmental stresses have led to a security crisis is Sudan.

According to the *Arab Human Development Report*:

> In the western provinces of Darfur and Kordofan, competition over pastureland erupted into tribal warfare on a scale that led to international intervention. Among the foremost factors aggravating the fighting in this region were scarcity of rainfall, the population explosion, and changes in the prevailing social system from nomadic shepherding or crop gathering to sedentary agriculture.[40]

Climate change can also contribute to national and human insecurity, for example, in increased pressure on government health facilities. Changes in the environment also have enormous impacts on global health. Increased migration, due to environmental or other reasons, can facilitate the spread of disease, and larger numbers of people living in close proximity in cities also makes disease transmission more likely. Climate change can also result in disease migration, like an increased prevalence of diseases previously found in tropical regions throughout the world; malaria is becoming much more common, for example.[41] An increased burden of disease can simultaneously lead to an increased demand for government health services and decreased government revenue as fewer people are able to work, putting the country in a very difficult position.

The consequences of 'poor, fragile states' like the ones Brainard et al., refer to have been studied under the label of weak and failing states. Council on Foreign Relations president Richard N. Haass defines a weak state as 'one that cannot defend itself, maintain internal peace, or address many of its most pressing challenges without outside help.' Like former US Under Secretary for Policy Eric Edelman, Haass sees weak states as the preeminent security challenge of the twenty-first century.[42] The best-known measure of failed states is that done by the Fund for Peace and *Foreign Policy Magazine*. The annually published Failed States Index uses a range of indicators to measure state vulnerability or risk of violence. The risk factors include demographic variables and environmental degradation. Specifically, the 12 indicators include extensive corruption and criminal behavior, inability to collect taxes or otherwise draw on citizen support, large-scale involuntary dislocation of the population, sharp economic decline, group-based inequality, institutionalized persecution or discrimination, severe demographic pressures, brain drain, and environmental decay.[43] In the 2009 rankings Uganda was twenty-first and a brief portrait of the country shows how all of these variables come together to create instability. There, '80 percent of the population is rural and depends on rainfed agriculture, while inherent fluctuations in the climate leave the country vulnerable to floods and droughts.' Uganda's high fertility rate has been described as 'a primary cause of poverty' due to shortages of land compounded through each successive generation. Researchers from Population Action International argue that underlying demographic and environmental trends bode poorly for the state's ability to adapt to coming challenges, including climate change. They say:

> The future impacts of climate change are not yet fully known, but those who will feel them most directly—women, children and the poor—are already the most affected by the consequences of rapid population growth and unintended fertility, low education, and limited access to social services that will in turn diminish Uganda's capacity to adapt to climate change.[44]

States can also tip from weak to failing because of external events, such as large natural disasters. However, demographic trends, particularly urbanization and rising population density, can increase vulnerability to natural disasters. As one researcher notes:

> Risk comes from increasing poverty and inequality and failures in governance, high population density, crowded living conditions and the siting of residential areas close to hazardous industry or in places exposed to natural hazard (including the modification of environments which generates new hazard, e.g. through the loss of protective mangroves to urban development, or subsidence following ground water extraction).[45]

Political security

A third line of research on population and national security focuses on a state's ability to remain in power, what we term political security, also known as regime security. Nazli Choucri has described regime security as 'the ability of the government and its institutions to discharge formal responsibilities and also to protect itself from domestic disorder, revolt or dissention.'[46]

Some population trends are more likely to be correlated with political insecurity than others. Three of the population variables regime security scholars have focused on are youth, urbanization, and gender. All of these are what Goldstone refers to as 'population distortions,' defined as population changes that 'make it difficult for prevailing economic and administrative institutions to maintain stable socialization and labor force absorption.'[47] The previous section on youth and rebellion looked at ways youth may be used as tools to bring dangerous leaders to power; but often, youthful rebellions are seen as positive forces for democracy, including the two previously cited examples of the US and China. Of course, rebellion of any flavor is a regime security threat from that country's perspective, even if it leads to democracy. Even dictators want to stay in power. But just as the previous section examined the correlations between youth and civil war, we must consider the possibility that youth could be a force for democracy, and the more countries there are with youthful age structures, the more transitions to democratic rule there could be.

The revolutions of the Arab Spring in 2011 provide strong support for the connection between age structure and revolution. Youth in Tunisia, Egypt, and Libya were part of a large and crowded young adult cohort, with limited opportunities for meaningful employment, political participation, and even marriage. Age structure and context set the arena as one ripe for revolution, but age structure can help us understand what comes after revolution as well. Cincotta posited that as the population of young adults reaches 40 percent or less of the total adults a country has a 50 percent chance of transitioning to some form of democracy.[48] Perhaps, he argues, it is too soon. Youthful age structures are too volatile, but as they mature the likelihood that the country will transition to democracy increases. In a different article, Cincotta suggests that the recent leveling off in measures of global democracy is temporary and will rise again as youthful demographic profiles mature. Latin America, North Africa, and Asia are all likely to see the emergence of new and more stable liberal democracies before 2020. It will take many years before we know what forms the regimes of Tunisia, Egypt, and Libya take but Cincotta's research suggests that we may eventually see an increase in democratic governance if fertility rates continue to fall worldwide.

Crime

Moving to less intense violence, we consider crime as a political security issue because it is part of a continuum of violence and is, at the most basic level, a challenge to the state's monopoly over the legitimate use of force. Though so far we have explored connections between 'youth' and violence, in actuality, researchers tend to focus more on young men than young women, as men are more prone to violent crime. Researchers at Population Action International reported that 90 percent of arrests for homicide in almost all countries surveyed were committed by men and three-quarters of these by young men between the ages of 15 and 34.[49] A young age structure implies an abundance of young men that could be involved in activities that undermine the ability of the state to keep order and a monopoly over the use of violence.

The experience of Haiti illustrates the connections between population and security. Almost 70 percent of Haitians are under age 30 and many are growing up in a very unstable environment. Demography and insecurity in Haiti are tightly coupled. There have been ongoing episodes of armed violence in Haiti's capital, where gangs are a significant problem. Gangs draw upon an abundance of young, poor males for members as they vie to control turf in the city and raise revenue through kidnapping and drug trafficking. According to the authors of a report from Population Action International, Haiti's gang violence is driven in part by the country's 'rapid urbanization process compounded with political instability, young age structure, and poor employment prospects for youth.'[50] These factors point to severe challenges for establishing order in Haiti. If the United States plans to continue to support Haitians, the military should expect continued demands for military involvement in the country.

While the growing urbanization of the world, and especially of developing countries, is often described as an issue falling under population studies, the environment also has an important role to play in the willingness of individuals to move to cities. But urbanization not only affects the environment in terms of the substantial environmental impact of millions of people living in a relatively compact area, but it also is affected by changes in the environment. As mentioned, environmental scarcity does not usually lead to conflict, but it does lead to increased migration from rural to urban areas. As environmental degradation makes it more difficult to earn a living from traditional means like agriculture, fishing, and the like, rural inhabitants are 'pushed' to live in cities.[51] But urban problems are difficult to solve; if the government tries to improve conditions for city dwellers to attempt to reduce problems like crime and poor public health, city living becomes more attractive and more rural citizens will be tempted to move to urban areas.[52] Furthermore, the developing countries that are increasingly facing these problems have a hard time finding the funds necessary to solve all grievances, leading to increased public dissatisfaction with the government.[53]

Gender imbalance, though related to youth, is another demographic trend that threatens regime security through its connection to crime. Gender imbalance results when parents engage in sex-selective abortion and/or female infanticide because of a cultural preference for male children, though infanticide is waning. In some areas of India—one country in which the practice of sex-selective abortion takes place—ratios of males to females has been recorded as high as 113:100.[54] Valerie Hudson and Andrea den Boer have introduced the idea that gender imbalance—particularly 'surplus males,' or 'bare branches' to the Chinese—could be a security concern partly because they link the treatment of females in society to the likelihood of that society engaging in conflict in general; the security of women is linked to the security of states.[55] Hudson and den Boer argue that in India and China, bare branches are dangerous because there may be a link between internal violence (especially against women) and violence between societies. Bare branches, in their view, are not necessarily the catalyst for this violence, but can be the fodder. They support their argument with evidence that contemporary bare branches are predisposed to substance abuse, violent crime, and collective aggression. While this thesis has become popular in national security circles, not all scholars agree that the links between gender imbalance and regime security are so direct, so this is an area with potential for future research.[56]

Societal security

Closely related to political threats are societal threats (real or perceived), which concern the identity or ethnic make-up and coherence of any given state. Analysts who study conflict and war might not think of these as security threats, but to the states *perceiving* impending demographic change that will work through institutions to displace an ethnic, religious, or other majority, the threat may seem just as real. In the following paragraphs we examine two examples: Western Europe and Russia.

A handful of Western European states have had a notable backlash to increasing proportions of immigrants, particularly of a Muslim religious background. Far right, anti-immigrant parties in Austria (Freedom Party) and Italy (the Northern League), as just two examples, have resurged in popularity. In fall 2008, two of Austria's far right parties received the biggest share of the vote—29 percent—'ever won by rightists in modern Western Europe.'[57] Of course, many sensational stories of clashes between immigrants and majority populations have made media headlines. The murder of controversial Dutch filmmaker Theo Van Gogh in 2004, whose film about violence against Muslim women sparked anger among immigrant groups, was one such incident that brought clashes over identity to light. The riots in the French *banlieues* in 2005 and in 2007 were an even more striking example of inequalities between immigrant and native groups that spilled over into heated discussions over class, race, and citizenship. Though mostly isolated, such incidents are reminders that there is at least the perception of cleavage within many European societies, and an underlying tension that could erupt into violence if only given the right spark. Some of the tension on the native side is over perceived higher birth rates of Muslim immigrants, and a fear of decline of the 'Western European culture.' In Austria in 2001, the birth rate for Catholics was 1.3 children per woman while the rate for Muslim women was 2.3.[58] This difference of one child, the former rate well below replacement level of 2.1 and the latter above it, will change the proportions of the population. No doubt such data, and perceptions of difference, have fueled the resurgence of Austria's far right.

Russia's declining Orthodox population has encouraged similar fears of takeover by Muslim minorities and of an invasion of Chinese into the Russian Far East. Some of the fear within Russia is because the country's ethnic groups have been growing at different rates. The leadership views this as a regime threat. Russia's 2009 national security strategy says that 'Social cohesion can be improved by fostering the "spiritual unity of the Russian Federation's multiethnic people", by such means as resisting orientation to "the spiritual needs of marginal strata", which is a "primary threat to national security in the cultural sphere".'[59]

Economic security

Population trends can be both negatively and positively related to economic security. Though many statesmen and scholars once argued that a state's economic growth comes from a large population in general, we now understand that *which* segments of the population are large matters more, as do the characteristics of that population. Populations with large proportions of either young or old dependents may economically struggle because there are too few workers to support the children and elderly in society. Societies with high fertility rates face dual economic challenges. First, each family must support a large number of children. Second, the societies struggle to create jobs fast enough to keep pace with the large cohorts entering working ages each year. There are many examples where high fertility or youth bulges have led to a lack of economic security and ultimately threatened national security. Iran's high fertility rate and struggling economy in the 1970s contributed to the revolution there in 1979 and more recently, the Arab Spring stemmed in part from frustrations over the lack of economic opportunities for youth.[60]

On the other end of the spectrum, societies with large numbers of elderly face economic insecurity as well. While countries with a lot of youthful dependents tend to be less developed, aging states are more often advanced, industrial democracies. In these societies, elderly dependents are typically more expensive to support than children because of high healthcare costs and generous entitlements. As larger and larger proportions of the population retire out of the workforce, there are fewer workers to support the elderly, creating severe economic strains. Europe has been struggling with the economic strains of population aging for over a decade and some of these problems came to a head in 2011 and 2012 when the European financial crisis hit.

Demography can bring economic benefits as well, though. Demographers often refer to the time along the demographic transition when a society has more workers than dependents as the 'window of opportunity.' In theory, states with this age structure can see higher growth in income per capita because both the household and the state can invest more per child, particularly in education.[61] East Asian states provide the most well-known examples of states that have taken advantage of their demographic bonus. Between 1965 and 1990, East Asia's working-age population grew nearly 10 times faster than its dependent population, and scholars have argued that demography can explain one-third to one-half of East Asia's economic 'miracle.'[62] Asia's economic growth has increased national security, as the rise of China as a world power has shown.

Demography can also bring economic benefits through migration. An influx of people can provide receiving states with specialized skills or labor. In aging states, particularly, migration can make up for shortfalls in the working-age population, delaying economic strains that would have occurred through the growing number of dependents. Part of the reason the United States and Canada have relatively less of an aging problem compared with Europe or Japan is the high levels of migration to those North American states. Countries more accepting of migrants may thus see a brighter future of economic growth.

A final point on population and economic security is that the number of people of dependent or working ages matters but other qualities of the population, such as education and health, matter as well. A state with large proportions of poorly educated and sick workers is not likely to see the same demographic dividend as a state that has invested in policy reforms to capitalize on the skills of their population. Though East Asia provides an example of a region that took advantage of its window of opportunity, many states in Latin America let their opportunity slide by. Although during the heyday of its demographic bonus East Asia experienced a per capita annual growth rate of 6.8 percent between 1975 and 1995, during that same time Latin America's growth rate was only 0.7 percent, though many Latin American states were experiencing a dividend then as well.[63] Thus, there is no guarantee that demography will translate into economic security, but with targeted policies to address literacy, skills, and health, states can turn their demographic bonuses to economic security, which is a vital component of national security.

The environment is closely tied to the economy, as nearly every activity, from agriculture to truck driving to utilizing electricity for lighting and climate control, relies on natural resources. As previously mentioned, environmental degradation can seriously upset both the political and economic situations in a state; African states like Sudan provide a timely example of this kind of trend but reducing environmental impact has proven difficult. The World Wildlife Foundation found that in 2008, the global consumption came to a 50 percent deficit; that is, it would take 1.5 Earths one year to replace the resources consumed in 2008.[64]

The debate about global responsibility for reducing consumption began at the Rio conference in 1992 and continues today. Developing countries argue that industrialized countries had to pollute a great deal to develop; indeed this pollution is one of the main causes of the current environmental struggle. Consequently, less developed countries feel equally entitled to concentrate on their economies and not on reducing their ecological footprint. Industrialized countries counter that since developing countries have rapidly growing populations, they must contribute to the environmental cause or success is impossible.

With the exception of those working on sustainable development, the current standard is to view the environment and the economy as separate areas of concentration; if a state focuses on one, it must necessarily neglect the other. Bob Giddings, Bill Hopwood, and Geoff O'Brien suggest that it would be more productive to see the economy, society, and the environment as interconnected and especially to see the economy as dependent on the latter two.[65] But as Denis Pirages points out, our way of life is so entrenched and dependent on fossil fuels. It would be very difficult to continue to support economic growth while restricting usage of petroleum and natural gases.[66]

Conclusion

When most people think of the environment, they tend to focus on climate change and its effect on the future of our planet. Similarly, people tend to spend more time discussing certain issues, like the rapid population growth in some countries, while neglecting other population issues. This chapter has argued that both population and the environment are complex, multi-faceted issues that affect many aspects of a state, from the location of the population to how they earn their living and beyond.

Population and the environment are closely interconnected. One only has to look at migration patterns following natural disasters—Hurricane Katrina is a good example—to discover how the environment can change population trends. People can also change the environment, as seen in cases of over-farming. Some scholars disaggregate population and the environment. They hope that by studying one aspect of International Relations they will be able to see trends more clearly; this is, in fact, often the case. But just as it is important to recognize factors beyond the environment and population, like the role of the government in the Iranian revolution, it is also vital to remember the connection between the two factors. While including multiple factors can sometimes make analysis more difficult, it can often present a clearer causal chain. If population and the environment are only discussed separately, these linkages may not be made.

A similar argument can be made for including studies of population and the environment in International Relations. While realism has long been the hegemonic theory in the discipline, it views each country as a 'black box'; that is, nothing taking place within the country is considered worthy of consideration. This chapter has presented numerous cases where environmental, population, or both factors have fundamentally changed the nature of a state. Cases ranging from the Asian Tigers' demographic bonus to the resource-cursed African states demonstrate that these factors can indeed have a significant impact on international politics. Not every scholar can focus exclusively on the environment or population, but neglecting these issues entirely risks losing a vital and fascinating facet of International Relations. As the post-Cold War era progresses, policymakers and scholars alike should note the ubiquity of these issues and ensure they are given their proper due in both publications and politics.

Notes

1   Nell Greenfieldboyce, 'An alien view of earth', 2010. Online: http://m.npr.org/news/front/123614938?page=0 (accessed August 2012).

2   B. Buzan, *People, States & Fear: An agenda for International Security Studies in the Post-Cold War Era*, second edition, Hemel Hempstead: Harvester Wheatsheaf, p. 116.

3   The five sectors of security are military, environmental, economic, political, and societal or identity security; see Buzan, *People, States & Fear*.

4   See for example, 'National Defense Strategy of the United States of America', US Department of Defense, June 2008; K. Giles, *Review: Russia's National Security Strategy to 2020*, Rome, Italy: NATO Defense College, 2009; *An Initial Long-Term Vision for European Defence Capability and Capacity Needs*, European Defence Agency, 2006.

5   R. Jackson and N. Howe, *The Graying of the Great Powers: Demography and Geopolitics in the 21st Century*, Washington, DC: Center for Strategic & International Studies, 2008, p. 19.

6   These works include: K. Davis, 'Social Changes Affecting International Relations', in J.N. Rosenau (ed.), *International Politics and Foreign Policy: A Reader in Research and Theory*, New York: The Free Press of Glencoe, 1961; K. Davis, 'The Political Impact of New Population Trends', *Foreign Affairs* 36 (2), 1958; A. F. K. Organski, *World Politics*, ed. V.O. Key, Borzoi Books in Political Science, New York: Alfred A. Knopf 1958; A. F. K. Organski, J. J. Spengler, and P. M. Hauser (eds), *World Population and International Relations*, The National Institute of Social and Behavioral Science, vol. II, *Symposia Studies*, Washington, DC: George Washington University, 1960; K. Organski and A. F. K. Organski, *Population and World Power*, New York: Alfred Knopf, 1961.

7   R. L. Tammen *et al.*, *Power Transitions: Strategies for the 21st Century*, New York: Chatham House Publishers, 2000, p. 9.

8   H. Sprout and M. Sprout, *Foundations of National Power*, Princeton, NJ: Princeton University Press, 1945.

9   N. W. Chamberlain, *Beyond Malthus: Population and Power*, New York: Basic Books, 1970; P. R. Ehrlich, *The Population Bomb*, New York: Ballentine Books, 1968; R. N. Gardner, 'The Politics of Population: Blueprint for International Cooperation', in L.K.Y. Ng and S. Mudd (eds), *The Population Crisis: Implication and Plans for Action*, Bloomington and London: Indiana University Press, 1965; P. M. Hauser, 'Demographic Dimensions of World Politics', in S. Mudd (ed.), *The Population Crisis and the Use of World Resources*, The Hague: Dr. W. Junk, 1964; J. C. Hurewitz, 'The Politics of Rapid Population Growth in the Middle East', *Journal of International Affairs* 19 (1), 1965.

10   S. Yoshihara, 'American Demographic Exceptionalism and the Future of U.S. Military Power', in S. Yoshihara and D.A. Sylva (eds), *Population Decline and the Remaking of Great Power Politics*, Washington, DC: Potomac Books, 2012, pp. 138–73.

11   See Jackson, and Howe, *The Graying of the Great Powers*, pp. 93–115.

12   M. L. Haas, 'A Geriatric Peace? The Future of U.S. Power in a World of Aging Populations', *International Security* 32 (1), 2007, 123.

13   'Lisbon Action Plan', Brussels: Commission of the European Communities, 2005.

14   N. Eberstadt, *Drunken Nation: Russia's Depopulation Bomb*, Washington, DC: American Enterprise Institute for Public Policy Research, 2009.

15   Quoted from Section IV.3.52 in Keir Giles, 'Review: Russia's National Security Strategy to 2020', in A.C. Monaghan (ed.) *Russia Review Series*, Rome, Italy: NATO Defense College, 2009, p. 10.

16   H. Urdal, 'The Demographics of Political Violence: Youth Bulges, Insecurity, and Conflict', in L. Brainard and D. Chollet (eds), *Too Poor for Peace? Global Poverty,*

Conflict, and Security in the 21st Century, Washington, DC: Brookings Institution Press 2007, p. 96.

17   E. Leahy, 'Beginning the Demographic Transition: Very Young and Youthful Age Structures', *Environmental Change and Security Program* 13, 2009.

18   R. A. Easterlin, 'Easterlin Hypothesis', in J. Eatwell, M. Milgate, and P. Newman (eds), *The New Palgrave: A Dictionary of Economics*, New York: Stockton Press 1987; D. J. Macunovich, 'Relative Cohort Size: Source of a Unifying Theory of Global Fertility Transition?', *Population & Development Review* 26, 2000, 236.

19   R. P. Cincotta, R. Engelman, and D. Anastasion, *The Security Demographic: Population and Civil Conflict after the Cold War*, Washington, DC: Population Action International, 2003, p. 44.

20   Urdal, 'The Demographics of Political Violence', p. 93.

21   Population Reference Bureau, World Population Data sheet 2012, p. 10. Online: www.prb.org/pdf12/2012-population-data-sheet_eng.pdf (accessed August 2012).

22   J. Perlez, 'Taliban Seize Vital Pakistan Area Closer to the Capital', in *The New York Times*, 22 April 2009.

23   *World Population Prospects: The 2008 Revision Population Database*, United Nations, 2009.

24   B. Daumerie and K. Hardee, 'The Effects of a Very Young Age Structure on Haiti: Country Case Study', in *The Shape of Things to Come Series*, Washington, DC: Population Action International, 2010, p. 1.

25   In J. A. Goldstone, 'Demography, Environment, and Security', in P.F. Diehl and N.P. Gleditsch (eds), *Environmental Conflict*, Boulder, CO: Westview, 2001, p. 85.

26   Wenche Hauge and Tanja Ellingsen, 'Causal Pathways to Conflict', in Diehl and Gleditsch (eds) *Environmental Conflict*, pp. 36–57.

27   See C. H. Kahl, *States, Scarcity, and Civil Strife in the Developing World*, Princeton, NJ: Princeton University Press, 2006; and C. H. Kahl, 'Demography, Environment, and Civil Strife', in L. Brainard and D. Chollet (eds), *Too Poor for Peace? Global Poverty, Conflict, and Security in the 21st Century*, Washington, DC: Brookings Institution Press, 2007, pp. 60–72.

28   B. Hartmann, 'Population, environment, and security: a new trinity', *Environment and Urbanization* 10 (2), 1999, 125.

29   Ibid, 126–27.

30   D. C. Pirages and T. M. DeGeest, *Ecological Security: An Evolutionary Perspective on Globalization*, Lanham, MD: Rowman & Littlefield, 2004, p. 20.

31   N. Poku and D. T. Graham, 'Redefining Security for a New Millennium', in N. Poku and D.T. Graham (eds), *Redefining Security: Population Movements and National Security*, Westport, CT: Praeger, 1998, p. 3.

32   R. S. Frey and I. Al-Mansour, 'The Effects of Development, Dependence, and Population Pressure on Democracy: The Cross-National Evidence', *Sociological Spectrum* 15 (2), 1995; J. A. Goldstone, *Revolution and Rebellion in the Early Modern World*, Berkeley: University of California Press, 1991; E. N. Luttwak, 'Where Are the Great Powers? At Home with the Kids', *Foreign Affairs*, 1994; J. M. Taw and B. Hoffman, *The Urbanization of Insurgency: The Potential Challenge to U.S. Army Operations*, Santa Monica, CA: RAND, 1994; A. Zolberg, A. Suhrke and S. Aguayo, *Escape from Violence: Conflict and the Refugee Crisis in the Developing World*, Oxford: Oxford University Press, 1989.

33   D. Pirages, *The New Context for International Relations: Global Ecopolitics*, North Scituate, MA: Duxbury Press 1978, p. 8.

34   M. Renner, 'Security Redefined', in L. Starke (ed.), *State of the World 2005: Redefining Global Security*, New York: W.W. Norton & Company, 2005, p. 5.

35   Pirages and DeGeest, *Ecological Security*, p. 19.

36   Renner, 'Security Redefined', p. 3.

37   Goldstone, 'Demography, Environment, and Security'; C. H. Kahl, *States, Scarcity, and Civil Strife in the Developing World*; Pirages and DeGeest, *Ecological Security*.

38   Renner, 'Security Redefined', p. 3.

39   L. Brainard, D. Chollet and V. LaFleur, 'The Tangled Web: The Poverty–Insecurity Nexus', in L. Brainard and D. Chollet (eds), *Too Poor for Peace? Global Poverty, Conflict, and Security in the 21st Century*, Washington, DC: Brookings Institution Press 2007, p. 1.

40   'Arab Human Development Report 2009: Challenges to Human Security in the Arab Countries', New York: United Nations Development Programme, 2009, p. 41.

41   K. O'Neill, *The Environment and International Relation*, Cambridge: Cambridge University Press, 2009, p. 43.

42   R. N. Haass, 'The Weakest Link', Newsweek, 2010; U.S. Senate Committee on Foreign Relations, Testimony Statement: Defense Undersecretary for Policy Edelman Urges Strengthening of Civilian Capacities, 31 July 2008.

43   *The Failed States Index 2009: FAQ & Methodology*. Online: www.foreignpolicy.com/articles/2009/06/22/2009_failed_states_index_faq_methodology (accessed August 2012).

44   B. Daumerie and E. L. Madsen, 'The Effects of a Very Young Age Structure in Uganda: Executive Summary', in *The Shape of Things to Come Series*, Washington, DC: Population Action International, 2010, p. 2.

45   M. Pelling, 'Urbanization and Disaster Risk', in *Population-Environment Research Network Cyberseminar on Population and National Hazards*, Population-Environment Research Network, 2007, p. 1.

46   N. Choucri, 'Migration and Security: Some Key linkages', *Journal of International Affairs* 1, 2002, 97–122, p. 99.

47   J. A. Goldstone, 'Flash Points and Tipping Points: Security Implications of Global Population Changes', *Environmental Change and Security Program* 13, 2009, 3.

48   R. P. Cincotta, 'Half a Chance: Youth Bulges and Transitions to Liberal Democracy', *Environmental Change and Security Program*, 13, 2009, 10.

49   Cincotta, Engelman, and Anastasion, *The Security Demographic*, p. 44.

50   Daumerie and Hardee, 'The Effects of a Very Young Age Structure on Haiti: Country Case Study', p. 6.

51   C. H. Kahl, *States, Scarcity, and Civil Strife in the Developing World*, p. 34.

52   Ibid, p. 41.

53   Ibid, p. 40.

54   V. M. Hudson, and A. M. den Boer, *Bare Branches: Security Implications of Asia's Surplus Male Population*, Cambridge, MA: MIT Press, 2004, p. 65.

55   V. M. Hudson *et al.*, 'The Heart of the Matter: The Security of Women and the Security of States', *International Security* 33 (3), 2009.

56   R. Cincotta, 'Review of Bare Branches: The Security Implications of Asia's Surplus Male Population', *Environmental Change and Security Program* 11, 2005, 70–73.

57   R. Nordland, 'Charging to the Right', *Newsweek*, 2008.

58   M. M. Kent, *Do Muslims Have More Children Than Other Women in Western Europe?*, Population Reference Bureau, 2008. Online: http://prb.org/Articles/2008/muslimsineurope.aspx (accessed August 2012).

59   Giles, 'Review: Russia's National Security Strategy to 2020', p. 7.

60   On Iran, see S. S. Lieberman, 'Prospects for Development and Population Growth in Iran', *Population & Development Review* 5 (2), 1979, 293–317; J. G. Scoville, 'The Labor Market in Prerevolutionary Iran', *Economic Development and Cultural Change* 34 (1),

1985, 143–55. On the Arab Spring, see R. Cincotta, 'Tunisia's Shot at Democracy: What Demographics and Recent History Tell Us', *The New Security Beat*, 25 January 2011.

61   A. Mason, 'Capitalizing on the Demographic Dividend', in United Nations Population Fund (ed.), *Population and Poverty, Population and Development Strategies*, New York: United Nations Population Fund, 2003, pp. 39–48.

62   D. E. Bloom, D. Canning and P. N. Malaney, 'Demographic Change and Economic Growth in Asia', *CID Working Paper*, Boston: Center for International Development at Harvard University, 1999, p. 3.

63   K. Wongboonsin, 'Labor Migration in Thailand', presented at *International Conference on Migrant Labor in Southeast Asia*, Armidale, Australia: UNE Asia Center and School of Economics, University of New England, 1–3December 2003.

64   World Wildlife Foundation (WWF), *The Living Planet Report 2012: Biodiversity, biocapacity and better choices*. Online: http://awsassets.panda.org/downloads/1_lpr_2012_online_full_size_single_pages_final_120516.pdf (accessed August 2012).

65   B. Giddings, B. Hopwood, and G. O'Brien, 'Environment, Economy, and Society: Fitting Them Together into Sustainable Development', *Sustainable Development* 10, 2002, 187–96.

66   D. C. Pirages and T. M. DeGeest, *Ecological Security: An Evolutionary Perspective on Globalization*, Lanham, MD: Rowman & Littlefield, 2004.

12

ENVIRONMENTAL SECURITY AND SUSTAINABLE DEVELOPMENT

Bishnu Raj Upreti

Introduction

In simple terms, environmental (in-)security becomes an issue when the routine practices and institutions of a society exploit natural resources in ways that are not sustainable. Unsustainable societies extract natural resources too quickly for them to be replenished, generate waste on a scale that cannot be assimilated by the environment, and disrupt the normal operations of ecosystems along with their capacity to adapt to internal and external stress. In today's world, many, and perhaps all, societies aspire to be sustainable, but they must contend with a host of insecurities related to the fact that they are not yet sustainable. This poses considerable challenges to governance, as the various forms of financial, social and technical capital needed to support the transition to sustainability are diverted to meet the more immediate demands of crisis and conflict. This, sadly, is a cycle that may be intensifying and speeding up in many of the poorer parts of the planet.

During periods of relative stability, however, and also in the aftermath of conflict and crisis, opportunities to build capacity and advance the long-term agenda of sustainability do emerge, but governments and other actors do not always have the knowledge base needed to take advantage of these opportunities. In this chapter I argue that more attention must be given to ensuring that the practitioner communities of the developing world have sufficient, appropriate knowledge to lay the foundations for and advance towards conditions of environmental security and sustainable development whenever a window of opportunity opens.

I begin by defining the key terms and by briefly examining the history behind these terms. I next explore the requirements for generating sufficient and appropriate knowledge for effective governance before concluding with a few final thoughts.

Environmental security and sustainable development

For the purpose of this chapter, environmental security is defined – following Jon Barnett – as 'the process of peacefully reducing human vulnerability to human-induced environmental degradation by addressing the root causes of environmental degradation and human insecurity'.[1] In other words, if environmental security is to be achieved, people must have both a healthy natural environment and some capacity to adapt to change, which is a constant in all complex systems. I follow common practice by defining 'sustainable development' in accordance with the Brundtland Commission's report (1987), which defined the term as follows:

> Sustainable development is development that meets the needs of the present without compromising the ability of future generations to meet their own needs. It contains *within* it two key concepts.
>
> • The concept of 'needs', in particular the essential needs of the world's poor, to which overriding priority should be given; and
> • The idea of limitations imposed by the state of technology and social organisations on the environment's ability to meet present and future needs.[2]

Insofar as one is satisfied with these definitions, I would like to suggest that sustainable development is focused in large measure on reconciling human needs with natural resource management and conservation. It is concerned with extraction rates, waste management and ecosystem integrity as they search for enduring, effective and efficient ways to alleviate poverty and encourage economic growth. Environmental security, on the other hand, is a complementary concept that investigates the heightened vulnerability people have to certain environmental stresses – which may be due to natural processes and phenomena as well as to unsustainable social activity. Although the two concepts are clearly distinct, it seems to me that in promoting one agenda, one would also be promoting the other.

When examining the security connection between environmental security and sustainable development, it is useful to differentiate between environmental security at the macro and the micro level. At the macro level, environmental security is primarily concerned with the effects of planetary processes such as global climate change, which encompasses wide-scale desertification, deterioration of ecosystems, rising sea levels and rising global temperature. At the micro level, environmental security is about local environmental problems including prolonged drought, flooding, drying up of fresh water sources, air and water pollution, metal (e.g., arsenic) contamination, emissions from power plants, immediate resource scarcity and poorly managed waste. Though the level and intensity of sensitivity and vulnerability varies in relation to the economic status of people and the ecological position of the country, the poor and marginalised people in the global South are vulnerable to both macro and micro stressors.

Developing countries suffer more from environmental insecurity than developed countries because of lower levels of adaptation capacity. This is in part why climate change is widely believed to 'be a massive threat to development'.[3] Environmental insecurity in developing countries is amplified further by natural disasters. Due to a lower level of adaptation to changing weather patterns as well as diminished resources to rebuild devastated areas after natural disasters have occurred, the poor in the developing world suffer greater hardships than their counterparts in the north. The 2000 floods in Mozambique, the tsunami of December 2004 affecting several countries around the Indian Ocean and the 2005 earthquake in Pakistan are just some of the most recent examples of natural disasters, which took their greatest toll on the vulnerable poor.

Environmental insecurity at the micro level is also high in developing countries because of its connection to poverty and bad governance. I understand good governance as a set of values, policies, technologies and institutions through which a society manages its political, economic, social and environmental resources in a transparent, accountable and sustainable manner. If governance is good, it provides an adequate framework for interlinking development and the environment. Bad governance, however, causes increased vulnerability in the poorer sections of society, which in turn places stress on natural resources and contributes to environmental degradation, consequently creating environmental insecurity. For a long time, the nexus between poverty and environmental degradation was not recognised in terms of conceptual orientation, strategic planning and implementation. This is changing and it is increasingly being realised that there is a direct and critical linkage between the quality of the environment and the incidence of poverty.[4] Poor people are most vulnerable to environmental degradation as they depend directly on natural resources for their livelihoods and wellbeing. Poverty is a complex and multi-faceted problem in developing countries, as it acts as both cause and effect of the degradation of environmental resources. Poor people, especially from rural communities, rely primarily on natural resources for their survival, while their investment in environmental resources is minimal, resulting in low levels of land fertility, high rates of deforestation, and poor watershed and water management. Hence, addressing poverty is one of the fundamental conditions for achieving environmental security.

The relationship between development and environmental security is complex. Environmentally insensitive development practices such as structural adjustment have contributed to numerous environmental challenges and problems such as soil erosion, deforestation, land degradation, water and air pollution, harmful chemical contamination, elite capture of resources, and, consequently, inequality, desertification and the disruption of ecosystems and biodiversity.

In turn, various studies[5] have suggested that environmental stress has contributed to, or aggravated, civil war and violent conflict in many developing countries, and that such stress often has been related to access, external interference in indigenous systems, environmental degradation, and the depletion of renewable resources such as water, land, forest, gas and oil. Scholars such as Thomas Homer-Dixon, Richard Matthew, Colin Kahl and and others have significantly contributed to understanding the links between human-induced environmental stress and social tension and conflict in many developing countries. They have established that there may be significant relationships between water scarcity and conflict, rapid urbanisation and urban violence, demographic changes and social tension, and environmentally induced displacement and ethnic tension. Their work (which includes both case studies and conceptual and methodological contributions) in different places such as Rwanda, South Africa, Bangladesh-Assam, Bihar (India), Pakistan, China, the Philippines, Indonesia, Nicaragua, Chiapas (Mexico), Gaza, and the Jordan and Senegal river basins provides the best evidence to date of likely linkages between environmental stress and conflict.

The complex relationship between development and human security has been summarised neatly by Frances Stewart[6] who identifies three connections between security and development. The first connection is related to the impact of security or insecurity on development achievement and, consequently, on the wellbeing of people; the second is the role of security in development; and the third is the effects of development on security. When development is seen mainly as economic growth and efforts are concentrated on achieving a high rate of economic growth, environmental concerns are sidelined. The consequence of an economic growth-centric approach is environmental insecurity and social tension. Pragmatically, sustainable development and environmental security can only be realised by linking human wellbeing to a healthy environment.

Historical background

The now well-publicised nexus between environment and development first gained international attention at the 1972 United Nations Conference on the Human Environment in Stockholm, Sweden. Under the leadership of Maurice Strong, that conference launched a way of thinking about development and the environment, whereby development was no longer to have unrestrained priority over environmental concerns, but rather development needed to be reconciled with environmental imperatives and constraints. The Conference's declaration reads as follows:

> In the developing countries most of the environmental problems are caused by under-development. Millions continue to live far below the minimum levels required for a decent human existence, deprived of adequate food and clothing, shelter and education, health and sanitation. *Therefore, the developing countries must direct their efforts to development, bearing in mind their priorities and the need to safeguard and improve the environment.* For the same purpose, the industrialized countries should make efforts to reduce the gap themselves and the developing countries. In the industrialized countries, environmental problems are generally related to industrialization and technological development.

Some 11 years later, in 1983, the UN created the World Commission on Environment and Development (WCED) to grapple with these problems. In 1987 that Commission, under the leadership of former Norwegian Minister for Environmental Affairs (1974–97) and Prime Minister (1981) Gro Harlem Brundtland, published its recommendations under the title *Our Common Future*. The Commission as well as its report were deliberately conceived in the spirit of the earlier Stockholm conference, epitomised by the concept of 'sustainable development'. The Brundtland report, as it is more commonly known, also laid the groundwork for the United Nations Conference on Environment and Development (Earth Summit) in Rio De Janeiro in the summer of 1992. At that conference, some 178 countries endorsed the Rio Declaration, consisting of 27 stewardship principles, many of which reflect the principles of effective environmental governance, including the need for transparency, accountability and public participation in environmental decision-making, the need to decentralise the management of natural resources, and the importance of environmental impact assessment. Above all, Rio endorsed Agenda 21 (popularly known as the Sustainable Development Agenda)[7], which focuses on the relationship between the environment and development. Its 27 principles are indicative of this focus. Principles three and four, for instance, read as follows:

Principle 3: The right to development must be fulfilled so as to equitably meet developmental and environmental needs of present and future generations. Principle 4: In order to achieve sustainable development, environmental protection shall constitute an integral part of the development process and cannot be considered in isolation from it.[8]

Principle 1, in turn, highlights the relationship between sustainable development and environmental security as human security. It reads: 'Human beings are at the centre of concerns for sustainable development. They are entitled to a healthy and productive life in harmony with nature.'[9]

In the evolution of the nexus between the environment and development, three conservation organisations have played a crucial role: United Nations Environment Programme (UNEP), The World Conservation Union (IUCN) and World Wildlife Fund (WWF). The Earth Summit's Agenda 21 is based on various important conceptual documents by these organisations: *Caring for the Earth: A Strategy for Sustainable Living;*[10] *Conserving the World's Biological Diversity;*[11] and *Global Biodiversity Strategy.*[12] These documents argue that improving the quality of human life is possible while maintaining the carrying capacity of supporting ecosystems.

A more recent international initiative is the commitment to achieving the Millennium Development Goals (MDGs). The MDGs were adopted by 189 nations and signed by 147 heads of state and governments during the UN Millennium Summit in September 2000 through the Millennium Declaration. There are eight goals to be achieved by 2015, which are broken down into 21 quantifiable targets measured by 60 indicators. The eight main goals are:

- Goal 1: Eradicate extreme poverty and hunger
- Goal 2: Achieve universal primary education
- Goal 3: Promote gender equality and empower women
- Goal 4: Reduce child mortality
- Goal 5: Improve maternal health
- Goal 6: Combat HIV/AIDS, malaria and other diseases
- Goal 7: Ensure environmental sustainability
- Goal 8: Develop a Global Partnership for Development[13]

All of these goals are interrelated and aim to meet the development challenges faced by the world, particularly by developing countries. Goal 7 is directly related to concerns of environmental security and development. A key factor in the realisation of the MDCs is 'good governance'. The latter is an integral part of the Millennium Declaration which reads:

Success in meeting these objectives depends, *inter alia*, on good governance within each country. It also depends on good governance at the international level and on transparency in the financial, monetary and trading systems. We are committed to an open, equitable, rule-based, predictable and non-discriminatory multilateral trading and financial system.

Through this 40-year history, 'good governance' has emerged alongside sustainable development as a key issue in development literature. Proponents argue that in developing countries environment and development related problems are directly related to governance.[14] They argue that environmental security and thus sustainable development can only be achieved by institutionalising environmental governance at the local level, where development projects and programmes are implemented, and at the higher level, where policies are negotiated. For this, it is imperative that national and local authorities have the same understanding of the relationship between environment and development. Authorities of both kinds need to revert to decentralised planning and implementation, a favourable regulatory environment, the promotion of indigenous knowledge and technologies, learning from best practices, promoting collective responsibility, identifying the potentials and limitations of environmental resources and services, and resolving environmental conflict through a collaborative approach. Building capacity for inclusive, equitable, productive and environmentally sensitive development is key to linking environmental security and sustainable development.

The World Summit on Sustainable Development (WSSD) in 2002 held in Johannesburg reaffirmed its commitment to the Rio Declaration on Environment and Development. The summit's political declaration clearly acknowledges that while the right ideas to achieving sustainable development exist, much more needs to be done in practice, especially if the world is to cope with the new challenges to both environmental security and development, namely globalisation and climate change.

> The global environment continues to suffer. Loss of biodiversity continues, fish stocks continue to be depleted, desertification claims more and more fertile land, the adverse effects of climate change are already evident, natural disasters are more frequent and more devastating and developing countries more vulnerable, and air, water and marine pollution continue to rob millions of a decent life. ... Globalization has added a new dimension to these challenges. The rapid integration of markets, mobility of capital and significant increases in investment flows around the world have opened new challenges and opportunities for the pursuit of sustainable development. But the benefits and costs of globalization are unevenly distributed, with developing countries facing special difficulties in meeting this challenge.[15]

Knowledge and environmental governance

Considering the complex relationship between environmental security and sustainable development, conventional and narrow disciplinary approaches cannot offer the knowledge, skills and learning[16] required for a holistic understanding needed to promote the integration of research, policy and practice. A transdisciplinary research approach is required to address problems related to sustainable development (balancing and promoting social, economic, political and ecological conditions in a given context) and to integrate environmental concerns into development policy and practices. The transdisciplinary approach is defined as:

different academic disciplines working together with practitioners to solve real world problems. It is cognitive and social cooperation across disciplinary boundaries and the direct application of scientific knowledge to both political decision making and societal problem solving by engaging non-scientific stakeholders within the research process.[17]

Social learning is an important element in understanding the relationship between the environment and development. It refers to an action-oriented form of learning for dealing with the complex problems related to the environment and development. It uses a participatory process and encompasses a positive belief in a potential social transformation based on critical self-reflection and effective communication.[18] Social processes such as the creation of a common platform for collective action, interactive goal setting, accommodation, shared learning, vision building from multiple realities, dialogue and negotiation, leadership development, resource mobilisation and concerted action are important elements of social learning that enable the better understanding of the relationship between environmental security and sustainable development. Social learning comprises learning through observation and interaction within the social context, learning by social aggregates, learning pertaining to environmental issues, and learning that results in recognisable social entities that make collective decisions and take concerted action towards realising the importance of environmental security in development. It also focuses on the procedures and incentives that promote dialogue among international actors, planners, policy makers, researchers, politicians, managers and resource-users to minimise conflict and to sustain the ecological capacity of natural resources to achieve sustainable development. In short, social learning produces knowledge and changes human behaviour to achieve the integration of environmental security elements into development approaches.

Social learning has pragmatic merit in understanding the links between environment and development as it focuses on the expanding knowledge network and creation of a platform from which to examine the dynamics. This process is called *soft-system methodology*.[19] The analysis of human activity is at the centre of the soft-system methodology, which explores messy, complex problematic situations to achieve purposeful action for improvement.[20] Soft-system methodology brings people together for collective learning, joint decisions and concerted action through which they can understand the problems of unsustainable development and environmental insecurity. Social learning through the soft-system methodology can mobilise human agency for collective action through shared, self-reflective understanding, to integrate environmental security concepts into development interventions.

Development and environment debates mainly revolve around four viewpoints: holism, reductionism, objectivism and constructivism. A constructivist perspective evolved with a holistic framework is essential to study the complexity of, and interrelations between, the environment and development. This perspective adopts an epistemological position in which multiple realities emerge from the interaction, realisation and construction of knowledge in dealing with environmental insecurity and developmental challenges. These constructs are socially and experientially based. The constructivist approach assumes multiple and conflicting realities as the product of human intellect and adaptation.[21]

Professor of Innovation and Communication Studies Niels G. Röling explains the constructivist perspective as follows:

> Over time, groups of people, through discourse, develop an inter-subjective system of concepts, beliefs, theory and practices that they consider to be reality. Based on their intention and experience, people construct reality creatively with their language, labour and technology. The same people change their reality during the course of time in order to adjust to changing circumstances.[22]

Multiple actors are involved in the environment and development, and these actors construct multiple realities. Hence, the constructivist perspective helps to look at things on the basis of potential multiple realities constructed by people through negotiation and agreement. A constructivist perspective emphasises the construction of visible linkages between development and the environment. To tackle the conflicting goals of the various actors, especially when there is conflict over ecological services and environmental issues related to natural resources, people need to agree, based on collective learning, to interact and act according to negotiated and shared goals. Adaptation to change through local initiatives, using various platforms for negotiation is required to achieve this joint learning and concerted action.[23]

In Figure 12.1, the quadrants are partitioned by two orthogonal axes: the lower part of the vertical axis contains the reductionist perspective, the holistic perspective is presented in the upper part, the right side of the horizontal axis presents the objectivist or positivist perspective, and the left side presents the constructivist perspective. In understanding the interrelationship between environmental security and sustainable development, holistic and constructivist perspectives are vital.

The soft-system approach helps in understanding the intricacies and interwovenness of development and the environment. This approach was developed as an alternative way of thinking in response to the failure of the hard-system approach in dealing with societal problems such as environmental insecurity and unsustainable development.[24] Hard-system thinking is based on reductionist and positivist assumptions. In contrast, soft systems are constructs. According to the soft-system approach, development and the environment are structured as whole entities or systems, with each system having properties different from the sum of its own parts, as well as from other systems around it.[25] The soft-system approach provides a perspective on the dynamics of the environment and development. According to Röling, soft systems are constructs with arbitrary boundaries that emerge as a result of collective learning and action through a cognitive process (developing collective agency through perception, emotion and action).[26] The interrelation of cognition (the process of perception, emotion and action) is a basis for the organisation of human action to achieve environmental security and sustainable development. The notion of cognition provides opportunities for the purposeful collective redesigning of human interactions and behavioural modification in pursuit of sustainable development. As cognition involves being emotional, perceptive and reflective, when people perceive environmental insecurity in their society, they assess it with emotion and act or react accordingly. Consequently, such reaction or action forces change in the situation. Therefore, the cognition based soft-system approach is a cyclic process of

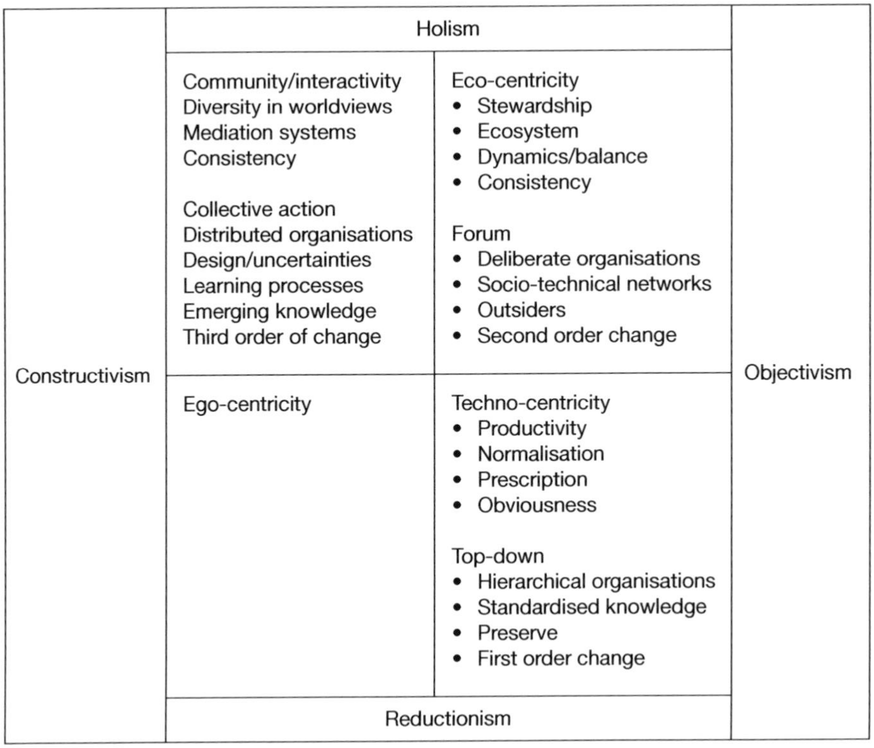

	Holism		
Constructivism	Community/interactivity Diversity in worldviews Mediation systems Consistency Collective action Distributed organisations Design/uncertainties Learning processes Emerging knowledge Third order of change	Eco-centricity • Stewardship • Ecosystem • Dynamics/balance • Consistency Forum • Deliberate organisations • Socio-technical networks • Outsiders • Second order change	Objectivism
	Ego-centricity	Techno-centricity • Productivity • Normalisation • Prescription • Obviousness Top-down • Hierarchical organisations • Standardised knowledge • Preserve • First order change	
	Reductionism		

Figure 12.1 Holistic and constructivist perspectives.

Source: Figure adapted from Hadorn, G.H., Hoffmann-Riem, H., Biber-Klemm, S., Grossenbacher-Mansuy, W., Joye, D., Pohl, C., Wiesmann, U. and Zemp, E. (eds) (2008) *Handbook of Transdisciplinary Research*. Bern: Springer Science, p. 117.

learning and action to achieve environmental security through purposeful cognitive human interaction.

Values, context, action and theory are vitally important in understanding environmental security and consequently contributing to sustainable development. From the social learning perspective, existing development practices and environmental security efforts are valuable sources of learning for developing shared goals for further action. In this context, communicative rationality is an important concept in developing shared goals.[27] It helps to converge multiple realities for concerted action.[28] Communicative rationality is a deliberate strategy for interaction through dialogue to identify the problems of unsustainable development and environmental insecurity. Collective decision making and concerted action are the subsequent steps based upon shared understanding and commitment. Jürgen Habermas has argued that in communicative action participants are not primarily orientated towards their own individual goals, but harmonise their plan of action on the basis of common or shared goals.[29] Communicative rationality emphasises people's ability to solve the problems of unsustainable development and environmental insecurity on the basis of negotiation for concerted action. Appropriate platforms for negotiation are necessary to achieve this goal. A platform can appear in various forms. It may already exist as social capital,[30] or be newly created. The aim of a platform is to create favourable circumstances to negotiate collective decisions and concerted action.

The effectiveness of platforms for collective action are demonstrated by different initiatives, such as the Landcare Movement in Australia, Chipko Movement in India, Integrated Arable Farming in the Netherlands, and Community Forestry and Ecotourism in Nepal.[31] The creation of a platform is an interactive process that arises from multiple perspectives and goals and communicative action. A platform helps to bring different development actors together to understand the social dynamics, actors' networks and ability, and consequences of environmental insecurity and unsustainable development. It focuses on negotiation between conflicting interests and accommodating differences. The environment and development encompasses both hard systems (in which natural laws define outcomes) and soft systems (in which social processes govern outcomes). The interface[32] between hard and soft systems helps to create platforms for interaction, action and negotiation. Platforms can demonstrate the impact of human actions on environmental resources and the social interdependence of the environment and people, as well as encourage collective decisions, facilitate a joint learning process and promote concerted action. When people feel the negative consequences of development they search for solutions, create new concepts and ideas, and negotiate a course of collective action. The interfaces between people with different interests, incompatible goals, and differential access to power and information bring them towards a minimum common understanding through appropriate platforms for facilitating adaptive management in implementing development and environmental security.

Adaptive management is the outcome of negotiation in platforms. Adaptive management focuses on designing an interface between society and the environment.[33] It moves away from a dominant linear-economic-goal seeking management strategy to social learning about complex systems. The release of human opportunities requires flexible, diverse and redundant regulation, monitoring that leads to corrective action, and experiential probing of the continually changing reality of the external world. The essential point is that evolving systems require policies and actions that not only satisfy social objectives but also achieve continually modified understanding of the evolving conditions and provide flexibility for adapting to surprise.[34] In relation to the relevance of learning in adaptive management, Lee states, 'adaptive management is an approach to natural resource policy that embodies a simple imperative: policies are experiment, learn from them'.[35] The pluralistic nature of human behaviour, motivations, interactions and daily practices affects environmental management and development. Hence, these practices need to be examined from a pluralistic perspective. The notion of legal pluralism is useful here. Legal pluralism refers to the existence of different types of laws (national laws, folk law, customary law, indigenous law, religious law, international law)[36] and is useful in understanding the diversity in the cultural, social and normative practices of actors engaged in the environment and development fields. The environment and development do not operate in a vacuum. Human behaviour determines activities, and, therefore, contradictions, confusions and conflict are common in the area of environment and development. The actions and behaviour of people towards development and the environment are not guided by a single comprehensive set of rules, but shaped by various local norms, practices, beliefs and customary laws, as well as legal regulations. These rules are different, and even contradictory, and can be a source of conflict. Hence, they have to be negotiated from appropriate platforms.

The preceding discussion indicates the need for different methodological approaches to understand the dynamics and interrelationship between environmental security and sustainable development. Hence, the need for an inter- and transdisciplinary approach that involves the diverse actors engaged in the environment and development (i.e., planners, politicians, technocrats, scientists and individuals).

Conclusions

Only during the later part of the twentieth century did poverty and the deterioration of environmental resources receive global attention. Unfortunately, these issues have been used by powerful countries as political tools in negotiations to fulfil their own political aims. Hence, poverty and environmental degradation have not received adequate attention at the international and national levels, but have merely been paid lip service. Despite the commitment of national governments and the international community to alleviate poverty and address environmental degradation, these issues have not been made a priority or translated into national policy and action. However, there have been a number of steps forward at the international level, for example various international events and instruments, including the various world summits that gave impetus to concepts such as sustainable development, the Millennium Development Goals and the importance of good governance.

This chapter has argued that if environmental security as human security is to be achieved then we require better environmental governance. Environmental governance can be understood as a set of values, principles, policies, technologies, institutions and procedures through which communities and societies manage environmental and natural resources in a transparent, accountable, participatory and equitable manner overcoming the barriers to sustainable development. It was further argued that sustainable development and environmental security are interdependent and mutually reinforcing. The relationship between environmental security and sustainable development is frequently interpreted as a tangible outcome. However, it is also a process of building relationships between people, and between people and their environment, and of facilitating planned change through concerted efforts and social learning. Social learning is crucial to cope with the challenges related to environmental insecurity that can lead to unsustainable development. Society has to develop the capacity to adapt to change by improving decision making through the sharing of information, communication and collective understanding. This is possible only by developing societies as learning societies. Environmental security and sustainable development are the outcomes of social learning and the concerted efforts of all relevant actors using innovative approaches. Environmentally friendly policy, based on principles of environmental security and environmental governance, greatly contribute to ecosystem management, gender equity, respect for human rights, conflict resolution and building community resilience. Preparing for future surprises or potential risks and new challenges requires adaptive management to build resilience in both social and ecological systems. Efforts should concentrate on improving the quality of life of people, without using natural resources beyond the capacity of the environment to supply them. Achieving environmental security and sustainable development, including the integration of diverse interests, empirical knowledge and different theoretical perspectives into a research process, requires a transdisciplinary approach. Actors in development have different and conflicting perspectives, values, objectives and knowledge systems. These differences are reflected in their behaviour. Therefore, a holistic and transdisciplinary approach that involves the diverse actors engaged in the environment and development is essential.

Notes

1   J. Barnett, *The Meaning of Environmental Security*, London: Zed Books, 2001, p. 129.

2   J. H. Brundtland *et al.* (ed.), *Our Common Future – The World Commission on Environment and Development*, Oxford: Oxford University Press, 1987, p. 43, emphasis added.

3   United Nations, Human Development Report 2007/08, *Fighting Climate Change: Human Solidarity in a Divided World*, New York: UNDP. Online: http://hdr.undp.org/en/media/HDR_20072008_EN_Overview.pdf (accessed August 2012).

4   See for detail: S. Hqu, H. Reid and L. A. Murray, *Climate change and development links*, IIED Gatekeeper series no 123, London: International Institute for Environment and Development, 2006.

5   For details see: R. Matthew and B. Upreti, 'Environmental Stress and Demographic Change in Nepal: Underlying Conditions Contributing to Decade of Insurgency', *Environmental Change and Security Programme Report*, Washington, DC: Woodrow Wilson International Centre of Scholars, Issue 11, 2005, pp. 29–39; See also the work of R. Matthew, for example, 'In Search of Environmental Leadership', *Georgetown Journal of International Affairs*, Winter/Spring 2000, 107–14; 'In Defence of Environment and Security Research', *Environmental Change and Security Project Report* 8, Summer 2002, 109–24; and 'Environment, Population and Conflict: New Modalities of Threat and Vulnerability in South Asia', *Journal of International Affairs* 56 (1), Fall 2002, 235–54. See also C. Kahl, 'Population Growth, Environmental Degradation, and State-Sponsored Violence: The Case of Kenya, 1991–93', *International Security*, 23 (2), Fall 1998, 80–119; B. R. Upreti, *The Price of Neglect: From Resource Conflict to the Maoist Insurgency in the Himalayan Kingdom*, Kathmandu: Brikuti Academic Publications, 2004; G. Dabelko, K. Conca and A. Carius, 'with Ken Conca and Alexander Carius', *State of the World 2005: Redefining Global Security*, New York: W. W. Norton, 2005, pp. 144–55; D. Buckles (ed.), *Cultivating Peace: Conflict and Collaboration in Natural Resource Management,* Ottawa/Washington: IDRC/World Bank Institute, 2005; F. Fouinat, 'A Comprehensive Framework for Human Security', *Conflict, Security and Development*, 4 (3), 2004, 289–97; T. Homer-Dixon, J. H. Boutwell and G. W. Rathjens, 'Environmental Change and Violent Conflict: Growing scarcities of renewable resources can contribute to social instability and civil strife', *Scientific American*, February 1993, 38–45.

6   F. Stewart, *Development and Security*, Working Paper 3, Oxford: Centre for Research on Inequality, Human Security and Ethnicity, University of Oxford, 2004.

7   Agenda 21 is a non-binding sustainable development strategy for action adopted by 178 countries.

8   Report of the United Nations Conference on Environment and Development (June 1992), *Annex 1 Rio Declaration on Environment and Development*. Online: www.un.org/documents/ga/conf151/aconf15126-1annex1.htm (accessed August 2012).

9   Ibid.

10   UCN, *Caring for the Earth: A Strategy for Sustainable Living*, Gland, Switzerland: IUCN, 1991.

11   J. A. McNeely, K. R. Miller, W. Reid, R. Mittermeier and T. B. Werner, *Conserving the World's Biological Diversity*, Washington, DC: World Resource Institute, 1990.

12   WRI, IUCN and UNEP, Global Biodiversity Strategy, Washington, DC: World Resource Institute, 1992.

13   Online: www.un.org/millenniumgoals/ (accessed August 2012).

14   See, for example, Indra de Soysa, chapter 3 in this volume.

15   UN World Summit on Sustainable Development Draft Political Declaration submitted by the President of the Summit. Online: www.un.org/esa/sustdev/documents/WSSD_POI_PD/English/POI_PD.htm (accessed August 2012).

16   Learning is a complex activity, which manifests itself in a change in people's behaviour. It is rooted in the human capacity to improve understanding and skills on the basis of day-to-day experiences, external knowledge and the surrounding environment.

17   See P. Burger and R. Kamber, 'Cognitive Integration in Transdisciplinary Science: Knowledge as a Key Notion', Issues in Integrative Studies 21, 2003, 43–73 for details.

18   See N. Röling and A. Wagemakers (eds), *Facilitating Sustainable Agriculture*, Cambridge: Cambridge University Press, 2008.

19   P. Checkland, and J. Scholes, *Soft Systems Methodology in Action*, Chichester: John Wiley, 1990.

20   For details see: P. Checkland, *Systems Thinking, Systems Practice*, Chichester: John Wiley, 1981; P. Checkland and J. Scholes, *Soft Systems Methodology in Action*, Chichester: John Wiley, 1990.

21   For details see: G. E. Guba, and Y. S. Lincoln, *Fourth Generation Evaluation*, London: Sage Publications, 1994.

22   See N. Röling, 'The Soft Side of Land', *ITC Journal* 3 (4), Special Congress Issue, 1997, 248–62; and N. Röling, 'Creating Human Platforms to Manage Natural Resources: First Results of Research Programme', in A. Budelman (ed.), *Agricultural R&D at the Crossroads. Managing Systems Research and Social Actor Approaches*, The Hague: Royal Tropical Institute, 1996.

23   For details see: N. Röling, 'Modeling Soft Side of Land: The Potential of Multiagent System', in C. Leeuwis (ed.), *Integral Design: Innovation in Agriculture and Resource Management*, Mansholt Studies 15, Mansholt Institute, Wageningen University, 1999, 73–97; and N. Röling and J. Jiggins, 'The Ecological Knowledge System', in N. Röling, A. Wagemakers (eds), *Facilitating Sustainable Agriculture: Participatory Learning and Adaptive Management in Times of Environmental Uncertainties*, Cambridge: Cambridge University Press, 1998, Ch. 16, pp. 283–307.

24   For details see: C. S. Holling, 'What Barriers? What Bridges?' in L. H. Gunderson, C. S. Holling and S. Light (eds), *Barriers and Bridges to the Renewability of Ecosystems and Institutions*, New York: Columbia University Press, 1995, Ch. 1, pp. 3–37; P. Checkland, *Systems Thinking Systems Practice*; Röling and Jiggins, 'The Ecological Knowledge System'.

25   For details see: R. Bawden, 'On the Systems Dimension in FRS', *Journal for Farming Systems Research-Extension* 5 (2), 1995, 1–18.

26   For details see: H. R. Maturana and J. F Varela, *The Tree of Knowledge, the Biological Roots of Human Understanding*, Boston, MA: Sambala Publications, 1992; F. Capra, *The Web of Life: A New Synthesis of Mind and Matter*, London: Harper Collins Publishers, 1996; N. Röling, 'Creating Human Platforms to Manage Natural Resources: First Results of Research Programme', in A. Budelman, (ed.), *Agricultural R&D at the Crossroads. Managing Systems Research and Social Actor Approaches*, The Hague: Royal Tropical Institute, 1996.

27 Habermas explains instrumental, strategic and communicative rationality. For details see: J. Habermas, *The Theory of Communicative Action*, Vol. 2, Boston: Beacon Press, 1989.

28 For details see: J. Jiggins and N. Röling, 'Adaptive Management: Potential and Limitations for Ecological Governance', article accepted for publication in the *International Journal of Agricultural Resources, Governance and Ecology* 1 (1), 2000, 28–43.

29 Habermas, *The Theory of Communicative Action*.

30 Social capital defined by Uphoff (p. 1) is 'an accumulation of various types of social, psychological, cultural, cognitive, institutional and related assets that increase the amount (or probability) of mutually beneficial co-operative behaviour'. For details see: N. Uphoff, 'Understanding Social Capital: Learning from the Analysis and Experience of Participation', paper presented at discussion forum on 13 September 2000, Wageningen.

31 For details of these achievements see: B. R. Upreti, *Management of Social and Natural Resource Conflict in Nepal: Realities and Alternatives*, New Delhi: Adroit Publishers, 2002; and B. R Upreti, *Conflict Management in Natural Resources: A Study of Land, Water and Forest Conflict in Nepal*, Published PhD Dissertation, Wageningen: Wageningen University, 2001.

32 For details see: N. Long (ed.), *Encounters at the Interface: A Perspective on Social Discontinuities in Rural Development*, Wageningen Studies in Sociology 27, Wageningen: Wageningen Agricultural University, 1989; N. Long and J. D. Van der Ploeg, 'Demythologising Planned Intervention', *SociologiaRuralis* 29 (3/4), 1989, 226–49; N. Long and A. Long (eds), *Battlefield of Knowledge, the Interlocking of Theory and Practices of Social Research and Development*, London: Routledge, 1992.

33 For details see: N. Röling, *Gateway to the Global Garden: Beta/Gamma Science for Dealing with Ecological Rationality*, Eighth Annual Hopper Lecture, University of Guelph, Canada, 24 October 2000.

34 C. S. Holling, 'What Barriers? What Bridges?, pp. 3–37.

35 For details see: K. N. Lee, *Compass and Gyroscope: Integrating Science and Politics for the Environment*, Washington, DC: Island Press, 1993, p. 9.

36 For details see: F. Benda-Beckmann, K. von Benda-Beckmann and H. L. J. Spiertz, 'Local Law and Customary Practices in the Study of Water Rights', in R. Pradhan, F. Benda-Beckmann, K. Benda-Beckmann, H. J. L. Spiertz, S. K. Khadka, and H. Azharul (eds), *Water Rights, Conflict and Policy*, proceedings of a workshop held in Kathmandu, Nepal, 22–24 January 1996, pp. 221–42.

13

ENSURING FOOD SECURITY

Meeting challenges from malnutrition, food safety, and global environmental change

Bryan McDonald[1]

Introduction

This chapter explores the ability of the now global food system to ensure 'food security,' by contemplating the various threats and vulnerabilities inherent to that system. Although many definitions of food security exist, most tend to emphasize that food security exists when *all people at all times have access to sufficient, safe and nutritious food necessary to lead active and healthy lives.*[2] By analyzing the issues of malnutrition, food safety, and global environmental change as they relate to the global food network, the chapter considers how processes of globalization are reshaping food systems in ways that have significant impacts on human security. It is argued that while a globalized food network provides a variety of new opportunities for improving health and well-being by connecting humans around the world through one of their most basic needs, a globalized food network also contains significant risks to individual and social health and safety, personal and national economies, and local, regional, and global environments. Unless we strive to fully understand the risks and opportunities posed by such a network we cannot use it to its full potential and run the risk of missing opportunities to address pressing global problems.

The ways people gain access to food has long been a defining feature of human civilizations; throughout history, changing methods of producing food have marked important shifts in the character of human societies.[3] In a time when economies are increasingly industrialized and focused on information and technology, food production remains an important sector of the global economy.[4] Globalization's acceleration of connections between the world's places and peoples has highlighted the important links between how humans produce, distribute, and consume foodstuffs with pressing social, political, and environmental issues. Changes brought by globalization—combined with reduced capacity of governments to address pressing issues, and an increasing role of non-state actors in national and international politics—have also produced a considerable amount of turbulence in world affairs.[5] This turbulence has led to transformations in the landscape of global security threats. Many current security challenges are transnational:

they cross state borders, but generally cannot be linked to foreign policies or behaviour of other states. Rather than being created and controlled by national governments, these threats are situated in a complex, dynamic, and global web created by modern communication, transportation, and information technologies.[6]

Recognition of the changing landscape of security threats has prompted a broadening of the referent object of security to include beside nation states also individuals. The concept of human security was the focus of the United Nations Development Program's *1994 Human Development Report*, which encompassed two main aspects, 1) 'safety from such chronic threats as hunger, disease and repression,' and 2) 'protection from sudden and harmful disruptions in the patterns of daily life.'[7] In short, freedom from want and freedom from fear. Food security is a fundamental component of human security's effort to ensure safety from both chronic threats and sudden disruptions, and is an integral component of meeting the Millennium Development Goals (MDGs).[8]

In the twenty-first century food security must address challenges that arise from a global food system where all people are experiencing some form of threat and vulnerability from a range of sources including hunger, obesity, disease, conflict, and terrorism. As a result of the emerging complexity of food security, other dimensions of the global food system are being recognized as relevant to security. This chapter examines three such aspects. First, it considers the challenge of nutrition. Second, it discusses ways in which the increasing speed and scale of the global food system's connections between places and peoples can serve as a vector for the transmission of accidental and intentional health threats from infectious diseases and biological weapons. Third, it reviews how better understandings of the process of global environmental change show that the food system both contributes to processes of global change and will experience significant impacts as a result of those processes. Finally, the chapter discusses the methods available for researchers to study the global food system. The chapter closes with thoughts on the need to develop a global food system that meets the nutritional needs of human populations in ways that foster the long term ecological health and security of the planet.

Ensuring nutrition

Providing sufficient food to all people has been a major challenge to human societies throughout history. Already in the nineteenth century the Reverent Thomas Malthus discussed what he foresaw as inevitable social problems based on inherent tensions between the different rates of increase between human populations and food supplies.[9] While his predictions were not borne out, Malthus' notion that human populations could, under certain conditions, increase at a geometric rate, while food supplies, under optimal conditions, could only increase at an arithmetic rate has led to considerable discussion about the factors that have allowed human populations to increase food production and questions about whether improvements in technology and methods have allowed humans to surpass or merely extend the limits to growth that Malthus predicted. In contrast to the perspective of Malthusian thought, so-called cornucopian thinkers such as Julian Simon and Bjørn Lomborg contend that limitless stocks of human ingenuity mean there are few, if any, limits on population growth and resource consumption and that properly functioning markets provide incentives that encourage conservation and substitution of resources and the development of new types of resources and technologies.[10]

In the twentieth century 'the right to food' has been a cornerstone of many of the foundational agreements of global governance.[11] In 1948, the United Nations General Assembly adopted the *Universal Declaration of Human Rights*, which included in Article 25 the statement that 'everyone has the right to a standard of living adequate for the health and well-being of himself and his family, including food.'[12] The 1966 *Covenant on Economic, Social and Cultural Rights* declared, 'the fundamental right of everyone is to be free from hunger.'[13] The importance of finding ways to meet the needs of growing human populations in a way that contributes to environmental sustainability and rural development was highlighted in chapter 14 of *Agenda 21* which was adopted by more than 175 Governments at the 1992 United Nations Conference on Environment and Development.[14] The 1996 *Rome Declaration on World Food Security* affirmed 'the right of everyone to have access to safe and nutritious food, consistent with the right to adequate food and the fundamental right of everyone to be free from hunger.' The evolving definitions of food security stress the fact that it is not a target to be met (such as ensuring all people receive 2,000 calories of food per day), but that it is a progressive goal of ensuring access to food that is adequate, safe, and nutritious.

The impact of processes of global change on the food system has been the creation of a global food network filled with pockets of both abundance and scarcity. This food network does not map clearly onto many of the traditional state-based or North/South models of the world; there are undernourished people in developed countries and over-nourished people in developing countries.[15] Malnutrition remains a significant threat to human health and well-being. 'Hunger and malnutrition kill more people every year than AIDS, malaria, and tuberculosis combined, and more people die from hunger than in wars.'[16] Tackling the challenge of malnutrition involves addressing the needs of people who do not consume sufficient calories each day but also of people who consume too many calories, as well as the many people who consume a sufficient amount of calories but receive inadequate nutrition from their diets. According to a 2006 report from the World Bank:

> Over one-fourth of all children in developing countries are either underweight or stunted. One third of the world's population (almost 2 billion people) suffers from various forms of iodine deficiency disorders (IDD). The same numbers have iron deficiency, which leads to anemia . . . Simultaneously, the proportion of people who are overweight or obese is growing, often in the same countries where undernutrition and micronutrient deficiency are concentrated . . . Some 1.1 billion adults are overweight, and 300 million are obese.[17]

The issue of ensuring proper nutrition for all of the world's people remains one of the most significant challenges facing human societies today.

Deciding how to best address questions of food insecurity involves exploring not just whether people have access to sufficient food, but must also consider how to actualize the standard of sufficient, safe, and nutritious food. Questions of hunger and malnutrition are questions about what kinds of food people have access to and how they access food. For many undernourished people in the world, their hunger comes not from a lack of sufficient food, but from the fact that they are too poor to access the food they need.[18] In March 2008, UN Secretary General Ban Ki-moon cited and number of causes, 'from rising demand in major economies such as India and China to climate- and weather-related events such as hurricanes, floods and droughts that have devastated harvests in many parts of the world,' as drivers of record highs in the price of staple crops such as wheat, corn and rice and the fact that prices had risen by as much as 50 percent in just six months, in a call for recognition of the changing nature of the challenge of hunger in the twenty-first century.[19] He recognized that while the causes of new challenges to hunger were diverse, so too must be the solutions, involving short term improvements in food aid systems but also long term changes to help develop a global food system that is more safe, resilient, and productive.[20]

Addressing natural and nefarious biological threats to food safety

While the global food system largely provides safe and wholesome food, it is a major connector between many different peoples and places. The food system itself can be disrupted by infectious diseases, such as Foot-and-Mouth disease or Bovine Spongiform Encephalopathy ('Mad Cow') disease, and it can also serve as a vector for the transmission of disease by actors with nefarious intent. A number of factors—including the 2001 Anthrax incidents in the United States, the rapid emergence and global spread of severe acute respiratory syndrome (SARS) in 2003, and the ongoing attention to the potential danger of a global influenza pandemic—have raised awareness of the need to ensure food safety.[21] Often discussed for their role in causing illness and death to individuals, infectious diseases can have significant and widespread impacts on societies.[22] 'The ability of infectious agents to destabilize populations, economies, and governments is fast becoming a sad fact of life. The prevention and control of infectious diseases are fundamental to individual, national, and global health and security.'[23] In addition to the problems directly caused by lack of proper nutrition, 'malnutrition magnifies the effect of every disease, including measles and malaria.'[24] While a number of factors are responsible for the spread and impact of infectious disease, the nature of threats from infectious disease has been accelerated by development of an increasingly interconnected world and processes such as increased flows of people and goods and a considerable reduction in the time required to travel from one part of the world to the next.[25]

Contamination of food supplies can have significant health impacts. At the global level, the World Health Organization estimates that approximately two million children die each year from diarrheal illnesses caused by contaminated food and water, and that one out of every three people in an industrialized country suffers from a foodborne illness each year.[26] Within the United States, the Centers for Disease Control and Prevention (CDC) estimates that approximately one out of every four Americans will develop a foodborne illness each year (or 76 million cases per year), and that these cases result in 325,000 hospitalizations and 5,000 deaths.[27] Contamination of food supplies can sicken large numbers of people over a relatively short period of time. For instance, in the US in 1994, an ice cream pre-mix that was contaminated with salmonella caused illness in over 220,000 people in 41 states.[28] More recently, contamination of fresh spinach by E. coli O157:H7 in 2006 and peanut butter with salmonella in 2007 and in early 2009 have attracted media coverage and focused attention on concerns about the safety of the food supply.

The food system provides a major means of interaction between humans and animals, and diseases can often transfer to humans through processes involved in the killing, processing, and consumption of animals.[29] Over 60 percent of identified infectious diseases can affect both humans and animals.[30] Food system contamination can also have a significant impact on the availability of food and the economic health of agricultural systems. In 2003 for example, one-third of global meat exports (6 million tonnes) were affected by an animal disease outbreak, causing an estimated $10 billion in losses.[31] Avian influenza A (H591) is only one of a series of livestock disease outbreaks that have caused losses of more than $100 billion (not including HIV/AIDS) over the past fifteen years.[32] Contamination of food supplies by diseases or other substances can cause high levels of wastage and spoilage of foods that lead to an increase in prices and a failure to recoup resources invested to bring crops or livestock to market.

Modern food systems are designed to rapidly move large numbers of goods to a large number of people and they could be an ideal target for disruption or for use as a delivery system for a biological agent by actors with nefarious purposes such as warfare, terrorism, or criminal endeavors.[33] Awareness of the possibility of intentional use of biological agents to cause harm was increased after the events of September 11, 2001 and Amerithrax anthrax incidents in the fall of 2001; following the incidents, a significant amount of attention was given to the vulnerability of citizens, livestock, and the food supplies to intentional attacks using biological weapons.[34] There is a general sense that advances in technology, along with increased mobility have created the conditions for different forms of threat from biological weapons than in prior years: 'Advances in biotechnology, coupled with difficulty in detecting nefarious biological activity, have the potential to create a much more dangerous biological warfare threat.'[35] This view of a looming threat from nefariously employed biological sources is not without its critics.[36] While concerns about the threat from nefarious uses of biological weapons have received a great deal of policy and media attention, it is important to note that as of 2009 instances of terrorist or criminal activity involving biological agents have caused mortality (death) and morbidity (illness) in relatively low numbers of people.[37] Efforts to ensure human security from malnutrition and food safety threats must address both natural infectious diseases and potential nefarious uses of biological agents. Yet such efforts must also be mindful that more people are impacted on a daily basis by chronic infectious disease threats and problems like malnutrition and a lack of clean water.

Environmental change and the global food system

The global food system is also facing a set of challenges related to processes of global environmental change. The way in which people satisfy their need for food, including farming, fishing, and raising livestock, is a significant contributor to the impact that human populations have on the environment. Unsustainable agricultural practices have driven many of the worst instances of human-induced environmental change, including the deforestation of China's uplands around 3000 BC, the deforestation of Mediterranean vegetation, and the North American Dust Bowl, a series of dust storms that caused significant environmental damage and hardship in the United States and Canada during the 1930s.[38] The environmental impacts of food continue to play an important role in global issues today. Since the early 1970s, efforts to increase food production have been the main driving force behind land clearance and resulting pressure on land resources; agricultural practices are closely linked to problems such as degradation and increased salinization of soils, increased stresses on water resources, impacts on water quality from agricultural run-off, and the development of antibiotic-resistant microbes.[39] Many current agricultural practices reduce the ability of ecosystems to provide goods and services such as carbon sequestration and the ability of soil to retain and absorb water.[40] Agricultural practices also play a role in using renewable natural resources, such as soil, at rates that are not renewable, with one estimate finding that soil is being depleted at a rate one to two orders of magnitude (between 10 and 100 times) greater than it is being replenished.[41] Agriculture not only plays a role in localized environmental changes, but also contributes to large-scale processes of global change. Agricultural practices 'account for around 13 percent of global greenhouse gas emissions, about the same as transport.'[42] This includes gases such as carbon dioxide from land clearance and deforestation, but more importantly gases like methane, nitrous oxide, and ammonia that result from crop and livestock production.[43]

Since the fourth assessment report by the Intergovernmental Panel on Climate Change in 2007 there is now general agreement that human activities have contributed to changes in the Earth's climate system. Reviews of the potential impacts of climate change have identified a number of threats from climate change to the basic component of human livelihoods, including: melting glaciers, declining crop yields, ocean acidification, rising sea levels, and a variety of impacts on ecosystems and species.[44] Based on these expected impacts, prominent global leaders have called for recognizing climate change as one of the highest priority issues facing the international community. Former UN Secretary-General Kofi Annan stated, 'Global climate change must take its place alongside those threats—conflict, poverty, the proliferation of deadly weapons—that have traditionally monopolized first-order political attention.'[45] The impacts of climate change are expected to be widespread and felt in a range of systems and sectors including water resources, health, and food production. Yet the impacts from climate change will not be equally distributed. Africa and Asia will face increasing levels of water scarcity, compromised food production, and impacts to coastal areas that could reduce fisheries and tourism revenues; regions like North America could see a 5 to 20 percent increase in agricultural yields, at least in the near term.[46] While global models about the impacts of climate change continue to be refined, there remain significant uncertainties about how, where, and when the impacts of climate change will impact food systems and the resulting effects such impacts will have on food security.

Climate change could also have indirect impacts on food systems. One mechanism of such impacts could be the increasing frequency, intensity, and duration of extreme weather events.[47] Events such as hurricanes, monsoons, and landslides disrupt social, economic, health, and sanitation systems, and can have devastating impacts on agricultural systems and fisheries.[48] Already, the impacts of climate change are linked to increasing severity of weather patterns such as the 2003 European heat wave that claimed 35,000 lives.[49] Climate change could also introduce new health threats by altering the range of pathogens and hosts for infectious diseases.[50] The impacts from climate change will also not be limited to human populations, likely increasing biodiversity loss through its impacts on plant and animal populations.[51] In addition to changing the global climate system, other human-induced environmental changes have had considerable and widespread effects on the global environment. Land clearance and deforestation, for instance, have had a significant impact on land resources; the UN estimates that such human activity has negatively affected the productivity of almost a quarter of arable land.[52] Human activities have been equally impactful with regard to forests, oceans, and populations of animal and plants species. Biodiversity loss can have significant and unexpected impacts on ecosystems, but can be especially noteworthy if loss impacts species that are vitally important for the food system.

The impact of environmental changes on agricultural systems will be diverse and widespread, and could include changes such as shifts in the ecological and economic viability of raising crops and animal species in a given environment, but also much more subtle changes such as impacts on species of soil bacteria that help fix nitrogen or encourage water intake or through loss of species such as honeybees and song birds that play key roles as pollinators. The industrialization of agriculture has resulted in significant increases in global food production, but has also led to major declines in the biological diversity of key food sources:

> Already three-quarters of biodiversity in crops has been lost in the last century, according to the United Nations Food and Agriculture Organization. Eighty percent of maize types that existed in the 1930s are gone, for example. In the United States, 94 percent of peas are no longer grown.[53]

Recognition of the impact of biodiversity loss has prompted renewed effort to create a global network of plant and seed banks to store samples and information about genetic resources of plant species.[54] Human actions also impact the food system in indirect ways, such as the loss of farmland to development as urban and peri-urban areas expand into previously rural areas. By 2050, global population is expected to increase from 6.7 billion to 9.2 billion people; by the end of 2008 half of the world's population will be living in urban areas and urban areas will absorb almost all the projected global population increase through 2050.[55] Expected growth of human populations and changes in dietary patterns suggest that the need to provide sufficient and safe food in an environmentally sustainable manner will likely remain a challenge to the global food system for many years to come.

Studying food security

As the global food system is a web of relations that involves producing, harvesting, processing, transporting, and consuming food, the range of subjects relevant to food security provides entry points for students and scholars from a variety of perspectives and disciplines. Timothy W. Luke writes that to understand pressing global issues 'we need to focus on hybridities of Nature/Society at sites which intermix the natural and the social, like the "built environment," "natural history," or "social ecology." '[56] Involving many thousands of years of interactions and intermingling and between humans and nature, the global food system is an excellent example of a site that is a hybrid of nature and society. Examinations of the food system can be a thematic focus of courses, projects, and lectures all their own, or they can be combined with other efforts to examine the political, social, and economic impacts of humans on the environment at a variety of levels of analysis from global to personal.

There is now a great wealth of information available about the food system at global and national levels of analysis. International organizations, national governments, research institutes, and non-profit organizations are developing increasingly sophisticated means of aggregating data about the global food system and food security. Advances in information and data storage capabilities are making it easier for users to access and utilize this information, as much of it is available on the internet. International organizations and UN agencies are excellent sources for global aggregate and national level information.[57] Many governments maintain helpful sites about food and food security including the United States, the United Kingdom, and the European Commission.[58] There are many other sources of information about the global food system, and global and national level information is only one means of examining the food system.

There is also a need for process tracing efforts that consider how food is produced and brought together with people and examines the costs, threats, and vulnerabilities involved in these connections. These efforts can examine a particular food or commodity chain. For example, a 2005 study examined the vulnerability of the milk supply in the United States and found that a few grams of a substance such as botulism toxin could enter the milk system through a variety of pathways and be delivered to hundreds of thousands of consumers.[59] In *The Omnivore's Dilemma*, Michael Pollan provides another example of process tracing as he uses four meals as a way to examine the various food chains that provide the ingredients for the meals.[60] Process tracing can be useful to identify key areas of vulnerability as well as opportunities to improve the functioning, safety, and sustainability of food systems.

There are also opportunities to better understand the global food system through fieldwork. It is now possible to access more and better data about large-scale impacts of global problems such as climate change. Yet even the most comprehensive global data set can provide only limited information about what such processes will mean for specific places. There is a need for field research to provide fine-grained understandings of how processes like climate change will impact certain people and places. For instance, at Thistle Hill Farm in North Pomfret, Vermont, John and Janine Putnam make traditional alpine cheeses whose flavor reflects the unique mix of flowers and grasses their cattle graze on during the year; climate change has the potential to impact their cheese production and product, from changing how much milk cows drink to the types of plants they consume.[61] For farmers in Vermont and around the world, climate change could have impacts like more rain or less rain, hotter summers, or cooler summers. Field investigations can provide important information, especially about which areas will be hardest hit by problems and what sorts of strategies are most effective at helping people mitigate and adapt to challenges.

Addressing food insecurity involves roles for actors at multiple levels. Some challenges, such as addressing the impact of subsidies, will require actions by states and international actors. Other challenges, such as improving local or regional food security, involve roles for actors at the provincial or metropolitan level. It is also important to recognize the increasing role that non-state actors, such as charities and food banks, play in addressing food security at all levels of activities. Finally, there is a set of activities related to food security that involve individual actions and food choices, but which aggregate to have significant impacts. The daily interactions between people and the food system provide opportunities examine such impacts through the integration of examinations of personal food systems into courses or projects. One approach is to encourage the keeping of a food diary in which students write down information about their food choices and interactions with the food system. They should be encouraged to record information such as what they eat, how much it costs them, and where it comes from. Even one day's record can provide an entry point into discussions about hunger, malnutrition, dangers posed by contaminated food, food prices, and the contribution that food makes to a person or family's ecological footprint.

Conclusion: toward a sustainable global food system

The global food system faces a number of challenges in the twenty-first century. Some of these challenges have received a great deal of policy and media attention, such as the potential danger from biological weapons and the threat from contaminated food from China and the impact of biofuels on food security, while others have received less attention, such as the continuing threat of malnutrition from micronutrient deficiency and the impacts of climate change on the food system. The transition of the food system from a localized system of production and consumption to a global system where millions of tons of crops, livestock, fish, and food products travel between countries and continents each year has added layers of complexity to this problem. Efforts to provide people with sufficient food have been, and continue to be, a major driver of environmental changes. These changes are often localized, such as cutting down forests, increasing erosion of topsoil, and reducing water quality through agricultural run-off. However, local changes can have regional and national impacts that contribute to problems such as toxic dead zones in rivers and oceans or global climate change. In addition to providing sufficient, safe, and nutritious food, the key challenge to the global food system is to find ways to meet these goals that at least have the lowest possible impact on the environment and at best can help with efforts to mitigate and adapt to environmental change.

Interest in improving the environmental sustainability of the food system has grown considerably in recent years. This interest has been driven by a variety of factors, including interest in food quality and healthfulness, concerns about technologies like genetic modification, focus on human health, and questions about the impact of agriculture on the environment.[62] The concept of sustainable agriculture refers to a range of techniques and practices intended to 'meet the dual goals of increased productivity and reduced environmental impact. They do this through diversification and selection of inputs and management practices that foster positive ecological relationships and biological processes within the entire agro-ecosystem.'[63] A global food system that is optimized around the goal of sustainability could help boost soil fertility and reduce erosion, improve local water quality and reduce run-off, and aid in efforts to mitigate and adapt to climate change by providing buffer zones, sinks for greenhouse gases, and the provision of energy generated from current biological sources.

To be certain, the development of a sustainable global food system will not be without its own set of complex challenges, and if changes are adopted without full consideration, efforts to promote sustainability can also have unintended consequences that cause threats and vulnerabilities to human health, security, and well-being. The global effort to promote biofuels provides an excellent example of this. Thus, efforts in the past few years to promote the use of biofuels have resulted in the unintended and sometimes significant increases in prices of key food crops such as corn leading to food insecurity in developing countries, while it is unclear whether biofuels actually cause more greenhouse gas emissions than they prevent.[64] The substitution of fuels made from current biological sources could prove important, but such a substitution must be renewable and done in a way fully aware of total impacts.

The efforts of past centuries have led to the creation of a global food system, and we must now turn our attention to the goal of optimizing that food system in an ecologically sustainable way that yet provides for human needs. Each of the sets of concerns discussed above is a component of the need to find a way to provide sufficient amounts of food to all people on the planet. Developing a sustainable global food system will also involve confronting ecological limiting factors—simply put, some people live in places that are very conducive to food production while others live in places where food cultivation is more difficult. There remains a great deal of discussion about the best way to accomplish the goal of ensuring food security; much of the discussion centers around how one moves from the overarching goal of ensuring food security to particular solutions. In such discussions, there is need to be mindful that food is not merely a substance that provides energy (calories) and 'spare parts' for the growth and maintenance of the human body (nutrients), but is an important connector between humans and their environment with cultural, ecological, economic, and social meanings that are constantly being negotiated and renegotiated.

One of the great challenges of envisioning a sustainable food system is that it must be reconciled with the increasingly globalized, networked structure of our world.[65] While processes of global change may make the effort to develop a sustainable global food system more difficult, advances in technology and communication could also open up new pathways for the food system. Developing a sustainable food system will require rethinking some of the key assumptions about it. For instance, the development of a more urbanized population has the potential to problematize the old urban core/rural periphery divide and raises new possibilities about improving urban agricultural systems by developing urban gardens and perhaps even vertical farms housed in self-sustaining skyscrapers.[66] A useful guiding standard for a sustainable global food system comes from Aldo Leopold, who suggests we should 'examine each question in terms of what is ethically and aesthetically right, as well as what is economically expedient. A thing is right when it tends to preserve the integrity, stability and beauty of the biotic community. It is wrong when it tends otherwise.'[67] The challenge of creating a sustainable food system is by no means simple, for it must be mindful of human needs, sensitive to current and future ecological conditions, and able to navigate the eddies of a complex system of global governance in a time of great turbulence and change.

Notes

1   A version of this chapter was presented at the 47th Annual Meeting of the International Studies Association, San Diego, California, March 22–25, 2006. I would like to thank Richard Matthew, Marilyn I. McMorrow, George Shambaugh, Kenneth Rutherford, and Patricia A. Weitsman for their comments. An extended version of the arguments and issues disccused in this chapter can be found in B. McDonald, *Food Secuirty*, Cambridge: Polity Press, 2010.

2   *The Rome Declaration on World Food Security and the World Food Summit Plan of Action*, Rome, Italy, November 13, 1996. Online: http://www.fao.org/docrep/003/w3613e/w3613e00.htm (accessed August 2012).

3   G. J. Armelagos, 'The Viral Superhighway', *The Sciences*, 1998, 24–9; J. Diamond, *Guns, Germs and Steel: The Fates of Human Societies*, New York: W.W. Norton, 1999; K. F. Kiple, *A Movable Feast: Ten Millennia of Food Globalization*, Cambridge: Cambridge University Press, 2007.

4   See S. Flynn, *America the Vulnerable: How our Government is Failing to Protect Us from Terrorism*, New York: HarperCollins, 2004, p. 113; Central Intelligence Agency (CIA), 'World Economy', *The World Factbook*, 2007.

5   J. N. Rosneau, *Turbulence in World Politics: A Theory of Change and Continuity*, Princeton, NJ: Princeton University Press, 1990.

6   R. A. Matthew and G. E. Shambaugh, 'Sex Drugs and Heavy Metal: Transnational Threats and National Vulnerabilities', *Security Dialogue* 29 (2), 1998, 163.

7   United Nations Development Programme, *Human Development Report 1994*, New York: Oxford University Press, 1994, p. 23.

8   United Nations Development Programme, *Human Development Report 2003*, New York: Oxford University Press, 2003, pp. 27–8.

9   T. Malthus, *An Essay on the Principle of Population: A View of its Past and Present Effects on Human Happiness*; with an Inquiry into Our Prospects Respecting the Future Removal or Mitigation of the Evils which It Occasions, 6th ed., London: John Murray, 1826. Online: www.econlib.org/library/Malthus/malPlong.html (accessed August 2012).

10   For an excellent discussion of the impacts of Malthus's ideas, see Thomas F. Homer-Dixon, *Environment, Scarcity, and Violence*, Princeton, NJ: Princeton University Press, 1999, pp. 28–44. See also L. Bjørn, *The Skeptical Environmentalist: Measuring the Real State of the World*, Cambridge: Cambridge University Press, 2001; D. Meadows, *et al.*, *Limits to Growth: A Report for the Club of Rome's Project on the Predicament of Mankind*, New York: New American Library, 1977; and J. Simon, *The Ultimate Resource*, Princeton, NJ: Princeton University Press, 1981.

11   J. Shaw, *World Food Security: A History since 1945*, New York: Palgrave MacMillan, 2007.

12   United Nations (1948) *Universal Declaration of Human Rights*. Online: www.un.org/en/documents/udhr/ (accessed August 2012).

13   United Nations, *Covenant on Economic, Social and Cultural Rights*, 1966. Online: http://www.unhcr.org/refworld/docid/3ae6b36c0.html (accessed August 2012).

14   United Nations, *Agenda 21*, 1992. Online: www.un.org/esa/sustdev/documents/agenda21/english/agenda21chapter14.htm (accessed August 2012).

15   The World Bank, *Repositioning Nutrition as Central to Development: A Strategy for Large-Scale Action*, Washington, DC: The World Bank, 2006.

16   Shaw, *World Food Security*, p. x.

17   The World Bank, *Repositioning Nutrition*, p. 43.

18   Food and Agriculture Organization of the United Nations (FAO), *Reducing Poverty and Hunger: The Critical Role of Financing for Food, Agriculture and Rural Development*, World Bank, 2002. Online: ftp://ftp.fao.org/docrep/fao/003/y6265E/Y6265E.pdf (accessed August 2008), p. 9.

19   B. Ki-moon, 'The New Face of Hunger', *Washington Post*, Wednesday, March 12, 2008, A19.

20   Ibid.

21   On the October 2001 Anthrax incidents, see U.S. Centers for Disease Control and Prevention, 'Update: Investigation of Anthrax Associated with Intentional Exposure and Interim Public Health Guidelines', *Morbidity & Mortality Weekly Report* 50, 2001, 889–93; Federal Bureau of Investigation, *Amerithrax Investigation*, 2008. Online: www.fbi.gov/anthrax/amerithraxlinks.htm (accessed August 2012). On SARS, see U.S. Centers for Disease Control, *Basic Information about SARS*, 2005. Online: http://www.cdc.gov/sars/about/fs-SARS.html (accessed August 2012). On pandemic influenza, see L. Garrett, 'The Next Pandemic?', *Foreign Affairs* 84 (4), 2005, 3–23; and M. T. Osterholm, 'Preparing for the Next Pandemic', *Foreign Affairs* 84 (4), 2005, 24–37.

22   J. B.Tucker, *Scourge: The Once and Future Threat of Smallpox*, New York: Grove Press, 2001; World Health Organization (WHO), The World Health Report 2007: *A Safer Future: Global Public Health Security in the 21st Century*, Geneva: World Health Organization, 2007. On global health threats, see National Intelligence Council, *National Intelligence Estimate: The Global Infectious Disease Threat and Its Implications for the United States*, 2000. Online: http://www.wilsoncenter.org/sites/default/files/Report6-3.pdf (accessed August 2012); L. Garrett, *Betrayal of Trust: The Collapse of Global Public Health*, New York: Hyperion, 2001; J. Brower and P. Chalk, *The Global Threat of New and Reemerging Infectious Disease: Reconciling U.S. National Security and Public Health Policy*, Arlington, VA: RAND, 2003; K. Jones, N. Patel, M. Levy *et al.*, 'Global trends on emerging infectious diseases', *Nature* 451 (21), 2003, 990–93.

23   M. S. Smolinski *et al.* (eds), *Microbial Threats to Health: Emergence, Detection, and Response*, Washington, DC: National Academies Press, 2003, p. xi.

24   World Hunger Education Service, *World Hunger Facts 2006*. Online: www.worldhunger.org/articles/Learn/world%20hunger%20facts%202002.htm (accessed August 2012).

25   For an extended discussion of factors impacting infectious disease threats, see Smolinski *et al.*, *Microbial Threats to Health* and Brower and Chalk, *The Global Threat of New and Reemerging Infectious Disease*.

26   World Health Organization and World Trade Organization, *WTO Agreements and Public Health: A Joint Study by the WHO and the WTO Secretariat*, 2002.

27   P. Mead, 'Food Related Illness and Death in the United States', *Emerging Infectious Diseases* 5, 1999, 607–25.

28   T. W. Hennessy, *et al.*, 'A national outbreak of Salmonella senteritidis infections from ice cream', *New England Journal of Medicine* 334, 1996, 1281–86.

29   See W. B. Karesh and R. A. Cook, 'The Human-Animal Link', *Foreign Policy* 4, 2005, 38–50.

30   Ibid., p. 40.

31   Food and Agriculture Organization of the United Nations (FAO), *Animal disease outbreaks hit global meat exports: One-third of global meat exports affected – losses could be high*, 2004. Online: www.fao.org/newsroom/en/news/2004/37967/index.html (accessed August 2012).

32   Bio Economic Research Associates, *SARS Shows Weakness In Biosecurity; Strategies To Address Root Causes Recommended*, 2003. Online: http://cmbi.bjmu.edu.cn/news/0306/156.htm (accessed August 2012).

33   S. Whitby and P. Rodgers, 'Anti-Crop Biological Warfare: Implications of the Iraqi and U.S. Programs', *Defense Analysis* 13 (3), 1997, 303–18; M. Nestle, *Safe Food: Bacteria, Biotechnology, and Bioterrorism*, Berkeley: University of California Press, 2003; Flynn, *America the Vulnerable*; T. Homer-Dixon, *The Upside of Down: Catastrophe, Creativity and the Renewal of Civilization*, Washington, DC: Island Press, 2006.

34   U.S. Centers for Disease Control and Prevention, 'Biological and Chemical Terrorism: Strategic Plan for Preparedness and Response, Recommendations of the CDC Strategic Planning Working Group', 49 *Morbidity and Mortality Weekly Report, Recommendations and Reports* (RR-4) 5–8, 2000; U.S. Food and Drug Administration, *Risk Assessment for Food Terrorism and Other Food Safety Concerns*, 2003. Online: www.cfsan.fda,gov/~dms/rabtact.html (accessed August 2012).

35   Central Intelligence Agency, *The Darker Bioweapons Future*, 2003. Online: www.fas.org/irp/cia/product/bw1103.pdf (accessed August 2012).

36   See for example J. Stern, 'Dreaded Risks and the Control of Biological Weapons', *International Security* 27 (3), 2002, 89–123; and B. Hartmann *et al.*, *Making Threats: Biofears and Environmental Anxieties*, Lanham, MD: Rowman and Littlefield, 2005.

37   Tucker, *Scourge: The Once and Future Threat of Smallpox*; WHO, 2007.

38   D. Worster, *Dust Bowl: The Southern Plains in the 1930s*, Oxford: Oxford University Press, 1979, p. 4.

39   United Nations Environment Programme (UNEP), *Global Environmental Outlook 3: Past, Present and Future Perspectives*, London: Earthscan, 2002.

40   J. Clay, *World Agriculture and the Environment*, Washington, DC: Island Press, 2004, p. 2.

41   D. R. Montgomery, 'Soil erosion and agricultural sustainability', *Proceedings of the National Academy of Sciences* 104 (33), 2007, 13268–272.

42   G. Walker and Sir D. King, *The Hot Topic: How to tackle Global Warming and Still keep the lights on*, London: Bloomsbury Publishing, 2008, p. 112.

43   Food and Agriculture Organization of the United Nations (FAO), *World Agriculture: Towards 2015/2030*, London: Earthscan, 2003, p. 334.

44   See for example, Stern Review, *The Economics of Climate Change*. Cambridge: Cambridge University Press, 2007; J. F. C. Dimento and P. Doughman (eds), *Climate Change*, Cambridge, MA: MIT Press, 2007.

45   Kofi Annan, *Annan Stresses Climate Threat at UNFCCC Conference*, 2006. Online: www.unep.org/Documents.Multilingual/Default.asp?DocumentID=495&ArticleID=5424&l=en (accessed August 2012).

46   Intergovernmental Panel on Climate Change, 2007.

47   Ibid.

48   Brower and Chalk, *The Global Threat of New and Reemerging Infectious Disease*.

49   WHO, 2007, p. 8.

50   National Research Council, 2001.

51   C. D. Harvell, *et al.*, 'Climate Warming and Disease Risks for Terrestrial and Marine Biota', *Science* 296, 2002, 2158–162.

52   United Nations Environment Programme, 2002.

53   E. Rosenthal, 'Near Arctic, Seed Vault Is a Fort Knox of Food', *The New York Times*, 2008. Online: www.nytimes.com/2008/02/29/world/europe/29seeds.html (accessed August 2012).

54   Ibid.

55   United Nations *World Urbanization Prospects: The 2007 Revision*, 2008. Online: www.un.org/esa/population/unpop.htm (accessed August 2012).

56   T. W. Luke, 'World Health and the Environment: Globalization's Ambiguities', presented at the Third Annual Staff Development Conference, University of Wisconsin System Institute of Global Studies, Lake Geneva, Wisconsin, October 28–20, 2001, p. 2.

57   See for example: The Food and Agriculture Organization of the United Nations (FAO), www.fao.org/; and the World Health Organization (WHO), www.who.int/nutrition/en/.

58   The United States: See Nutrition.gov, www.nutrition.gov/; The Food and Drug Administration, http://www.fda.gov/; United States Department of Agriculture, http://www.usda.gov/. The United Kingdom: Department for Environment, Food and Rural Affairs, www.defra.gov.uk/foodrin/index.htm; The Food Standards Agency, www.food.gov.uk/. The European Commission: Agriculture, http://europa.eu/pol/agr/index_en.htm; Food Safety, http://ec.europa.eu/food/index_en.htm

59   L. M. Wein and Y. Liu, 'Analyzing a bioterror attack on the food supply: The case of botulinum toxin in milk', Proceedings of the *National Academy of Sciences*, 102 (28), 2005, 9985–989.

60   M. Pollan, *The Omnivore's Dilemma: A Natural History of Four Meals*, New York: Penguin, 2006.

61   K. Levine, *Cheesemakers Taste a Change in the Weather*, National Public Radio, 2007. Online: www.npr.org/templates/story/story.php?storyId=13981929 (accessed August 2012).

62   E. Schlosser, *Fast Food Nation: The Dark Side of the All-American Meal*, New York: Houghton Mifflin, 2001; Pollan, *The Omnivore's Dilemma*; B. Glassner, *The Gospel of Food: Everything You Think You Know About Food Is Wrong*, New York: HarperCollins, 2007; M. Pollan, *In Defense of Food: An Eater's Manifesto*, New York: Penguin, 2008.

63   FAO, *World Agriculture*, p. 304.

64   E. Rosenthal, 'Biofuels Deemed a Greenhouse Threat', *The New York Times*, 2008. Online: www.nytimes.com/2008/02/08/science/earth/08wbiofuels.html (accessed August 2012); Ki-moon, 'The New Face of Hunger'.

65   A. L. Barabási, *Linked: How Everything Is Connected to Everything Else and What It Means for Business, Science, and Everyday Life*, New York: Plume, 2003; Homer-Dixon, *The Upside of Down*.

66   L. Mougeot, *Agropolis: The Social, Political, and Environmental Dimensions of Urban Agriculture*, London: Earthscan, 2005; L. Chamberlain, 'Skyfarming', *New York Magazine*, 2007. Online: http://nymag.com/news/features/30020/ (accessed August 2012); The Vertical Farm Project, www.verticalfarm.com/ (accessed August 2012).

67   A. Leopold, *A Sand County Almanac: And Sketches Here and There*, Oxford: Oxford University Press, 1989, pp. 224–25.

14

CHALLENGING INEQUALITY AND INJUSTICE

A critical approach to energy security

Adam Simpson

Introduction

Energy security is a nebulous and contested concept. As demonstrated by the various issues addressed in this book, it is only one aspect of broader environmental security studies, but it has often been dealt with independently due to its central importance to the modern industrialized state. Unlike other environmental security issues energy security has a long history as a matter that has exercised both security analysts and national policy makers. Energy security concerns dominated global headlines following the oil shocks of the 1970s, but low fossil fuel prices for the subsequent two decades reduced its significance for both the public and political decision makers. With the return of high oil and commodity prices since the turn of the century, energy security as a concept is once more at the top of the global agenda: 'Lawyers, bankers, brokers, economists, geographers, geologists, engineers and journalists speak of energy security with the same confidence as generals, development workers, defence analysts or environmental activists.'[1]

The importance of energy to industrial society has meant that although water, for example, is far more essential to human survival, it has been the efforts to secure national energy supplies, largely from fossil fuel sources, that have been of central concern to most national governments and as a result, energy security as a concept has been traditionally approached from a state-centred national security perspective. It is only in extremely water-deprived areas, such as the Middle East, that water security is considered to be a similar national security concern.[2]

Most definitions of energy security include some combination of availability, affordability and reliability, with recent work adding concepts such as sustainability.[3] From the perspective adopted in this chapter, however, it is also important that energy security is not achieved at the expense of other aspects of environmental security, such as food or water security. Energy security could therefore be broadly defined as being achieved when there is sufficient energy available to satisfy the reasonable needs of the political community (the referent object) in an affordable, reliable and sustainable manner as long as pursuing it does not cause environmental insecurity to that or any other political community.

When considering the achievement of energy security, as with any security concept, there are three fundamental questions to ask: 'for whom'; 'from what threats'; and 'by what means'. Underlying most definitions of energy security – and Sovacool provides a list of 45[4]– is the assumption that it is a concept that applies primarily to the nation-state. It is clear that for traditional theorists and national security advisers the state is the referent object and, therefore, the answer to the first question. The second question could receive a variety of answers, even from traditional theorists. Realists might focus on restrictions in supply that arise due to military competition with great powers while liberals might focus on impediments to free flowing energy markets such as sanctions or trade restrictions. Similarly, in answer to the third question realists would be likely to emphasize the necessity of securing access to energy through increasing state military power and influence while liberals might focus on the need to establish bilateral or multilateral free trade agreements that promote free energy markets.

It comes as no real surprise that traditional discourses of energy security have focused particularly on fossil fuels and large-scale electricity projects due to their importance to the industrial development, and particularly, the military security of the nation-state even in the South. The quest for this sort of energy security has barely progressed since the nineteenth century and it is still the main reason for many modern imperial interventions and conflicts.[5] This interpretation of energy security coincides with academic or government research funding opportunities, but it also fits neatly with the predominant large-scale and hierarchical, top-down development paradigm of 'high politics' prescribed by international financial institutions and adopted by governments around the world.

Some environmental security approaches treat energy security in this traditional state-centric manner, linking environmental degradation to a looming 'energy gap' particularly in relation to fossil fuels,[6] a gap which is seen by some environmental security theorists as the most likely cause of future international conflict.[7] As noted in chapter 1, however, many scholars adopting an environmental security lens are likely to be informed by a broadly critical epistemology that challenges existing social structures, rather than attempting to solve problems within them. This critical approach to energy security challenges the existing economic, political and technical assumptions that underpin traditional debates on energy production and consumption, but it also challenges traditional notions of security that have the nation-state as their referent object. As chapter 1 demonstrates, the recent history of security studies and environmental security in particular, is characterized by the attempts of critical theorists and practitioners to challenge this state-centric assumption, but energy security, due in large part to its singular importance to industrial and military activity, has proved particularly wedded to both the traditional security architecture and a focus on fossil fuels.

Despite this enduring emphasis on fossil fuels, fuel wood remains the primary energy source for most households in the less affluent global South.[8] Unlike studies from the North most studies of states in the South, even if they adopt a national focus, recognize the dominance of biomass as a fuel.[9] Economic development in China and other emerging economies has reduced this dependence on biomass, but according to the International Energy Agency (IEA) over 1.4 billion people, many in Africa and South Asia, still live outside the electricity–fossil fuel nexus that the affluent North takes for granted.[10] As a result, in much of the South 'renewable' energy is not an 'expensive luxury item', but the only viable energy source on offer.[11] The difference in energy consumption patterns between the global South and North therefore highlights a pertinent contrast between renewable and non-renewable energy. Although a dominant view in the North sees poverty in the South as a significant cause of environmental degradation, the difference between energy consumption levels and the type of energy consumed means that energy security from small-scale renewable sources in the South can often be achieved more easily, and with less environmental impact, than with the fossil fuel or nuclear technologies deployed in the North. While this situation still characterizes much of the South, some emerging economies and China in particular are becoming significant global consumers of fossil fuel, nuclear and large-scale hydroelectric energies at great cost to human health and the environment.

Rather than focusing on the state, a critical energy security perspective relates more to the ability of individuals, particularly in marginalized or deprived communities, to secure sufficient access to energy for their personal needs. This approach draws on Jon Barnett's definition of environmental security that adopts a human security standpoint and focuses on the inequitable distribution of degradation resulting from unequal social structures: '[a] human-centred environmental security concept places the welfare of the disadvantaged above all else'.[12] Almost all textbook chapters that examine energy security do so, however, primarily from a state-centric US or European perspective, with developing states of the global South mentioned only cursorily or as 'supplier' states. China now appears more regularly as a 'consumer' state, but often this is discussed primarily in relation to how it will impact on the energy security of the North. While not ignoring the economically and politically powerful Northern centres this chapter focuses more on the critical approach that characterizes much environmental security literature and which prioritizes the needs of the most marginalized and disadvantaged in the global community. This critical approach shifts the referent object of energy security away from the state to the individual and community and shifts the focus away from the North to the South. The shift in the referent object does not, however, necessarily result in a lack of analysis of state decision making, but this analysis is undertaken only in relation to the impacts on the energy security of poor or marginalized individuals or communities rather than national energy security which may, for example, prioritize the provision of energy to the military or associated industries. As with the Aberystwyth School approach to security the state, if in a cosmopolitan guise, remains a potentially powerful agent of progressive emancipatory change, but it can also act as a regressive force causing greater insecurity.[13]

As energy security in this book is considered through the prism of environmental security it should be noted that energy security is not a desirable goal if it is to be only achieved at the expense of some other significant environmental insecurity. This potential trade-off, as with most environmental security issues, is most pertinent in the South where, for example, a project to build a large hydroelectric dam may supply electricity to local communities, but may submerge the forest from which the community derived their livelihood. In the South, which is often characterized by varying degrees of authoritarian governance, these projects are frequently undertaken with little public consultation or public interest and the energy is often exported or sent further afield to satisfy large business interests or the military. In the context of climate change, the burning of fossil fuels now undoubtedly brings adverse climate security issues later, although the impacts are often geographically displaced. Nevertheless, the pursuit of energy security by Chinese cities through increasing numbers of coal-fired power stations is a poor strategy if it results in catastrophic flooding or drought in the future, in addition to the short-term air pollution causing dire local impacts on human health.

Even considering the short history of environmental security literature, however, there have been relatively few attempts to explicitly adopt a critical approach to energy security.[14] Although it is an emerging area it appears that many environmental security scholars have preferred to focus on 'softer' issues such as food and water that have traditionally been more closely aligned with human security within the development studies discipline rather than International Relations (IR) or security studies. Authors such as Deudney[15] and Levy[16] have challenged the usefulness of securitizing the environment itself, but energy has always been considered a security issue, even if as a 'non-traditional' issue compared with military security. The dominance of state-centred approaches to energy security may have shifted interested critical scholars towards alternative analyses, including the use of 'energy poverty',[17] a relatively new concept primarily within the development studies discipline that considers the energy access of households. Although this concept is gaining broader acceptance, it is not necessarily possible to substitute it for energy insecurity, even from a critical perspective. Indeed, one of the characteristics of critical approaches to environmental security in general and energy security in particular, is that they are predominantly qualitative in nature whereas energy poverty, by its very nature, is distinctly quantitative.

Outside of security studies and IR there has been another stream of academic work focusing on energy issues that could be grouped together under the 'green politics' banner. Much of this work has advocated greater reliance on renewable energies, even before the emergence of the climate change discourse, and focused on questions of energy efficiency, energy sources and technologies, personal energy usage and lifestyle choice. Growing out of the environment movements of the 1960s and 1970s, this scholarship has examined local, national and global energy issues but has focused primarily on issues as they affect the North and has largely avoided using security terminology. Some of the green politics writings are linked to a body of work that examines energy security by focusing on the type of energy sources available and the extraction and employment technologies associated with them. The energy sources most commonly examined are fossil fuels, and oil takes precedence as the most crucial for industrial economies.

Climate change, however, is increasingly considered an existential threat to modern societies and contemporary analyses often include the relationship between pursuing energy security through fossil fuel combustion and the climate risks associated with increased greenhouse gas emissions.[18] Climate change from this perspective should be seen as primarily an energy systems problem rather than as an emissions problem.[19] The climate change discourse has seen an unlikely alliance between the nuclear industry and some prominent environmentalists – such as George Monbiot – who argue that the cataclysmic ramifications of climate change result in an argument for the extension of nuclear power to replace fossil fuels as a means of ensuring future energy security.[20] Most environmentalists, however, continue to argue that the rapid expansion of renewable energy sources such as wind and solar power, with perhaps natural gas as a transitional fuel, together with efficiency and conservation measures would be enough to meet the energy needs of the future.

There are many books and book chapters that examine energy security from a state-centric perspective and although this approach is not the focus of this chapter, the following section provides an overview of some of the state-centric literature, which is dominated by a focus on fossil fuels, to place in context the energy security issues that deprived communities face. The rest of the chapter provides an outline of an alternative model of energy security, beginning with an analysis of various energy sources and the intrinsic bias of the technologies that are employed to extract and use them. The final three sections examine different theoretical approaches for addressing energy security that are broadly critical in nature: a critical security studies approach; an energy poverty approach; and a green politics approach. All these approaches have merit and analysing them together provides an opportunity to reconsider the dominant state-centric approach to energy security.

The state-centred national energy security perspective

Historically, the state as the referent object has been most clearly associated with realist or mercantilist schools of thought, but a national energy security perspective also predominates within most liberal approaches. Traditionally, there also has been a focus on energy security in the US and Europe, but this has now shifted to Asia, and more specifically China and India.[21] Former US Secretary of State Henry Kissinger, an archetypal realist, argued that 'aside from military defense, there is no project of more central importance to national security and indeed independence as a sovereign nation than energy security'.[22] Moran and Russell agree, arguing that '"[e]nergy security" is now deemed so central to "national security" that threats to the former are liable to be reflexively interpreted as threats to the latter'.[23] As a result, they contend that energy security is the 'one area of international life' where large-scale conflict among developed states is possible. As can also be surmised from national news headlines and much extant energy security analysis, this state-centred energy security approach also results in a focus on fossil fuels, and primarily oil.

Michael Klare is a prominent realist whose work on energy security draws on his broader research, and that of Michael Ross, on the resource curse.[24] Klare argues that diminishing energy resources, and particularly oil, are likely to cause severe and enduring threats to energy security, and therefore national security from both regional competition for the control of energy resources and great power rivalry and conflict. He notes that from a US perspective the Carter Doctrine has dictated that any threat to the free flow of oil in the Persian Gulf is regarded as an assault on the vital interests of the US and would be repelled 'with any means necessary', including military force. The most obvious example of this policy was the first Gulf War in 1990 to oust the Iraqi army from Kuwait. He also notes that the US is placing greater emphasis on its 'regionally deployed forces' to provide protection for oil supply sea lanes, which may stimulate conflict particularly in the increasingly contested East Asian waters.

Stokes and Raphael similarly argue that US policy since the Second World War has been largely focused on ensuring energy security for itself and its allies through the promotion of stability and the free flow of oil through the Persian Gulf, regardless of the nature of the regimes in power.[25] They argue that counter insurgency aid and other foreign policies have supported oil-rich non-democratic regimes, protecting them from domestic uprisings or external threats in exchange for US energy security. From this perspective the Iraq invasion of 2003 was clearly about oil and energy security with the Iraqi nationalized oil industry restructured and essentially privatized through Production Sharing Agreements between foreign corporations and the Iraqi State.

A relatively liberal and more optimistic approach to energy security, although one still focused on state-driven industrial development, is adopted by Daniel Yergin, an oil industry insider whose 'reasoned confidence' is based on what he terms the 'globalization of innovation'. In 1991 – just as the first Gulf War erupted – Yergin's book, *The Prize*,[26] was published and became what is probably the most influential popular publication on the oil industry and its effect on global politics, winning him the Pulitzer Prize. As an encyclopaedic and historical analysis of oil politics it is not surprising that the focus of the book is on the 'epic quest' of states to secure their oil supplies – although looking back on it now it seems remarkable that there was no mention of China as a potentially significant consumer – but it also demonstrated the narrow state-centric view of energy security that permeated both academia and policy circles at the time.

Twenty years after its initial release Yergin released its successor, *The Quest*,[27] that documented the influence of oil in the intervening period – this time including three chapters on China – and also examined the prospects for renewable and alternative energies. Although noting the possibilities for interstate conflict, Yergin argues that mutual interest in energy security is more likely to predominate, with the promotion of privatized energy markets the priority in Iraq rather than 'a mercantilist 1920s style ambition to control Iraqi oil'. Fettweis makes similar arguments when he suggests that states are increasingly concluding that the costs of going to war over oil far outweigh the benefits.[28] Nevertheless, as can be seen from the earlier views regarding the invasion of Iraq, his argument that 'there has never actually been a war over fossil fuels' is highly contentious.

Liberal-influenced approaches extend, however, from this rather narrow view of economically rational decision making to include a greater emphasis on multilateralism that, while acknowledging the importance of states in driving global change, pursues the development of global institutions that transcend national interests as the means to achieving global energy security. As Lesage *et al.* note, 'as long as energy security is viewed through national prisms, the concept is impracticable as a driving force for global collective action'.[29] Müller-Kraenner concedes, in a realist tone, that '[t]he gloves are off in the battle for the last resources; securing national energy supply has become the tough realpolitik for every country',[30] but he then argues that a new world environment organization and a global agency for the promotion of renewables must be established as the only feasible route to energy security. Realists argue that such liberal projects have little chance of success but, regardless of the various institutional structures employed, the choice of particular energy sources and technologies can themselves influence the structure of society and constrain the possible outcomes from the pursuit of energy security.

Energy sources and technologies

As the previous section suggests, most energy security analysis focuses particularly on fossil fuels; primarily oil, which dominates the military and transport sectors, but also natural gas and coal. This is closely followed by nuclear energy, which is absent from some states, such as Australia, while others are highly dependent on it; France draws over three-quarters of its electricity from nuclear plants. Increasingly, renewable energies such as wind and solar are gaining importance in the global energy discourse due to the impacts of climate change and the security threats surrounding traditional fuel sources. Critical analysis of all these energy technologies, however, suggests that they underpin particular modes of thought in relation to the structure of society. As Kranzberg emphasized, technology is not neutral[31] and it often 'reflects and reinforces existing power relations'.[32] Any critical analysis of energy security, therefore, should include some discussion of the intrinsic bias attached to given energy technologies.[33]

Fossil fuel and nuclear energy technologies all favour large-scale industrial development and have centralizing political and economic consequences. Nuclear technology in particular requires a centralization of political and economic power that frequently results in a lack of democratic oversight and transparency in relation to the corporations that operate nuclear power stations. The close relationship that exists between the operators and the regulators, sometimes considered an example of regulatory capture, was evident in the wake of the triple meltdown at the Fukushima Daiichi nuclear plant following an earthquake and tsunami in March 2011. The operator, Tokyo Electric Power Company (TEPCO), consistently failed to maintain adequate safety measures for many years and the regulator, the Nuclear and Industrial Safety Agency (NISA), failed to act on these omissions.[34]

The concerns over these regulation issues, together with concerns over the safety of nuclear power plants, the proliferation of nuclear weapons and the storage of nuclear waste, resulted in a general downswing in global support for nuclear energy following the Chernobyl nuclear accident in 1986. In the first decade of this century, however, with global energy demand increasing dramatically and climate change becoming a significant political issue, nuclear energy became much more attractive to governments,[35] and a global 'nuclear renaissance' was predicted. Although China will continue building nuclear power stations to complement its other sources, the predicted 'renaissance' in the US and Europe was unlikely, even before the Fukushima accident. As Duffy argues, nuclear power has never been economic without significant subsidies and attempts to deal with climate change could be addressed by investments in other technologies, 'which pose far fewer problems and which yield quicker returns'.[36] Despite the emerging support of some environmentalists for nuclear power to supplement renewable technologies in replacing fossil fuels, the evolving international attitude to nuclear energy was more accurately captured in the title of a special report in the *Economist* a year after the Fukushima melt down, 'The Dream that Failed'.[37]

Although there are more immediate security concerns over nuclear technologies than fossil fuels technologies, due to their associated processes and products, the large-scale exploitation of fossil fuels ushered in the industrial revolution and a new scale of environmental destruction that was accompanied by the growth of energy companies on a similar scale; in early 2012 three of the four largest corporations in the world by market capitalization (along with electronics giant Apple) were energy 'super majors'. Most modern electricity production, from fossil fuels, nuclear technology or large-scale hydroelectric dams, favours large oil and gas, nuclear and construction companies. As I have argued elsewhere: 'the discourse of energy security is still employed by government and business elites to justify top-down investments in large-scale energy projects, which require significant initial capital injections and subsequent industrial-scale capital returns.'[38]

These projects provide governments and businesses with significant opportunities for profit and corruption and enhanced centralization of political and economic control. In assessing this route to energy security it should be noted that the criteria that are used to measure the success of particular technologies are socially constructed and are set by a social subgroup, usually comprising the engineers and technocrats that benefit from adherence to the technology.[39] The governments and corporations who employ the technocrats, who broadly design and assess the energy technologies, generally favour the resultant increased centralization of political and economic control.

All these large-scale centralizing technologies, however, have human and environmental costs that are rarely considered when calculating the cost of the energy produced, which can result in favourable comparisons with renewable and decentralized forms of electricity generation such as wind and solar power.[40] Even without consideration of these sorts of externalities, British economist Nicholas Stern argued in his review on the *Economics of Climate Change* that established technologies will have a price advantage over newer technologies and making widespread renewable technologies economically competitive requires targeted government support.[41] For the first time since the industrial revolution carbon pricing systems in Europe and Australia now add small amounts to the costs of burning fossil fuels in recognition of the impacts on climate change, but this does not take into account all the carbon historically emitted and the costs are kept exceedingly low due to political constraints. If provided with similar subsidies to earlier technologies the flexibility of renewable technologies such as wind and solar power generation, even if they are linked into national electricity grids, provide opportunities for the decentralization of both electrical and political power, reducing dependence on central governments and large corporations.[42] While natural gas is a fossil fuel it emits less greenhouse gases than other fossil fuels and is often seen as a 'transitional' fuel from coal and oil to a renewable energy future, particularly in combined heat and power (CHP) generation plants.[43] In contrast to nuclear in particular, it is amenable to small-scale power plants and can therefore promote a shift to decentralized energy production, as demonstrated in Azerbaijan,[44] although the pipelines that are often used to transport the gas can cause social and environmental insecurities.[45]

A de facto decentralized energy production system is already well established in the South because few states have extensive electricity grids that service all communities. While local biomass fuels are used extensively by households in poor communities across the South, in some areas, particularly those with low levels of forestation such as Ladakh in India on the Tibetan plateau, NGOs have promoted passive micro-solar technologies that have transformed living conditions and livelihoods for remote households and communities.[46] In both the North and the South various biomass energy systems are increasingly seen as an attractive replacement for fossil fuels.[47]

These biofuels, however, raise a crucial justice dimension to energy security and should only be employed if they do not use land or crops that would otherwise be utilized for food. If global markets are used to determine the allocation of resources then as oil returns to its 2008 peak of close to US$150 per barrel, farmers in the South are more likely to substitute biofuels for existing food crops, resulting in increased food insecurity. In addition, the growing of biofuels such as palm oil has caused widespread deforestation across Southeast Asia and Brazil, exacerbating climate change, causing flooding and increasing air pollution. From a critical perspective energy security should not be pursued at the expense of other aspects of environmental security, and food insecurity in the South is only likely to increase as populations grow and arable land declines. The issues of justice and sustainability are of central importance to critical scholarship and these concerns are further explored in the remaining sections by briefly analysing three broadly critical approaches to energy security.

Critical security studies

There is a broad range of potentially critical approaches to energy security, but as with many theoretical perspectives, its borders are fuzzy and ill defined. Shane Mulligan has written recently on a 'critical' approach to energy security, but he does not address issues specific to the South, where energy shortages are so prevalent and there is a focus in this work on fossil fuels, and peak oil in particular. Although there are critical aspects to Mulligan's analysis, in general these are not primarily critical concerns; as Simon Dalby notes, 'oil is not a resource that the marginalized peasantry of the Third World are directly fighting over; it's a matter of superpower competition'.[48] In one article Mulligan examines the history of connections between environmental security and energy security particularly in relation to fossil fuels. He argues that although with the rise of climate change in particular,

> fossil fuels are now decidedly a principal threat to, rather than a component of, 'the environment' . . . [i]nternational agreements addressing other environmental resources may offer . . . models for dealing with a future of declining energy and climate security, without resort to 'traditional' mechanisms of conflict among the world's states. But such a shift in the *practice* of energy security necessitates a shift in the *concept* of security that, instead of emphasizing state-centered and military aspects, is grounded in discourses of global and human security.[49]

In another article Mulligan examines the security literature more closely including the Copenhagen School's concept of the 'securitization' of issues, which may result in authoritarian responses by the state: 'Securitization is thus a tool that enables states to take exceptional measures, including repression or the suspension of the public freedoms considered normal in the West.'[50]

While the Copenhagen School is clearly constructivist, and has broadened the security agenda away from the purely militaristic national security approaches, it is not readily identifiable as 'critical' and Mulligan contrasts the implications of securitization for the Copenhagen School with the concept of 'emancipation' employed by Ken Booth and others in the Critical Security Studies (CSS) of the Aberystwyth School. Rita Floyd also examines the Aberystwyth School approach, surmising that 'true security refers to the emancipation of the poor and disadvantaged from the way they live today'.[51]

In applying this critical perspective to energy security the referent object shifts from the state to the individual and individual energy security becomes the goal, provided it is not at the expense of other aspects of environmental security or the environmental security of other members of the political community. As Matt McDonald notes, however, even with the referent object as the individual, communities still remain important: first, although the state can sometimes create insecurities it can also perform the role of a progressive security agent or provider; and second, communities can function as important sites for group identification and the pursuit of justice.[52] In the latter case, this identification of individuals with particular communities, particularly if they are marginalized, allows a critical approach to energy security to be expanded to include a marginalized cultural or political community as a broader referent object, provided that community does not replicate the inequalities and injustices that characterize the wider society.

In my own analysis of complex energy relationships in Southeast Asia I combine a broad critical international political economy approach with elements of critical security studies to ensure that energy security is a concept particularly focused on justice for marginalized individuals and communities in the South. Drawing also on Barnett's approach to environmental security I have examined the impacts of large-scale transnational energy projects in regions with marginalized and oppressed communities, particularly in Myanmar.[53] These communities are faced with severe energy deficits, but they often see no relief when an energy project is completed. While the discourse of energy security is used to justify the project, communities living in its vicinity may remain without electricity following its completion and have other elements of their security, such as food, water or livelihood, undermined. In this situation it becomes pertinent to ask what is actually being secured by the project. Unfortunately, despite government protestations to the contrary, it is often the financial security of governing and business elites that determines project decision making at the expense of local communities' environmental security. In addition, due to the transnational nature of these energy projects the professed pursuit of energy security in one relatively affluent state may cause environmental and energy insecurities in another.[54] This critical security studies focus on the most marginalized communities has some similarities with an emerging 'energy poverty' concept.

Energy poverty

From a security studies perspective Ciuta argues that, 'the story of energy security is relatively straightforward . . . energy insecurity is the product of the contradiction between a general trend of increasing energy consumption and a contradictory trend of decreasing energy reserves.'[55] This definition, however, assumes national or global concerns and a focus on the relatively affluent. In the case of many communities in the South 'increasing energy consumption' in their communities is not part of the problem at all. A critical approach to energy security could therefore link energy insecurity to 'energy poverty', which avoids the state-centric assumption through a direct focus on households. Although a fringe energy issue compared with national energy security for policy makers the energy poverty concept is gradually gaining acceptance in national and supra-national fora including at the EU level where, despite its overall wealth, member states such as Bulgaria face considerable problems at the 'energy affordability-social inequality nexus'.[56]

Energy poverty can be defined as a condition wherein a household is unable to access energy services at the home at a materially necessitated level and usually includes a lack of regular access to electricity and a dependence on the traditional use of biomass for cooking and heating. Almost one-third of the world's population uses this sort of biomass for cooking and energy access over the next two decades under business-as-usual assumptions suggests that the absolute amount of people in energy poverty is likely to remain static.[57]

Energy poverty as a concept appears more closely related to development studies than security studies or IR, but viewing environmental security from a critical perspective results in a more human-centred security approach that could clearly encompass energy poverty. As might be expected from its development studies lineage, however, energy poverty tends to be far more quantitative than most critical theoretical analysis; it provides specific daily energy consumption quantities, depending on social and environmental requirements, while critical analysis tends to focus more on challenging social and political structures. Nevertheless, energy poverty can be useful in highlighting areas where critical analysis might be most useful. For instance a study by Pachauri *et al.* estimated that at the turn of the century 30 per cent of the population of India was 'energy poor' – lacking access to modern energy sources and consuming amounts that only provide minimum services – but another 15 per cent was 'very energy poor', consuming less than enough to cook two meals per day and basic lighting.[58] Similar entrenched energy poverty is seen in Myanmar, which has an electrification rate of 13 per cent compared with its neighbour, Thailand, with 99.3 per cent.[59] In these cases energy poverty data and analysis can be employed by critical analysts to address entrenched social and political inequalities.

A green politics approach to energy security

The last approach briefly examined here, and one that has similarities with other critical approaches outlined above, is a 'green politics' approach to energy security. Jon Barnett has comprehensively elucidated a green politics approach to environmental security that, although not focusing specifically on energy security, critiques the assumptions that underpinned the post-Second World War notion that 'energy security is [simply] the theory and practice of securing energy for the nation-state'.[60] The four core pillars of green politics are social justice, nonviolence, participatory democracy and ecological sustainability,[61] and it is clear that the pursuit of a national energy security that encompasses large-scale industrial consumption of fossil fuels – which has resulted in wars fought over resources and the development of a military-industrial complex in many states – undermines these pillars, particularly in the South.

Although there has been limited academic writing on energy from a security perspective in the green literature, partially because of the militaristic and statist connotations that it traditionally evokes, there has been much written and done in the green and environmental movements more broadly that attempts to address issues of energy demand and supply while adhering to the four green pillars. Much of the recent writing in this area is couched in terms of mitigating climate change,[62] but there is also a vibrant global activist community that promotes energy efficiency and renewable energy sources as part of a broader shift towards an ecologically conscious lifestyle born out of the New Age and alternative community movements of the 1970s.[63]

While some environmentalists pursue ecologically friendly energy efficiencies and sources within their own private lives, the Centre for Alternative Technology (CAT) in Wales epitomizes the attempt to build a more cooperative community based on eco-friendly principles. Founded in 1973 in a disused slate quarry, CAT opened a visitor centre in 1975 to demonstrate its alternative technologies. It now receives 65,000 visitors every year and offers two onsite Master degrees on energy as well as a Doctor of Ecological Building Practices through the University of East London.[64] Much of this green alternative energy movement is influenced by the writings of E. F. Schumacher who promoted human-scale technologies instead of the large-scale industrial technologies identified above.

> Experience shows that whenever you can achieve smallness, simplicity, capital cheapness and nonviolence, or, indeed, any one of these objectives, new possibilities are created for people, singly or collectively, to help themselves, and that the patterns that result from such technologies are more humane, more ecological, less dependent on fossil fuels and closer to real human needs than the patterns (or lifestyles) created by technologies that go for giantism, complexity, capital intensity and violence.[65]

In his most influential book, *Small is Beautiful*, Schumacher argued for a Buddhist approach to economics that took into consideration the fundamental difference between renewable and non-renewable energy sources with non-renewables to be used only if they are indispensable and with 'meticulous care for conservation'.[66] While the drivers of the contemporary societal shift towards renewable energies and energy efficiencies are not primarily the alternative lifestyles, but more hard-headed national concerns such as economic efficiency, national economic security and global climate change, there is little doubt that the technologies and approaches that have facilitated this shift have emerged from within the environmental movements of the last four decades. As noted above, however, technologies are created by particular philosophies and while these philosophies usually infuse the society that develops the technology it is possible that technologies can change the society. Despite the current unsustainability of modern industrialized societies, by adopting these technologies they are gradually shifting to a potentially more sustainable and ecologically sound approach to energy security.

Conclusion

Energy security can be defined and interpreted in many ways but due to the economic and military importance of energy, and particularly oil, to the modern nation-state it has, until recently, been considered an issue about, and of concern to, primarily states. Unlike other aspects of environmental security, this lineage has ensured that energy security has been discussed widely within broader politics and IR discourses, particularly since the 1970s. This chapter has demonstrated, however, that a new and distinct approach to energy security based on broadly critical principles has emerged within the wider environmental security literature.

Although environmental security encompasses a wide variety of perspectives, these critical approaches to energy security that prioritize justice, equity and sustainability have provided an important antidote to the traditional definitions of energy security that are associated with militarism, wars and unsustainable, unnecessary and inappropriate levels of industrial development. In addressing the three fundamental questions relating to security – 'for whom', 'from what threats' and 'by what means' – this critical perspective shifts the referent object to the individual and highlights injustices in existing social and political structures that institutionalize inequality. The answer to the first question therefore focuses on the individual, the answer to the second is the threat to adequate energy supplies due to unjust social and political structures, and the answer to the third is the political and economic reforms that eliminate these structural inequalities.

Inevitably, the most critical energy security analysis focuses on the South, where insecurities and injustices are most prevalent, and the concept of energy poverty may be increasingly employed to supplement this sort of energy security analysis. In alleviating energy insecurity, however, the nature of energy technologies and sources must be assessed to ensure that the pursuit of energy security is not at the expense of other insecurities, such as food and water, and does not precipitate the displacement of environmental insecurities to other communities. While this approach to energy security is still in its infancy its importance is only likely to increase as awareness grows that the extraction and use of traditional sources of energy at current rates is socially, economically and ecologically unsustainable.

Notes

1   F. Ciuta, 'Conceptual Notes on Energy Security: Total or Banal Security?' *Security Dialogue* 41 (2), 2010, 123–44, at 123–24.

2   See R. Dannreuther, *International Security: The Contemporary Agenda*, Cambridge: Polity Press, 2007, Chapter 5.

3   J. Elkind, 'Energy Security: Call for a Broader Agenda', in C. Pascual and J. Elkind (eds), *Energy Security: Economics, Politics, Strategies, and Implications*, Washington, DC: Brookings Institution Press, 2008, pp. 119–48.

4   B. K. Sovacool, 'Introduction: Defining, Measuring, and Exploring Energy Security', in B.K. Sovacool (ed.), *The Routledge Handbook of Energy Security*, London and New York: Routledge, 2011, pp. 3–6.

5   D. Stokes and S. Raphael, *Global Energy Security and American Hegemony*, Washington, DC: Johns Hopkins University Press, 2010.

6   A. Dupont, *East Asia Imperilled: Transnational Challenges to Security*, Cambridge: Cambridge University Press, 2001.

7   T. Homer-Dixon, *The Upside of Down: Catastrophe, Creativity, and the Renewal of Civilization*, Washington, DC: Island Press, 2007.

8   P. Calvert and S. Calvert, *The South, the North and the Environment*, London and New York: Pinter, 1999, p. 24; C. M. Shackleton *et al.*, 'Urban Fuelwood Demand and Markets in a Small Town in South Africa: Livelihood Vulnerability and Alien Plant Control', *International Journal of Sustainable Development and World Ecology* 13 (6), 2006, 481–91, at 481.

9   K. Gunasekera and A. Najam, 'Energy and Security in South Asia: A Multiple Frameworks Analysis', in A. Najam (ed.), *Environment, Development and Human Security: Perspectives from South Asia*, Lanham, MD: United Press of America, 2003, p. 172.

10   International Energy Agency, *World Energy Outlook 2010: The Electricity Access Database*. Online: www.iea.org/weo/database_electricity10/electricity_database_web_2010.htm (accessed August 2012).

11   T. Doyle, 'Fire and Firepower: Energy Security in the Indian Ocean Region', in T. Doyle and M. Risely (eds), *Crucible for Survival: Environmental Security and Justice in the Indian Ocean Region*, New Brunswick, NJ and London: Rutgers University Press, 2008, p. 190.

12   J. Barnett, *The Meaning of Environmental Security: Ecological Politics and Policy in the New Security Era*, New York: Zed Books, 2001, p. 127.

13   K. Booth, *Theory of World Security*, Cambridge: Cambridge University Press, 2007, p.142.

14   See, for example, S. Mulligan, 'Energy and Human Ecology: A Critical Security Approach', *Environmental Politics* 20 (5), 2011, 633–49; A. Simpson, 'The Environment–Energy Security Nexus: Critical Analysis of an Energy "Love Triangle" in Southeast Asia', *Third World Quarterly* 28, (3), 2007, 539–54.

15   D. Deudney, 'The Case against Linking Environmental Degradation and National Security', *Millennium* 19 (3), 1990, 461–76.

16   M. A. Levy, 'Is the Environment a National Security Issue?', *International Security* 20, (2), 1995, 35–62.

17   M. Bazilian *et al.*, 'More Heat and Light', *Energy Policy* 38, (10), 2010, 5409–12.

18   L. Anceschi and J. Symons (eds), *Energy Security in the Era of Climate Change: The Asia-Pacific Experience*, New York: Palgrave Macmillan, 2011.

19   K. Shaw, 'Climate Deadlocks: The Environmental Politics of Energy Systems', *Environmental Politics* 20 (5), 2011, 743–63.

20   G. Monbiot, 'Why I Am Urging David Cameron to Act against Friends of the Earth', *The Guardian*. 15 March 2012. Online: www.guardian.co.uk/environment/georgemonbiot/2012/mar/15/david-cameron-friends-of-the-earth (accessed August 2012).

21   See, for example, B. D. Cole, *Sea Lanes and Pipelines: Energy Security in Asia*, Westport, CT: Praeger Security International, 2008; M. Li, 'Peak Oil, the Rise of China and India, and the Global Energy Crisis', *Journal of Contemporary Asia* 37, (4), 2007, 449–71; G. Luft and A. Korin (eds), *Energy Security Challenges for the 21st Century*, Santa Barbara, CA: Praeger Security International, 2009; M. Wesley (ed.), *Energy Security in Asia*, New York: Routledge, 2007; D. Yergin, *The Prize: The Epic Quest for Oil, Money, and Power*, New York: Pocket Books, 1993; D. Yergin, *The Quest: Energy, Security, and the Remaking of the Modern World*, New York: Penguin, 2011.

22   A. N. Stulberg, *Well-Oiled Diplomacy: Strategic Manipulation and Russia's Energy Statecraft in Eurasia*, Albany: State University of New York Press, 2007, p. 3.

23   D. Moran and J. A. Russell, 'Introduction: The Militarization of Energy Security', in D. Moran and J.A. Russell (eds), *Energy Security and Global Politics: The Militarization of Resource Management*, London and New York: Routledge, 2009, p. 2.

24   M. T. Klare, *The Race for What's Left: The Global Scramble for the World's Last Resources*, New York: Metropolitan Books, 2012; M. L. Ross, *The Oil Curse: How Petroleum Wealth Shapes the Development of Nations*, Princeton, NJ: Princeton University Press, 2012.

25   Stokes and Raphael, *Global Energy Security and American Hegemony*.

26   Yergin, *The Prize: The Epic Quest for Oil, Money, and Power*.

27   Yergin, *The Quest: Energy, Security, and the Remaking of the Modern World*.

28   C. J. Fettweis, 'No Blood for Oil: Why Resource Wars Are Obsolete', in G. Luft and A. Korin (eds), *Energy Security Challenges for the 21st Century*, Santa Barbara, CA: Praeger Security International, 2009, pp. 66–77.

29   D. Lesage, T. Van de Graaf, and K. Westphal, *Global Energy Governance in a Multipolar World*, Farnham: Ashgate, 2010, p. 38.

30   S. Müller-Kraenner, *Energy Security: Re-Measuring the World*, London: Earthscan, 2008, p. xi.

31   M. Kranzberg, 'Technology and History: "Kranzberg's Laws"', *Technology and Culture* 27, (3), 1986, 544–60.

32   G. Curran, *21st Century Dissent: Anarchism, Anti-Globalization and Environmentalism*, Basingstoke, UK: Palgrave Macmillan, 2006, p. 75.

33   J. Mander, 'Technologies of Globalization', in J. Mander and E. Goldsmith (eds), *The Case against the Global Economy and for a Turn toward the Local*, San Francisco: Sierra Club Books, 1996, pp. 347–48.

34   D. Kaufmann and P. Penciakova, *Japan's Triple Disaster: Governance and the Earthquake, Tsunami and Nuclear Crises*, Washington DC: Brookings Institution, 2012. Online: www.brookings.edu/opinions/2011/0316_japan_disaster_kaufmann.aspx. (accessed August 2012).

35   C. Mitchell, *The Political Economy of Sustainable Energy*, Basingstoke, UK: Palgrave Macmillan, 2008, Chapter 4.

36   R. Duffy, 'Déjà Vu All over Again: Climate Change and the Prospects for a Nuclear Power Renaissance', *Environmental Politics* 20, (5), 2011, 668–86, at 683.

37   *The Economist*, 'Nuclear power: The dream that failed', 2012. Online: www.economist.com/node/21549936 (accessed August 2012).

38   Simpson, 'The Environment–Energy Security Nexus', p. 543.

39   R. Kline and T. Pinch, 'The Social Construction of Technology', in D. MacKenzie and J. Wajcman (eds), *The Social Shaping of Technology*, Buckingham, UK: Open University Press, 1999, p. 114.

40   R. Douthwaite, *Short Circuit: Strengthening Local Economies for Security in an Unstable World*, Totnes, UK: Green Books, 1996, p. 184; D. Elliott, 'A Sustainable Future for Energy?', in D. Elliott (ed.), *Sustainable Energy: Opportunities and Limitations*, Basingstoke, UK: Palgrave Macmillan, 2010.

41   N. Stern, *The Economics of Climate Change: The Stern Review*, Cambridge: Cambridge University Press, 2007.

42   Mitchell, *The Political Economy of Sustainable Energy*, Chapter 5.

43   Elliott, 'A Sustainable Future for Energy?'

44   D. M. Sweet, 'The Decentralized Energy Paradigm', in G. Luft and A. Korin (eds), *Energy Security Challenges for the 21st Century*, Santa Barbara, CA: Praeger Security International, 2009, pp. 308–17.

45 A. Simpson, 'Gas Pipelines and Security in South and Southeast Asia: A Critical Perspective', in T. Doyle and M. Risely (eds), *Crucible for Survival: Environmental Security and Justice in the Indian Ocean Region*, New Brunswick, NJ and London: Rutgers University Press, 2008.

46 LEDeG, *Ladakh Ecological Development Group (Ledeg)*. Online: http://ledeg.org/ (accessed August 2012).

47 J. Scurlock, 'Biomass: Greening the Transport Sector', in D. Elliott (ed.), *Sustainable Energy: Opportunities and Limitations*, Basingstoke: Palgrave Macmillan, 2010, pp. 49–65.

48 S. Dalby, *Security and Environmental Change*, Cambridge: Polity Press, 2009, p. 75.

49 S. Mulligan, 'Energy, Environment, and Security: Critical Links in a Post-Peak World', *Global Environmental Politics* 10, (4), 2010, 79–100, at 85.

50 Mulligan, 'Energy and Human Ecology', p. 639.

51 R. Floyd, *Security and the Environment: Securitisation Theory and US Environmental Security Policy*, Cambridge: Cambridge University Press, 2010, p. 48.

52 M. McDonald, *Security, the Environment and Emancipation: Contestation over Environmental Change*, London and New York: Routledge, 2012, pp. 52–53.

53 A. Simpson, *Energy, Governance and Security in Thailand and Myanmar (Burma): A Critical Approach to Environmental Politics in the South*, Farnham, UK: Ashgate, 2013.

54 Simpson, 'The Environment–Energy Security Nexus', p. 540.

55 Ciuta, 'Conceptual Notes on Energy Security', p. 126.

56 S. Bouzarovski, S. Petrova and R. Sarlamanov, 'Energy Poverty Policies in the EU: A Critical Perspective', *Energy Policy* 40 (1) 2012, 480–89.

57 Bazilian *et al.*, 'More Heat and Light'.

58 S. Pachauri *et al.*, 'On Measuring Energy Poverty in Indian Households', *World Development* 32, (12), 2004, 2083–104.

59 International Energy Agency, *World Energy Outlook 2010: The Electricity Access Database*.

60 Barnett, *The Meaning of Environmental Security*, p. 34.

61 N. Carter, *The Politics of the Environment: Ideas, Activism, Policy*, 2nd ed., Cambridge: Cambridge University Press, 2007 p. 47.

62 See, for example, M. Diesendorf, *Climate Action: A Campaign Manual for Greenhouse Solutions*, Sydney: University of New South Wales Press, 2009.

63 See, for example, Douthwaite, *Short Circuit*; D. Pepper, *Communes and the Green Vision: Counterculture, Lifestyle and the New Age*, London: Green Print, 1991.

64 Centre for Alternative Technology (CAT): www.cat.org.uk/ (accessed August 2012).

65 E. F. Schumacher, *Good Work*, London: Jonathan Cape, 1979, p. 57.

66 E. F. Schumacher, *Small Is Beautiful: A Study of Economics as If People Mattered*, London: Blond and Briggs, 1973, p. 55.

15

CLIMATE CHANGE AND SECURITY

Richard A. Matthew

In 1896, the Swedish physicist and chemist Svante Arrhenius speculated that carbon dioxide emissions could lead to global warming through a greenhouse gas effect. In 2007 – 111 years later – the Intergovernmental Panel on Climate Change (IPCC), a global network of climate scientists set up by the United Nations, published its fourth report assessing research on this hypothesis, and concluded that Arrhenius was correct. Human activities, and especially anthropogenic carbon dioxide emissions, are forcing global climate change. The clarity and urgent tone of this report triggered a wave of speculation and research on the empirical and possible security implications of climate change. In this chapter, I briefly review the principal findings of climate science, and then discuss three important and emerging social science responses to this: (1) the argument that climate change poses a threat to national security, (2) the argument that climate change poses a threat to human security, and (3) the argument that climate change needs to be addressed explicitly in the context of peace and peacebuilding.[1]

Climate change science

Climate change refers to transformations in the statistical characteristics of weather over time. These transformations have, in fact, been a regular feature of the planet since it aggregated from space debris some four and a half billion years ago. Indeed, placed in the context of earth history, the observations of the past century are far less dramatic than some of the climate transformations that took place in the distant, pre-anthropocene past. But today's changes are significant, and they are unprecedented in that they are being forced by human activities that increase the concentration of carbon dioxide and other greenhouse gases in the atmosphere and that reduce carbon storage on the earth's surface through deforestation and other types of land modification. Scientists are in agreement that since the mid nineteenth century, coincident with the rise of the industrial age, the planet's average temperature has increased by one degree centigrade, and that it is because of this that weather patterns typical of the last several millennia are changing.[2]

This agreement has taken a long time to forge, with much of the research that supports it being quite recent.[3] The gathering of hard weather data, however, began in the 1600s, when the thermometer was invented, and was first formalized in global protocols and standards through the creation of the International Meteorological Organization in 1873.

By the 1930s, several scientists, such as G. S. Callendar, began compiling time series data on temperature, and speculating on the impacts of changes in atmospheric carbon dioxide levels. Today, it is "The high-accuracy measurements of atmospheric CO_2 concentration, initiated by Charles David Keeling in 1958, [that] constitute the master time series documenting the changing composition of the atmosphere."[4] According to Le Treut *et al.*:

> climate change science literature grew approximately exponentially with a doubling time of 11 years for the period 1951 to 1997. Furthermore, 95% of all the climate change science literature since 1834 was published after 1951. Because science is cumulative, this represents considerable growth in the knowledge of climate processes and in the complexity of climate research.[5]

The broad conclusions of this research are now well known. Global warming is expected to continue for decades and perhaps centuries. On the whole, this will lead to the dry areas of the planet becoming dryer and the wet areas becoming wetter. Severe weather events will increase in frequency and intensity. Ecological systems and global biodiversity will be affected in multiple ways. However, these patterns will not be uniform across the planet but rather will display considerable variety. Some changes will not be foreseen at all. This is partly because climate change is affected by phenomena that have proven impossible to model, at least so far, such as changes in cloud cover, as well as by unpredictable feedback responses from both natural and human systems. It is partly because complex adaptive systems, such as forests, display non-linear behavior and generate new properties. In short, while much is known about anthropogenic climate change, and broad patterns of future climate change are being predicted, there also are important areas of uncertainty.

Curiously, many people appear to have assigned high value to the areas of uncertainty and on this basis have called into question the broad areas of scientific agreement. For example, in 2010 a Populus poll conducted in the United Kingdom showed that only 26 percent of the public was convinced that climate change was real and caused by human behavior.[6] In both Europe and North America, in addition to concerns raised about the limits of climate change models, arguments have emerged that climate change was invented by liberals and radicals to advance a hidden agenda of wealth redistribution. A handful of scientists contend that climate change is real but is the result of natural phenomena such as solar variation and elliptical migration. While these types of explanation can never be dismissed by science, which evolves through paradigmatic shifts that can bring marginal views to the center very quickly, science-based counterarguments have so far garnered little support from the broad scientific community. Thus humankind is well advised at this point in time to focus attention and resources on adapting to a world characterized by far more severe floods, droughts, heat waves, storms, and other climate phenomena than people have experienced – and built infrastructure to withstand – in the past few millennia. And since human activities are driving these costly changes, acting to mitigate climate change is a prudent course of action.

The next IPCC report is scheduled for release in 2014. Current discussions of the draft texts suggest that the predominant message of the report will be that climate change effects are being widely experienced today, if not always acknowledged as such, and in the near future they will occur on a larger scale and with greater speed than previously anticipated.[7]

Climate change and security

One way of framing climate change impacts on human societies is as security issues. Since the publication of the 2007 IPCC report, known as the *Fourth Assessment Report* or *AR4*, this approach has become widespread in both the academic and practitioner communities. While research and debate in this area have been wide-ranging, three clusters of concern have predominated and will be addressed here: the implications of climate change for national security, human security, and peacebuilding.

It is perhaps worth noting that researchers and practitioners could and at times do choose to frame climate change in other ways, including as a development or ethical issue. The pros and cons of these alternative frames will not be discussed here. The use of one frame does not preclude the use of other frames, although as discussed in chapters 1 and 16, securitizing climate change can mitigate against certain goals, values, and forms of cooperation.

Climate change and national security

The Baron of Brentford, Sir Nicholas Stern, has been a key and highly visible figure in the process of framing climate change as a security issue. After a stint at the World Bank, Stern went to work for the British Treasury, and in late 2006 released his influential and controversial report, *The Economics of Climate Change*. Among its numerous gloomy forecasts was the assertion that as many as 200 million people would be displaced permanently by climate change effects by 2050, leading to social upheaval in many parts of the world.[8] Very quickly, the security implications of rising sea level, massive flooding, and long droughts attracted attention from other analysts around the world. Working for the consulting firm CNA, for example, a group of retired US military brass authored *National Security and the Threat of Climate Change*, predicting a future in which "climate change acts as a threat multiplier for instability in some of the most volatile regions of the world" and contributes "to tensions even in stable regions of the world."[9] At the same time, another influential group in Europe, the German Advisory Council on Global Change, released *World in Transition: Climate Change as a Security Risk*, predicting "Climate change will overstretch many societies' adaptive capacities within the coming decades."[10] International Alert's Dan Smith and Janna Vivekananda identified "46 countries – home to 2.7 billion people – in which the effects of climate change interacting with economic, social, and political problems will create a high risk of violent conflict."[11]

In later work, Vivekananda sought to specify the process through which climate change could generate instability and violent conflict:

> The basic trajectory by which climate change could combine with other variables and increase the risk of instability or violent conflict is determined by the role of governance. The impact of climate change will challenge and reduce the

resilience of people and communities to varying degrees. In some situations, it will cause extreme disruption with which people simply cannot cope, as it overwhelms them and renders their homes and livelihoods unviable. If the governance structures that the community regards as safeguards of their human security are not up to the task, climate change will weaken confidence in the social order and its institutions and damage the glue that holds societies together. In some contexts, this can increase the risk of instability or violence. This is a particular problem in conflict-prone or conflict-affected contexts where governance structures and institutions are often weak, regardless of climate change.[12]

He further noted, as an example, that:

[when] the Koshi River [in Nepal] burst its banks in the summer of 2008, 240 people were killed, crops and infrastructure destroyed and 60,000 people displaced in the worst flooding in five decades. These people were resettled among communities who were themselves struggling to survive. Tensions between the host communities and flood victims quickly escalated and were further fuelled by political groups who used flood victims' unmet expectations for clean water and shelter to feed antigovernment sentiments. The situation became violent and 200 policemen were called in to maintain order in the camps.[13]

These early reports by government officials, consultants, and non-governmental organizations created a platform for policy activity. Hence, Germany stated that during its tenure on the United Nations Security Council (UNSC), it would strive to have climate change acknowledged as a security issue "in the broadest sense."[14] In the US, the 2010 *National Security Strategy* describes climate change as a significant security threat, a position endorsed later that year in the *Quadrennial Defense Review Report*, which suggested that future conflict could be linked to climate change.

In large measure, these reports accept at face value those aspects of the 2007 IPCC reports that underscore the high sensitivity to climate change impacts of parts of South Asia, the Middle East, and sub-Saharan Africa.[15] The basic idea is that these regions are relatively more sensitive than other parts of the world by virtue of both their geography and their more fragile social institutions. In particular, they lack the capacity to dedicate substantial resources to prevention, adaptation, and response, and so they will tend to face greater hardships managing droughts, floods, storms, and heat waves. Weak public health systems, rapid population growth and urbanization rates, widespread poverty, a history of recent violent conflict, and a high dependency on agriculture all serve to deepen the vulnerability of these areas. Looming in their near future, then, are the prospects of rapid and extensive population displacement, economic development setbacks, public health losses, food insecurity, institutional failure, and violent conflict – all triggered or enhanced by climate change.

South Asia provides a particularly good – and alarming – example of the interaction of climate change effects with these types of social vulnerability. The IPCC forecasts that, in keeping with global trends, South Asia will experience significant warming over the next few decades; its dry areas will become drier; its wet areas will become wetter; glacial lake outburst floods (GLOFs) will cause extensive damage in mountain countries such as Nepal and Bhutan; the monsoon will change in terms of timing, location, and intensity; and severe weather events will increase. In some regards, these changes will be more dramatic than elsewhere – for example, warming could average 3.3 degrees centigrade, which would be higher than the global average. South Asian countries, poor and conflict-prone, may not be able to rally adequate responses to these stresses, opening the door to security problems.

While attentive to the IPCC reports, these reports and predictions have tended to neglect academic research in this area. In this arena, claims about the ways in which climate change will affect national security have been more cautious, building on arguments and counter-arguments developed over the past twenty years about how environmental change, such as natural resource scarcity, is related to violent civil war, state failure, and other forms of insecurity and violence.[16] Elsewhere I have organized the ways in which climate change could affect national security into three areas: effects that (1) weaken the elements of national power; (2) contribute to state failure; or (3) lead to, support or amplify violent conflict. I will summarize that argument here, addressing each area in turn.[17]

Climate change impacts on national power

National power is the sum of many variables.[18] From the classical writings of Thucydides, to the contemporary work of Richard Armitage and Joseph Nye, these variables typically have included environmental elements like geography and resource endowment; military capacity; intelligence capacity; population size; social cohesiveness; political regime type; and size and performance of the national economy.[19] Each one of these elements of national power could be affected negatively by climate change. For example, militaries might face challenges in projecting and exercising power if they have to operate in conditions of drought, or on flooded terrain, or during severe storms or heat waves. Global warming could reduce a country's national power by reducing the availability of water or disturbing its stock of other natural resources like fish or timber. Regime type could be compromised if governments activate extraordinary powers to deal with climate-based disasters, and are accused of corruption, incompetence, and neglect. Intelligence could fail if it has to contend with making sense of a domain characterized by high levels of uncertainty about social effects.

Perhaps the greatest foreseeable peril relates to the extent to which climate change could undermine economic development, especially in the world's poorer and more fragile countries. These are the countries that Paul Collier characterizes as trapped in conditions that obstruct development. Specifically, fragile countries tend to have a history of violent conflict, supplies of high value natural resources such as oil or diamonds that groups can compete to control, corrupt and ineffective governments, and unstable neighbors with similar problems.[20] Costly and hard to manage events such as floods, droughts, and fires tax the scant resources of these countries, burden their economies, and hence reduce national power.

Climate change impacts on state failure

It is not hard to imagine numerous climate-related pathways to state failure, a condition under which states are not able to provide basic services or adequately protect themselves and their citizens. Colin Kahl, for example, has argued that resource scarcity – a condition climate change could amplify – can lead to state failure or state exploitation (where a collapsing state acts to preserve itself by giving greater access to natural resources to groups it believes can prop it up).[21]

According to the 2009 report of the International Federation of the Red Cross and Red Crescent Societies (IFRC):

> The threat of disaster resulting from climate change is twofold. First, individual extreme events will devastate vulnerable communities in their path. If population growth is factored in, many more people may be at significant risk. Together, these events add up to potentially the most significant threat to human progress that the world has seen. Second, climate change will compound the already complex problems of poor countries, and could contribute to a downward development spiral for millions of people, even greater than has already been experienced.[22]

According to the 2011 report, the cost of climate-related disasters tripled between 2009 and 2010 to almost $110 billion. The costs of disasters are mounting dramatically, a trend that could well continue alongside climate change. In all countries, but especially in the poorer and more fragile ones, disasters draw scarce resources away from poverty alleviating activities such as building critical water, transportation, and energy infrastructure, and investing in education and job training. Diminishing resources could easily affect a government's functional capacity, and the trust and legitimacy it commands.

Climate change effects also can also affect other aspects of a society, such as its food and water availability, public health, urban development, and rural livelihood structure. Faced with deteriorating socioeconomic conditions, people may find themselves forced to migrate into marginal lands and hostile communities, attracted to the easy solutions of extremist ideology, or willing to turn to crime or war to survive. All of these social pressures can complicate governance, sharpen social cleavages, and foment distrust and anger in the civilian population.

For some states, climate change could prove to be an existential threat, causing states to fail because they can no longer function as their lands disappear under rising seas. This already is happening to the Maldives and some forty other island states. Glacial outburst floods might cause similar devastation in countries such as Bhutan and Nepal, and a reversal in the warming properties of the ocean conveyer could result in countries like the United Kingdom disappearing under meters of ice within a matter of years.

Climate change impacts on violent conflict

Finally, some analysts have suggested that climate-induced resource scarcities could become key drivers of violent conflict in the not too distant future. This prediction builds on research covered earlier in this volume, and most explicitly associated with scholars such as Thomas Homer-Dixon. Homer-Dixon was concerned above all with the social effects of natural resource scarcity, which he argued was likely to increase for many critical natural resources. Economic and population growth would continue to increase demand. Some resource stocks would decline. Some social groups would limit access to others. Developing countries, Homer-Dixon argued, were and would continue to be hard pressed to flourish and even survive under conditions of severe resource scarcity. Over time, patterns of resource capture (one group seizes control of the resource) or ecological marginalization (people are forced to move into resource-poor lands) would become more prominent, creating conditions conducive to diffuse violent conflict.[23] While some researchers reject arguments linking environmental stress to violent conflict, many others believe they offer important insights, and a number of policymakers and practitioners have integrated these arguments into their worldviews.

If not due to scarcity, violent conflict might erupt in response to other climate change effects. I noted above the prediction made by Sir Nicholas Stern that, by 2050, as many as 200 million people could be displaced permanently due to climate change. These vast flows of people could speed up rural to urban migration, and propel large groups across the ethnic, economic, and political boundaries that have divided them for generations. It is not hard to imagine this type of stress contributing to violent conflict, and other expressions of social discontent.

Stern's prediction, however, has been criticized by researchers like Henrik Urdal, who contends that the:

> potential for and challenges related to migration spurred by climate change should be acknowledged, but not overemphasized. Some forms of environmental change associated with climate change like extreme weather and flooding may cause substantial and acute, but mostly temporal, displacement of people. However, the most dramatic form of change expected to affect human settlements, sea-level rise, is likely to happen gradually, as are processes of soil and freshwater degradation.[24]

Whether there is strong historical and empirical support for the arguments of Homer-Dixon, Stern, and others has become a very controversial topic in the academic literature. A recent issue of the *Journal of Peace Research*, for example, offers a number of criticisms of the climate change–violent conflict thesis. In this issue, Rune Sletteback argues that:

> Despite climate change, economic and political variables remain the most important predictors of conflict. Rather than over-emphasizing conflict as a result of climate change, I would recommend keeping the focus on societal development, including building resilience against adverse effects of climate change. While this promises the possibility of alleviating the danger of climate change, it can also lead to strengthened societies in the face of natural disaster and civil war.[25]

In the same special issue, Eric Gartzke argues that:

> Interstate warfare is not generally inflamed by higher temperatures. Instead, economic development contributes to both global warming and interstate peace. Development creates nations that are no longer interested in territorial conquest, even if occasionally they continue to use force in punitive ways, or to police the growing global commons, coercing non-compliant states, groups, or leaders. In a somewhat ironic twist, the same forces that are polluting our planet and altering the climate also have beneficial effects on international conflict.[26]

In contrast, another contributor, Tor Benjaminsen *et al.*, concludes:

> When it comes to the effect of environmental variability on the onset of conflicts, we have observed two different and contrasting scenarios. First, the Sahelian droughts of the 1970s and 1980s led rice farming to move down the riverbed and encroach on the dry season burgu pastures. In this sense, a drought may play a role in causing confrontations between farmers and pastoralists, increasing intercommunal tensions and, quite possibly, escalating latent conflicts to the use of violence. Conversely . . . good rainfall years with generous flooding might also induce more conflicts as the zone of potential contestation is expanded to areas with less established norms of ownership and control. The key factor in avoiding both these scenarios would be better policies and laws recognizing the needs of pastoralists and generally improving the relationship between the government and rural populations.[27]

Clearly, the research on climate change as a driver of violent conflict and other national security issues is controversial, and the historical record, which may or may not be relevant to today's climate-changed world, is also contested. The disagreement can be organized along two parameters. On the one hand, there is disagreement over how best to interpret empirical evidence. Both case study and large N analyses, working with imperfect data, have shown a significant contribution or very little to no contribution of climate change to conventional national security issues. However, most of the focus of research over the past two decades has been on violent conflict as the dependent variable, itself a disputed term, and almost none has been given to how climate change is affecting state functioning or elements of national power. On the other hand, there is also disagreement over the normative role of the analyst. For scholars such as Urdal, the writings of analysts like Stern are alarmist, and this is most likely because of a normative desire to securitize climate change, lifting its political profile, attracting resources to its cause, and enabling the use of extraordinary measures to combat it.

This is a concern familiar to the subfield of environment and security. Since the early 1990s, as noted in chapter 1, scholars such as Daniel Deudney, Marc Levy, and Ole Wæver have expressed concerns about militarizing or securitizing the environment, and thus linking it to military and intelligence attitudes and tools. Such a linkage, they suggest, could prove to be a costly and inefficient way of promoting adaptation and mitigation, while undermining cooperation on other fronts, and perhaps encouraging misperceptions and mistrust at home and abroad.

Amplifying this attitude in a series of editorials in *Foreign Policy* magazine, Stephen Walt has argued that climate change is simply not a national security issue – at least not for the United States and similar countries.[28] And where it does have major destabilizing impacts, it is better framed as a humanitarian issue. For Walt and other realists, national security should only be about war and the survival of the state. He wonders about the value – in terms of analysis, strategy, and policy formulation – of expanding the concept of national security to include threats that are vague or unknown.

These lines of division and critiques are muted or non-existent in the research and policy activity that seeks to link climate change to human security, with the exception that here, too, Walt's concerns about diluting the very concept of security are germane.

Climate change and human security

While versions of the concept of human security may be traced back through centuries of the tradition of political thought, the term entered the contemporary field of international relations through the United Nations Development Programme's (UNDP) 1994 *Human Development Report*. The UNDP argument was that a narrow, state-centric and militaristic understanding of security had dominated research and policy for decades, but did not capture many of the ways in which the lives and welfare of most people on the planet actually are threatened. To fill this gap, the UNDP suggested attention be given to a set of global threats to "human life and dignity."[29] It organized these threats into seven categories: economic, food, health, environmental, personal, community, and political. The core of this approach often is summarized in terms of freedom from want and freedom from fear. Tariq Banuri offers an insightful justification for the term:

> security denotes conditions which make people feel secure against want, deprivation, and violence; or the absence of conditions that produce insecurity, namely the threat of deprivation or violence. This brings two additional elements to the conventional connotation (referred to here as political security), namely human security and environmental security.[30]

In response to the UNDP report, a research program on Global Environmental Change and Human Security was established in 1996. The members of this program argued that:

> human security is a variable condition where people and communities have the capacity to manage stresses to their needs, rights, and values. When people do not have enough options to avoid or to adapt to environmental change such that their needs, rights, and values are likely to be undermined, then they can be said to be environmentally insecure.[31]

In a similar way, Halsnæs and Verhagen have argued that:

> It is here demonstrated that there are numerous linkages between climate change impacts and the MDGs starting with the influence from climate change on livelihood assets and economic growth, and continuing with a number of serious

health impacts including heat-related mortality, vector-borne diseases, and water and nutrition. Specific gender and educational issues are also identified as areas that indirectly will be impacted.[32]

The argument that climate change impacts can and do threaten human security is straightforward and does not incite criticism. There is controversy over whether the term "human security" is useful for policy and analysis. Roland Paris, for example, argues that its inclusiveness can "hobble the concept of human security as a useful tool of analysis," although he concludes that:

> definitional expansiveness and ambiguity are powerful attributes of human security. . . . Human security could provide a handy label for a broad category of research . . . that may help to establish this brand of research as a central component of the security studies field.[33]

In any case, this critique should be familiar to the readers of this volume and will not be rehearsed further here.

Climate change and peacebuilding

In 1992 the United Nations Secretary-General Boutros Boutros-Ghali published *An Agenda for Peace*, which introduced a new post Cold War program for peacebuilding. The basic idea was to describe and valorize a process of state and civil society capacity building that ought to begin during the peacekeeping phase and could continue for years beyond that.[34] After a decade of peacebuilding activity, the Peace Research Institute of Oslo (PRIO) analyzed 336 peacebuilding projects funded over a ten-year period by Germany, the Netherlands, Norway, and the United Kingdom.[35] The study concluded that peacebuilding had organized itself into four mutually reinforcing programmatic areas:

1. Social, economic, and environmental capacity building[36]
2. Governance and political capacity building
3. Security capacity building
4. Truth and reconciliation capacity building.

Insofar as climate change can be linked to national and human security, integrating some level of climate sensitivity into peacebuilding operations, and building the capacity to prepare for and respond to climate change effects, has a prima facie attraction. There is already a precedent for this in the recent efforts to integrate natural resource management into peacebuilding.[37] The obvious strategy for doing this is to identify entry points in some or all of the programmatic areas.[38] In keeping with the logic of peacebuilding, the first step likely would involve the identification and assessment of climate sensitive sectors, and is likely to be more relevant to categories 1 and 2 above, than 3 and 4. Also early in the process, criteria for climate-sensitive general capacity building would need to be articulated, so that projects and other initiatives in each of the programmatic areas could be filtered though a climate lens. Depending upon the results of the initial assessment, a next step might be to develop climate change specific capacity building in line with the national framework for reconstruction. The most ambitious activity during a peacebuilding operation might involve capacity building requiring bilateral, regional, or global cooperation.

Climate-sensitive capacity building, climate-specific capacity building, and bilateral and regional climate capacity building also would tend to fall into the first two categories. For example, in the domain of politics and government:

> one way to integrate climate change adaptation into peacebuilding is to assist the new government to enhance its capacity for managing climate risk. For example, functioning meteorological services could support the collection and analysis of climate data – which could, in turn, help to establish early-warning systems to prepare for, and minimize, the impact of events such as storms, floods, disease outbreaks, and famine. Measures that encourage governments to create adaptation-related ministerial and departmental posts, establish interdepartmental coordination units, take a long-term perspective, and be flexible in planning and policy development (through measures such as review processes), can further develop adaptive capacity and encourage wider participation.[39]

There are also opportunities to build capacity for mediation and conflict resolution that can assist in dealing with climate-related tensions. For example, Bronkhorst writes that the Sudan Environmental Conservation Society's "Water for Peace project provides affordable water at strategic points along the livestock routes, and works towards separating water sources for humans and livestock, easing congestion and preventing conflict by creating alternative water sources in areas of high demand."[40]

Certainly, the largest set of opportunities will be in the area of socioeconomic programming. For example, resettling people and helping them to recover or develop new livelihoods is a large budget in which the decisions ought to be climate sensitive. This has not always been the case. "In Rwanda, for example, where refugees and IDPs numbered over 2 million in 1994, many families were resettled in marshes and steep hillsides, increasing their exposure to climate-related hazards such as flooding and landslides."[41]

> Another immediate need is for infrastructure. Inevitably, post-conflict societies want functional transportation, communication, and water and energy systems to be the focus of government policy and donor support. Although a strong argument can be made for climate-proofing critical infrastructure (e.g., by constructing higher bridges, wider drainage systems, and more efficient irrigation systems, and by ensuring that water sources are not at risk of salinization) and for introducing new infrastructure (such as sea walls) that will help manage climate risks, the sole example of such an approach has so far been on a small scale in Sierra Leone.[42]

Discussing the work of SOS Sudan, Bronkhorst describes how their:

> sand dams capture water and prevent rapid evaporation, as people need to dig a little through sand before getting to the water. There are a number of other

benefits, including a reduction in hygiene problems and helping to recharge and regenerate areas in terms of greenery.[43]

Infrastructure investment is often critical to the typical national priorities of stabilizing the economy and making it attractive to both local and foreign investors. Here again, there may be an opportunity to build capacity to assess infrastructure and other investments in terms of climate risk. Finally, depending on the local conditions and sensitivities, capacity building programming related to public health and education might also provide valuable entry points for the integration of climate change capacity.

Ultimately, the prospects for integrating climate change capacity into peacebuilding are promising. This is partly because:

> the two processes are similar in important ways: both focus on building capacity and resilience, and both promote the adoption of a longer-term perspective, while requiring enough flexibility to react to changing circumstances. Moreover, because both processes are context dependent, interventions need to be informed by context-specific conflict analysis, capacity assessments, vulnerability assessments, and scenario planning.[44]

Moreover:

> the development aspects of peacebuilding are often valuable to supporting the wider recognition of underdevelopment as a root cause of conflict. The development community has not always found the right balance between short-term, externally driven results and the less glamorous medium- to long-term capacity building. Peacebuilding provides an opportunity to rethink development, and integrating adaptation into peacebuilding can make both more sustainable.[45]

However, in an earlier study my co-author Anne Hammill and I argued that there also are:

> important differences between the two processes. In particular, what is good for peacebuilding may not always be good for climate adaptation, and vice versa. For example, settling people around Virunga National Park was critical to jump-starting livelihoods and stabilizing communities in the Democratic Republic of the Congo, as it permitted access to forest resources for construction, fuel, food, and medicinal needs. The resulting degradation of ecosystem services, however, may have undermined the longer-term adaptive capacity of the system. Similarly, putting money into climate forecasting may seem extravagant when people are struggling to meet daily needs. In sum, climate change adaptation (like development in general) involves trade-offs, some of which may directly conflict with peacebuilding initiatives. Further complicating matters, even if one wished to integrate climate change adaptation into peacebuilding, assessments of climate risk may be difficult or impossible to obtain in the time frame available, and it is always hard to plan and act under conditions of great uncertainty.[46]

Conclusion

There are pros and cons to linking climate change to national security, human security, and peacebuilding. As noted in chapter 1, disagreements on issues such as this often reflect deeper divisions over what security is and what the security analyst should be doing. These questions are taken up in detail in the concluding chapter of this volume by my co-editor, Rita Floyd.

Notes

1   This article is informed by earlier work including A. Hammill and R. Matthew, 'Peacebuilding and Climate Change Adaptation', *St. Antony's International Review* 5 (2), 2010, 89–112.5; A. Hammill and R. Matthew, 'Peacebuilding and Climate Change Adaptation', in C. Bruch, D. Jensen, M. Nakayama and J. Unruh (eds), *Post-Conflict Peacebuilding and Natural Resource Management*, London: Earthscan, 2012; R. Matthew, 'Environmental Change, Human Security and Regional Governance: The Case of the Hindu Kush-Himalaya Region', *Global Environmental Politics*, forthcoming 2012; R. Matthew, 'Environmental Security and US Politics', in M. Kraft and S. Kamienicki (eds), *The Oxford Handbook of U.S. Environmental Policy*, Oxford: Oxford University Press, forthcoming 2012; and R. Matthew, 'Is Climate Change a National Security Issue?', *Issues in Science and Technology* 27 (3), 2011, 49–60. Some of the ideas presented here have been derived from an unpublished work in progress by Anne Hammill, Richard Matthew and Dennis Tänzler. Research assistance for this chapter was provided by Jamie Agius.
2   S. Solomon *et al.* (eds), *The Physical Science Basis: Contribution of Working Group I to the Fourth Assessment Report of the Intergovernmental Panel on Climate Change*, Cambridge: Cambridge University Press, 2007. Online: www.ipcc.ch/publications_and_data/ar4/wg1/en/contents.html (accessed August 2012).
3   For an overview of the history of climate change science, see also S. Weart, *The Discovery of Global Warming: Revised and Expanded Edition*, Cambridge, MA: Harvard University Press, 2008.
4   H. Le Treut, R. Somerville, U. Cubasch, Y. Ding, C. Mauritzen, A. Mokssit, T. Peterson and M. Prather, 'Historical Overview of Climate Change' in S. Solomon, D. Qin, M. Manning, Z. Chen, M. Marquis, K. B. Averyt, M. Tignor and H. L. Miller (eds), *Climate Change 2007: The Physical Science Basis. Contribution of Working Group I to the Fourth Assessment Report of the Intergovernmental Panel on Climate Change*, Cambridge: Cambridge University Press, 2007, p. 100. Online: www.ipcc.ch/pdf/assessment-report/ar4/wg1/ar4-wg1-chapter1.pdf (accessed August 2012).
5   Le Treut *et al.*, 'Historical Overview of Climate Change', p. 98.
6   BBC News, 'Climate Skepticism "On the Rise", BBC Poll Shows', 7 February 2010. Online: http://news.bbc.co.uk/1/hi/8500443.stm (accessed August 2012).
7   Based on private communications with several authors of the fifth assessment report.
8   N. Stern, *The Economics of Climate Change: The Stern Review*, Cambridge: Cambridge University Press, 2007.
9   CNA, *National Security and the Threat of Climate Change*, 2007. Online: http://securityandclimate.cna.org/ (accessed August 2012); See also K. M. Campbell, *Climatic Cataclysm: The Foreign Policy and National Security Implications of Climate Change*, Washington, DC: Brookings Institution Press, 2008.

10   German Advisory Council on Global Change, *World in Transition: Climate Change as a Security Risk*, London: Earthscan, 2008.

11   D. Smith and J. Vivekananda, *A Climate of Conflict: The Links between Climate Change, Peace and War*, London: International Alert, 2007, p. 3. Online: http://www.international-alert.org/resources/publications/climate-conflict (accessed August 2012).

12   J. Vivekananda, 'Practice Note: Conflict-Sensitive Responses to Climate Change in South Asia', International Alert, 2011, p. 8. Online: www.international-alert.org/sites/default/files/publications/201110IfPEWResponsesClimChangeSAsia.pdf (accessed August 2012).

13   Ibid.

14   See, for example, UN General Assembly, 'Climate change and its possible security implications: report of the Secretary-General', 11 September 2009, A/64/350.

15   Intergovernmental Panel on Climate Change (IPCC), *Working Group II Report: Climate Change Impacts, Adaptation, and Vulnerability*, 2007. Online: www.ipcc.ch/ (accessed August 2012).

16   See, for example, T. Homer-Dixon, *Environment, Scarcity and Violence*, Princeton, NJ: Princeton University Press, 1999; C. H. Kahl, 'Contested Grounds', in D. Deudney and R. Matthew (eds), *States, Scarcity, and Civil Strife in the Developing World*, Princeton, NJ and Oxford: Princeton University Press, 2006; H. Urdal, *Demographic Aspects of Climate Change, Environmental Degradation and Armed Conflict*, United Nations Expert Group Meeting on Population Distribution, Urbanization, Internal Migration and Development, January 2008.

17   Matthew, 'Is Climate Change a National Security Issue?'.

18   For an excellent bibliography on the copious literature on the elements of national power, see National Defense University, *Elements of National Power*. Online: www.jfsc.ndu.edu/library/publications/bibliography/Elements_of_National_Power.pdf (accessed August 2012).

19   R. L. Armitage and J. S. Nye, Jr., *CSIS Commission on Smart Power: A Smarter, More Secure America*, Washington, DC: Center for Strategic & International Studies, 2007. Online: http://csis.org/files/media/csis/pubs/071106_csissmartpowerreport.pdf (accessed August 2012).

20   P. Collier, *The Bottom Billion: Why the Poorest Countries Are Failing and What Can Be Done About It*, New York: Oxford University Press, 2007.

21   Kahl, 'Contested Grounds'.

22   IFRC, *World Disasters Report 2009*. Online: www.ifrc.org/Docs/pubs/disasters/wdr2009/WDR2009-full.pdf (accessed August 2012).

23   Homer-Dixon, *Environment, Scarcity and Violence*.

24   Urdal, *Demographic Aspects of Climate Change*.

25   R. T. Slettebak, 'Don't blame the weather! Climate-related natural disasters and civil conflict', *Journal of Peace Research* 49, January 2012, 163–76.

26   E. Gartzke, 'Could climate change precipitate peace?', *Journal of Peace Research* 49, January 2012, 177–92.

27   T. A. Benjaminsen, K. Alinon, H. Buhaug and J. T. Buseth, 'Does climate change drive land-use conflicts in the Sahel?', *Journal of Peace Research* 49, January 2012, 97–111.

28   S. Walt, 'National Security Heats Up', *Foreign Policy*, 10 August 2009. Online: http://walt.foreignpolicy.com/posts/2009/08/10/national_security_heats_up (accessed August 2012); S. Walt, 'A Heated Rant About Climate Change', *Foreign Policy*, 11 May 2012. Online: http://walt.foreignpolicy.com/posts/2012/05/11/a_heated_rant_about_climate_change (accessed August 2012).

29   United Nations Development Programme, *Human Development Report 1994: New Dimensions of Human Security*, Geneva: UNDP, 1994, p. 22.

30   T. Banuri, 'Human Security', in N. Nauman (ed.), *Rethinking Security, Rethinking Development*, Islamabad: SDPI, 1996, pp. 163–64.

31   R. Matthew, J. Barnett, B. McDonald and K. O'Brien (eds), *Global Environmental Change and Human Security*, Cambridge, MA: MIT Press, 2009, p. 18.

32   K. Halsnæs and J. Verhagen, 'Development based climate change adaptation and mitigation – conceptual issues and lessons learned in studies in developing countries', *Journal of Peace Research* 12 (5), 2007, 665–84. Online: http://dx.doi.org/10.1007/s11027-007-9093-6 (accessed August 2012).

33   R. Paris, 'Human Security: Paradigm Shift or Hot Air?', *International Security* 26, 2001, 101.

34   See, for example, UN General Assembly, 'Report of the Secretary-General on Peacebuilding in the Immediate Aftermath of Conflict', 11 June 2009, A/63/881-S/2009/304.

35   D. Smith, 'Towards a Strategic Framework for Peacebuilding: Getting Their Act Together: Overview Report of the Joint Utstein Study of Peacebuilding', Oslo: PRIO, 2004. See also Peacebuilding Initiative, 'Introduction to Peacebuilding', International Association for Humanitarian Policy and Conflict Research, 2007–08. Online: www.peacebuildinginitiative.org/index.cfm?pageid=1776 (accessed August 2012).

36   For a discussion of capacity building, see UNDP, 'Capacity Development, Technical Advisory Paper 2', July 1997. Online: http://mirror.undp.org/magnet/cdrb/TECHPAP2.htm (accessed August 2012).

37   United Nations Environment Programme (UNEP), 'From Conflict to Peacebuilding: The Role of Natural Resources and the Environment', Nairobi: UNEP, 2009.

38   This discussion is based largely on the author's personal experience as a member of United Nations Peacebuilding missions in Rwanda and Sierra Leone.

39   Hammill and Matthew, 2012, page unavailable.

40   S. Bronkhorst, 'Climate Change and Conflict: Lessons for conflict resolution from the southern Sahel of Sudan', African Center for the Constructive Resolution of Disputes, 2011, p. 32. Online: www.accord.org.za/publications/reports/910-climate-change-and-conflict-lessons-for-conflict-resolution-from-the-southern-sahel-of-sudan (accessed August 2012).

41   Hammill and Matthew, 2012, page unavailable.

42   Hammill and Matthew, 2012, page unavailable.

43   Bronkhurst, 'Climate Change and Conflict', p. 35.

44   Hammill and Matthew, 2012, page unavailable.

45   Hammill and Matthew, 2012, page unavailable.

46   Hammill and Matthew, 2012, page unavailable.

16

WHITHER ENVIRONMENTAL SECURITY STUDIES?

An afterword

Rita Floyd[1]

It was the aim of this book to provide a comprehensive and up-to-date overview of the field of environmental security studies (ESS). To this end, *Environmental Security: Approaches and Issues* includes a set of chapters on the competing theoretical approaches and another set of chapters on the most pertinent issues arising in the context of the natural environment and national, human and biosphere security.

In chapter 1 of this volume I argued that the meaning of environmental security and hence how it is studied are contested. Scholars differ with regard to their views of what the term security signifies. Some adopt a very narrow view of the term, generally on the grounds that this allows scholars to function as a sort of epistemic community building more sophisticated models and theories over time, and that it also allows for a close relationship to the security community. Others embrace a much broader understanding of the term, widening it to include non-military threats such as disease and climate change to subnational and trans-national groups as well as to states. In addition, on the grounds that environmental stress is very often a site for cooperation and conflict resolution a third group of scholars within ESS rejects the focus on security altogether.

Scholars also differ with regard to the role they identify for the analyst and consequently the role they attribute to theory. Is it purely analytical and predictive, or is it Critical emancipatory, or is it some combination of analytical analysis and normative theorizing?

As a result one is left with many different competing *approaches* to environmental security. The approaches themselves are of course also not homogeneous and many nuanced versions of, for example, 'environmental security as human security' exist, with individual scholars suggesting different routes for how we can achieve this ideal type.

Throughout part one of the book, proponents of the different approaches have sought to make the case for the relevance of their preferred approach. In part two of the book scholars have shown how the insights of one or more theoretical approach speak to some of the most pertinent environmental security issues of our time, including water, conservation, population, sustainable development, food, energy and climate change. In this concluding chapter I do not seek to summarize and repeat the arguments made throughout this book; instead I want to speculate about the future of environmental security studies. Three things are particularly important in this context. First, I want to consider arguments against linking climate change with security. This is important because given the extent of adverse social effects related to climate change, climate security could become the definitive concern for all of ESS. Second, I examine if any of the research findings from within ESS have relevance for the discipline of security studies more generally. Third, by highlighting existing gaps in the field of ESS I aim to suggest topics and issues for future research.

Arguments against climate security

More than twenty years of environment and security research have helped shed light on the meaning of environmental security and on its practice. Today environmental security scholars are increasingly concerned with the security implications of climate change and/or with the meaning of climate security. The questions for ESS are not only whether climate change will lead to environmental conflict or cooperation, or indeed what climate security *should* mean. But also whether climate change does affect the various issue areas discussed. Bryan McDonald (chapter 13), for example shows that climate change will adversely affect global food security, while Bishnu Upreti (chapter 12) highlights that the intimate connection between sustainable development and environmental security as human security becomes even more important in the face of global warming.

Although, and as argued in chapter 1, many of the key debates within ESS will remain unchanged by the new focus, the nexus between 'climate' and 'security' is important and interesting to an audience beyond ESS. Among those new to 'climate security' some appear almost completely unaware of ESS, and one rationale for this book was to demonstrate that ESS makes an ideal entry point for climate security research.

The securitization of climate change is, however, not universally endorsed. Recent attempts to introduce climate change into the UN Security Council, for example, have been resisted by some developing countries, which prefer a framing that deepens and broadens the linkages to development and human rights. Even some scholars of environmental security have reservations against 'climate security', with the latter nearly always influenced by their knowledge of the history and practice of environmental security. My concern here is with the future of ESS. Given that climate security is fast becoming the focus of ESS an examination of the critiques of climate security is in order.

Three arguments have been advanced against 'climate security' in particular: 1) climate change requires long-term political solutions, not short-term security solutions; 2) climate security measures defy the logic of securitization; and 3) securitization can be abused. Before outlining these in more detail I should like to stress two things. First, this list is not comprehensive, but addresses my assessment of the most important arguments against climate security. Second, I do not ascribe much value to defending climate security against each and every one of these critiques. Instead I consider the work of critics an important component of environmental security studies and one that helps to advance the field.

Critique 1. Climate change requires long-term political solutions and not short-term security measures

One critique of environmental security is the thought that environmental problems require long-term political solutions not short-term security measures. While it is debatable that security measures are necessarily short term, there can be little doubt that it is precisely a desire for action on the causes of climate change that has made many policymakers and concerned individuals link climate change to security. After all, the concept of security with its narrative of threats and survival harbours within it a unique and unprecedented capacity to mobilize national elements of power. Critics, however, maintain that in order to deal with climate change we do not need short-term mobilization; instead we need to change our lifestyles.[2] In particular, we need to change how and for what purposes we consume energy. Governments can and in many countries already do play a major role in the switch over to green sources of energy. In the UK alone there are numerous governmental funding opportunities incentivizing industry, business, public sector organizations and householders to switch to sources of renewable energy.[3]

While governments may be able to incentivize change, ultimately what is needed is a change of behaviour at the individual level.[4] The philosopher Peter Singer has long maintained that even our choice of recreation is not ethically neutral from an environmental standpoint:

> At present we see the choice between motor car racing or cycling, between water skiing or windsurfing, as merely a matter of taste. Yet there is an essential difference: motor car racing and water skiing require the consumption of fossil fuels and the discharge of carbon dioxide into the atmosphere. Cycling and windsurfing do not. Once we take the need to preserve our environment seriously, motor racing and water skiing will no more be an acceptable form of entertainment than bear-baiting is today.[5]

Arguably, such change in individual behaviour cannot be brought about by the scaremongering and negativity inherent to security narratives. E. H. Carr, though in the wildly different context of achieving a stable peace after the Second World War, saw this clearly when he argued that a stable peace could only be achieved if the new order was construed 'in positive rather than negative terms, striving for the achievement of good rather than for the avoidance or suppression of evil. ... A positive and constructive programme is the first condition of any effective moral purpose.'[6]

Some contemporary authors writing on the issue of climate change appear to share this sentiment; they hold that a new and greener world order cannot be built by focusing on negatives alone. Anthony Giddens, for example, in his recent book on *The Politics of Climate Change* (2009) explicitly speaks out against linking climate change with security.[7] 'Fear and anxiety are not necessarily good motivators, especially with risks perceived as abstract ones, or dangers that are seen as some way off.'[8] Especially if a securitizing move (a speech act that declares a referent object of security existentially threatened, cf chapter 1) is not followed by subsequent security practice, the language of security (dramatization and scaremongering) can lead to 'attention fatigue'[9] and achieve nothing. The very recent history of climate action in the UK appears to confirm Giddens' warning. Here the height of the general public's climate awareness, concern and willingness to act came just before the unsuccessful Copenhagen climate summit in December 2009. And here as elsewhere the urgency for immediate climate action was fostered and underlined by what Lene Hansen calls 'visual securitizations'.[10] In this case abundant (in public places, on websites etc.), large, visual countdowns depicting the time left to act on climate change, which was equal to the time left until the Copenhagen summit. Retrospectively, the problem with this securitizing move was that despite the fact that nothing was achieved at Copenhagen, bluntly put, no one (in the UK at least) has died as a result of climate change and things continue as before. What has changed, however, is that the British public is no longer as receptive to the message of climate change in the same way as they were before the summit. And especially in light of the Euro crisis and governmental austerity measures climate change has been allowed to move into the background. In this case then the language of security has triggered nothing other than attention fatigue.

Critique 2. Contemporary climate security measures defy the logic of securitization

A related second major argument against climate security is that it is simply not clear what is gained from securitization as opposed to a politicization of climate change. This is a pressing issue, because by and large proposed climate security measures defy the logic of securitization (cf below), which holds that security responses are considered distinct from normal politics; indeed securitization explicitly permits the breaking of established rules and, if necessary, the use of exceptional measures.[11] In contrast to this, however, proponents of climate security generally advance as a solution a global environmental regime that imposes binding emissions cuts on all governments, and not short-term extraordinary security measures.[12] In response to this claim two things can be said. First, regardless of the actual form securitization ends up taking (i.e. extraordinary emergency measures or more ordinary measures), calling something a matter of security, which is to say identifying an existential threat to some referent object and stating 'a point of no return' unless we act now,[13] *can* still mobilize actors into action. And arguably that is what matters. Thus the most important problem as regards climate action we face is not that humanity does not know what we ought to do to curtail anthropocentric climate change; the problem is rather that climate change is a collective action problem. Convergence on the logic of climate security might just prove instrumental in overcoming the latter; however, there can be no guarantee of success as there is always the danger of attention fatigue.

Second, it is a little too early to argue that climate security excludes extraordinary measures. For the West at least climate change is primarily regarded as a future problem. Politicians in turn are concerned with the here and now and with re-electability. The use of exceptional measures to deal with a problem most of us have yet to see and experience could cost politicians their own political future. In other words, exceptional measures might in time become part of climate security. In particular some form of geo-engineering is not outside the realm of possibility. Already private investors, business, governments and militaries everywhere are investing into research in this area. At the same time there is growing evidence that climate change impacts already are extensive, and that the pace and extent of these impacts are growing rapidly.

Critique 3. Securitization can be abused

Finally, one of the longest standing and strongest arguments against securitization in general and environmental security in particular is the fact that (environmental) security can be abused by powerful actors. Abuse can take different forms and it is facilitated by the contested nature of environmental and climate security.[14] One of the earliest critiques of environmental security was that the concept is counterproductive because it can lead to the militarization of the environment. In Robin Eckersley's words in a review of Jyrki Käkönen's (ed.) *Green Security or Militarized Environment* (1994) this refers to the worry:

> that enlisting the military in the task of environmental protection is likely to confer key decision making authority on undemocratic elites who are more likely to act in ways that protect the 'military-industrial complex' – which is part of the problem rather than the solution.[15]

One look at the Clinton administrations' environmental security policy suggests that this worry was very much justified, as it was mostly the national security establishment and above all the Department of Defense that benefitted from US environmental security policies, and not the American people, who were the declared referent object of such policies.[16]

As Matthew has shown in chapter 15 of this volume, the United States government is turning once again to the issue of environmental security, if this time under the label 'climate security'. However, this is not necessarily a positive development; Betsy Hartman has argued that just as environmental security benefitted the US military, the hyperbole of climate-induced resource scarcity conflicts bolsters the Pentagon. By depicting the poor and marginalized people of the developing world as dangerous villains, who when having outlived the carrying capacity of their own lands, will turn to violent conflict and migrate en masse towards the Western world, the US gains a new enemy.[17]

Perhaps more worrying still is that in the context of the war on terrorism, human security has evolved from a 'vision of integrating existing aid networks into a coordinated, global system of international intervention able to complement the efforts of ineffective states in securing their citizens' into a foreign policymaking tool serving homeland security.[18] Post 9/11 US development aid became increasingly tied to eradicating the roots of terrorism. The Bush administration's foreign aid initiative the Millennium Challenge Account, for example, was not about magnanimous development goals, but rather it was concerned with failed states and how to prevent them, because failed states could harbour terrorists.[19] These new notions of human security serving the interests of the homeland are apparent also with regards to climate change. Hartman, for example, points out that the military's increasing role and interest in post-environmental disaster relief increases the military's budget to such an extent that there is now talk of an 'aid-military complex'.[20]

A related critique is that once securitized, climate change could attract an excessive focus, while other issues are pushed into the background. Climate change has already become an explanation for a whole host of problems while other causes are ignored. The World Bank's 2010 development report, for instance, discusses poverty predominantly in relation to climate change, while little attention is given to other causes.[21] Similarly, some environmentalists are worried that the excessive focus on climate change will take attention away from other pressing environmental problems. The internationally renowned ecologist Edward O. Wilson, for example, has stated that 'the problem of biodiversity loss had been "eased off centre stage" because of the focus on climate change.'[22] While Johan Röckstrom *et al.* state that climate change is one of *three* areas where humanity has exceeded planetary boundaries – biodiversity loss and the disruption of nitrogen and phosphorus cycles being the other two.[23]

The contested nature of security and its consequent potential for abuse is something many advocates of climate security overlook. Anthony Giddens, somewhat surprisingly given his rejection of the securitization of climate change discussed above, promotes 'energy security' as a positive strategy towards solving climate change. In his view energy security is to be achieved not by scaring people that we are running out of fossil fuels and so on, but 'by emphasising the advantages of having homes that are snug, protected against the elements and which also save money'.[24] While it seems then that much can be said for Giddens' strategy of focusing on political and economic convergence, his assessment of energy security leaves much to be desired, as it is naïve as regards the beneficiaries and consequences of different types of energy security.

Just like environmental security, and as Adam Simpson has pointed out in chapter 14, energy security means different things to different people and far from every interpretation of energy security is conducive to climate action and indeed to safeguarding the environment. Energy security might mean no more than a government securing fossil fuels so that it can continue to live in the same environmentally and climate inconsiderate way as it has done so far. The rush to the Arctic, which contains about 30 per cent of the world's undiscovered gas and 13 per cent of the world's undiscovered oil,[25] sadly confirms this. Energy security, like environmental security at times, can be a dangerous concept that must be promoted with care.

This brief analysis of the outright critiques of climate security has shown that climate security is approached with considerably more caution and scepticism than was the concept of environmental security when it emerged. In part this is a result of environmental security research itself, with many of those who have studied the practice of environmental security now warning against climate security. The securitization of climate change might potentially hold the answer to curtailing detrimental climate change – for example, by helping to enact a binding global climate change regime, or by funding geo-engineering, while it could also lead to attention fatigue, the security dilemma, inaction and possibly even wars. Ultimately only time will tell. In any case, outright criticism of climate security is not detrimental to ESS; to the contrary these insights make for an altogether stronger subfield.

Environmental security studies: lessons for security studies

This book has shown that ESS is a rich and diverse field. Many environmental security analysts have a background in security studies, while others come from different disciplines including anthropology, geography, environmental studies, demography, political science and political theory. What is clear is that anyone working in environmental security studies needs a firm grounding in the different theories of security and to a lesser extent also in International Relations theory. Thus it is vital to understand the changed nature of security after the end of the Cold War – both in theory and in practice. It further is important to recognize the different epistemological positions a security analyst can inhabit, that is, whether they consider themselves value-neutral, or if they embrace their own politics as observers. It is important to understand that security is contested, that it can be a positive concept, tantamount to human well-being, while it can also be regarded as a negative development, a closing of debate, potentially bringing with it conflict triggering the security dilemma. In this section, however, I want to say no more about what the environmental security analyst can, or indeed must, learn from security studies. Instead I am interested in: what, if anything, can security studies learn from environmental security studies? Quite a lot it seems, hence in the following I will argue that research from ESS has implications for 1) the nature of threats; 2) the condition of security; 3) the practice of security; 4) the understanding of desecuritization; and finally 5) the ethics of securitization and the ethics of security.

Lesson 1. The nature of the threat

Traditionally in security studies threats are measured (insofar as this is possible) in terms of an aggressors' intentions and their capabilities. By focusing on these two factors we can hope to answer questions such as: does an actor really want to harm me – or by extension someone else or something else? Or in other words, is the threat objectively present? This logic makes sense in all sectors of security bar the environmental sector (which includes health security)[26] because in each one of the other sectors of security (military, economic, societal and political)[27] we are dealing with the conflictual relationships of different groups of people. The environmental sector is the exception; here the threat can be seen to stem from an ill-functioning environment without anyone doing the threatening.[28] Indeed, the fact that intentional behaviour structures the field of security has meant that many critics have dismissed threats from environmental stress and degradation as outside the security logic.[29] Those who hold, however, that 'threats without enemies'[30] are an oxymoron and favour instead the language of danger, risk or vulnerabilities for such issues are not necessarily right to do so. The Oxford English dictionary for one defines threat (noun) as: '(1) a stated intention to inflict injury, damage, or hostile action on someone; (2) a person or *thing* likely to cause damage or danger, [and also] 3) the possibility of trouble or danger.'[31] I would also suggest that labelling threats from environmental change anything other than threats risks downgrading them in importance vis-à-vis other threats, for example, terrorism.

In the absence of intentional behaviour, how then can we measure whether threats in the environmental sector of security are objectively present? It seems to me that what we are hoping to uncover is whether the environment itself is now so damaged/degraded that it poses an objective existential threat[32] to our (which is to say humanity's) survival.[33] I would propose that for this assessment we rely on peer reviewed research from the natural sciences. Of course, as Saleem Ali and Mary Watzin argue in chapter 10, even scientific research is at times not entirely free from biases.[34] Plus even that research will not be able to conclusively predict the socio-economic consequences of, for example, climate change. Indeed, for now it remains unable to tell us by how much exactly global temperature will increase, hence much of the Intergovernmental Panel on Climate Change's (IPPC) work is concerned with mapping different scenarios for different temperature increases.

A possible way forward here would be to develop thresholds that determine at what point climate change would constitute an objective existential threat to different referent objects of security, similar to the way the Uppsala Conflict Data Program (UCDP) differentiates between, for example, minor armed conflicts, intermediate armed conflicts and war judging by the number of battle deaths in any given year. Thus it could be argued that climate change poses an objective existential threat to any given state if, for instance, 'x per cent of the population are dying from climate change related diseases and disasters per calendar year', or when there is 'y amount of net loss of economic output due to climate change', or 'z amount of arable land lost' and so on. Naturally in all these cases it will be difficult to single out climate change as the independent variable. Moreover, given the ample comparative research on state failure it might be easier to work out/agree on thresholds that determine at what point states can be said to be existentially threatened by climate change. Indeed Joshua W. Busby (2008) has already made important inroads into this kind of thinking, when he identified a number of conditions for when climate changes poses a 'direct threat' to (US) national security:

1. Climate change threatens the existence of the country.
2. Climate change could decapitate the seat of government.
3. Climate change threatens the country's monopoly on the use of force.
4. Climate change could disrupt or destroy critical infrastructure.
5. Climate change could lead to such catastrophic short-run loss of life or general well-being as to undermine the government's legitimacy.
6. Climate change could cause these effects on neighbors to spur refugee crises.
7. Climate change could alter the territorial borders or waters of the country.[35]

A similar exercise for human security is likely to get caught up in the usual lengthy debates about rights versus needs, while it will also be difficult to fix how many humans exactly would have to be objectively existentially threatened, the latter an important criterion for just securitization theory.[36] Is one enough, or would it have to be a group, but, how large is a group? These are just some of the difficult questions that will need answering in this context. Indeed the difficulty of this is such that states could once more become the most written about referent object of (climate) security.

Lesson 2. The condition of security

In chapter 1 of this volume I draw a distinction between, on the one hand, those environmental security scholars with a *primary* interest in the condition of environmental security, and those whose *primary* interest is with the practice of environmental security, on the other. It was also argued that these two distinctions are apparent in the discipline of security studies more generally; where some writers are concerned *primarily* with the condition of security, while others are *primarily* concerned with how it is practised.[37] Security, for those with an interest in the condition of security, is a positive value to be cherished and protected. It usually has something to do with the physical safety of people from (external) aggression and from the threat of violence. In more comprehensive formulations it concerns not only the absence of violence, but also the presence of certain basic human capabilities or needs.[38]

What sort of thing is either perceived as – or is objectively – threatening changes over time and with new developments. Cyber-security, for example, was quite obviously not an issue before the invention and near ubiquity of information and communication technology. Economics, geopolitics, politics and technology are among the dominant forces that play a part in what sort of thing is (objectively or perceived) threatening to people (of course people are affected differently depending on where they live, on their social mobility etc.).

The 'natural' environment, argues Simon Dalby, is another such force, and one that needs to be acknowledged much more fully if true security is to be achieved.[39] Following a number of earth-system scientists Dalby has repeatedly argued (including in chapter 6 of this volume) that we now live in the era of the Anthropocene, that is, the era in which human beings have become a major force in altering the biosphere.[40] Anthropocentric climate change epitomizes this new era.

In the age of the Anthropocene, the traditional aspiration of national security facilitated by military machinery is built on an unsustainable fossil fuel culture that is at the heart of insecurity.[41] Drawing on political ecology Dalby argues that we must realize that the world is interconnected. The developed world's consumer culture and throw-away society has direct security implications for the poor and marginalized in the developing world, who are already feeling the effects of climate change today. Insecurity is a product of the way we live our lives and not something external that can be defended against. Security will require 're-imagining ourselves as within a changing biosphere rather than on an earth we can dominate and control'.[42]

Lesson 3. The practice of security (securitization)

ESS does not only hold valuable lessons for the condition of security, it also holds lessons for how security is practised, or in other words for the nature of securitization. Although I work with a generic definition of 'securitization' here as opposed to Ole Wæver's version, when something is a matter of security, and has been securitized, for most people this means that the issue is of primary importance and that it will be addressed accordingly, which is to say with emergency measures that may be extraordinary in nature.[43] Dalby's rendition of securitization, which does not refer back to Wæver, is a case in point. He argues:

> The key point about the operation of securitization is precisely that it refers to pressing and immediate situations that normal political life cannot address.

In this regard, the invocation of emergencies in present times is important to deal with the climate-change issue and the larger matters of ecological disruption.[44]

Empirical examples, however, have shown that in practice environmental security does not always play out that way. M. Julia Trombetta, for example, has argued that the practice of environmental security constitutes a challenge to the established conception of security as exception. This is so, she argues, because environmental security measures often take the form of 'prevention, risk management and resilience'.[45] It is of course possible to draw the conclusion – as some have – that in the absence of extraordinary emergency measures there simply is no case of environmental security.[46] However, the more likely conclusion is that just as the practice of security changed from a focus on military threats towards including environmental and other non-traditional security threats after the end of the Cold War, it has changed again. If this is so then either our existing security theories need to adapt (insofar as this is possible) in order to accommodate this change or we will need new theories. One such new theory is offered by risk security writers who argue that threats posed by climate change are 'uncertain, diffuse, difficult to quantify and yet potentially catastrophic [and thus reflect] the logic of a risk society'.[47]

Another comparatively new theory is offered by the so-called Paris school[48] of security studies, who theorize security practice below the level of exception and who hold that:

> Security is in no sense a reflection of an increase of threats in the contemporary epoch – it is a lowering of the level of acceptability of the other; it is an attempt at insecuritization of daily life by the security professionals and an increase in the strengths of police potential for action.[49]

Utilizing the Paris school's framework for analysis Angela Oels has argued that we are currently experiencing what she calls the 'climatization of the security field', the process whereby 'existing security practices are applied to the issue of climate change and . . . new practices from the field of climate policy are introduced into the security field'.[50] One manifestation of this is the news media's representation of 'climate refugees' as threatening to Western industrialized states, which foregoes any systematic examination of the causation of climate change induced migration and which does not grant a voice to alleged climate refugees, many of whom find the term offensive and deliberately marginalizing.[51]

Lesson 4. The understanding of desecuritization

Desecuritization, or in Jef Huysmans' words, the 'unmaking of securitization',[52] is one of the most underdeveloped areas of security studies. Although some notable work[53] has been done in this area, the focus for most security analysts is either the practice of security or the condition of security, and not the unmaking of security. Yet, studies into how securitization is unmade hold important lessons for conflict resolution and peace-building, including, importantly, who can desecuritize. Is it only the securitizing actor, or can social movements, scientists and even security analysts play a (significant) role? Under what circumstances is desecuritization most likely to occur? What, for example, are the preconditions for conflicting parties to engage in dialogue? Are there conditions that will lead to a consistent peace between conflicting parties? If yes, what are they and how can they be promoted?

Empirical evidence such as that collected by Achim Maas and Alexander Carius (chapter 5), Patrick MacQuarrie and Aaron Wolf (chapter 9) and Saleem Ali and Mary Watzin (chapter 10), offers some tentative answers to at least some of these questions. Thus the environmental peace-building and conflict resolution literature suggests that although securitizing actors need to be willing to engage in conflict resolution, other actors can contribute to the likelihood of this occurring in various ways. Even security analysts, as long as they have the right kind of access, can, by providing examples of (environmental) conflict resolution elsewhere, 'nudge' practitioners in the right direction.

The environmental peace-building literature also suggests, as have Ken Booth and Nicholas Wheeler in the context of overcoming the security dilemma,[54] that the elusive concept of trust is elementary in peace-building and conflict resolution. Different theories of trust and on how trust between conflicting parties can be achieved might then be instrumental towards explaining and bringing about processes of desecuritization. The latter could potentially serve to build further bridges between two leading European schools of security, as Booth is a leading member of the Aberystwyth school, while desecuritization is associated with the Copenhagen school.[55]

Lesson 5. The ethics of securitization and the ethics of security

While there is little or no concern for the ethics of securitization (i.e. what form security practice *should* take, who or what *should* be protected and against what threats) and/or the ethics of security (i.e. reflections on the concept of security itself and what it *should* be) among traditional security scholars it is an intrinsic component of non-traditional security studies, also known as critical security studies.[56] The normative commitment within non-traditional security studies takes different forms. For some it rests with a commitment to desecuritization on the grounds that it equates to dialogue and democratization, thus ultimately on the fact that liberal democracy is valued as the best form of government. Others (radial poststructuralists) have a normative commitment to celebrate insecurity, while Critical Theorists advance a vision of security as emancipation, whereby 'emancipation seeks the securing of people from those oppressions that stop them carrying out what they would freely choose to do'.[57] It is a well-known problem that each one of these ideas suffers from a lack of specification.[58] Chris Browning *et al.* have recently argued the ethical commitments within all critical/non-traditional security theories 'are insufficiently developed to provide a genuine account of what constitutes ethical commitment regarding security', and they suggest that the latter is an important future research agenda for critical/non-traditional security studies.[59]

In the same way as insights from the environmental peace-building and conflict resolution literature hold valuable lessons for desecuritization, environmental security in theory and practice can serve as an important stepping stone for developing more systematic ethical theories of securitization. This is so for at least three different reasons.

First, a case can be made that the natural environment should survive and be defended, if necessary, by recourse to extraordinary measures because the natural environment is intimately connected with human well-being. Any claims to the survival of the natural environment for its own sake, as some proponents of ecological security would have it, can be dismissed on the grounds that, and as Dennis Pirages holds in chapter 7, we as human beings cannot escape our own anthropocentric nature, and that the natural environment is ultimately valued from a human perspective. While arguments for the survival of states vis-à-vis a threatening environment ignore the close interdependence between nature and human well-being. Not to mention the fact that on ethical grounds, it is impossible to endorse the survival of one group (say the wealthy Global North) at the expense of another (the poor Global South). For the ethics of securitization more generally this suggests that human well-being is an important unit of value. Indeed, more systematic theories of the ethics of securitization could flow from the question: what features of our world are conducive to human well-being and as such *should* be defended, if necessary by recourse to extraordinary emergency measures?[60]

Second, the practice of environmental security and also the many debates about it readily show that security means different, even opposing things, to different people. With reference to environmental issues Matt McDonald, for example, has argued that security practice is best understood as 'a site at contestation over a group's core values' and that the nature of these core values, as well as the ways to preserve or advance them, determines the ethical value of securitization.[61] In other words, from a moral point of view not all cases of securitization are the same.

Related to that, third, the practice of environmental security shows that the language of security is easily abused (cf above). For example, a security policy can be launched with the intention to benefit the securitizing actor and not the stated referent object of security, and/or security language can be used to declare commitment to an issue when in reality nothing much is done to follow up on that language. All of these and more are *among* the factors that need to be taken into consideration when thinking about the value of any given securitization, and when developing theory concerned with the ethics of securitization more generally.

In addition, environmental issues and especially climate change have influenced our thinking on the concept of security itself and what it *should* entail. Environmental threats have played an important role in highlighting the by now commonly held view that security is about more than the absence of military threats, and also that security cannot be achieved by military means. Echoing Dalby, however, we still need much more widespread acceptance that genuine security will require redressing the throw-away consumer society that has become established in the West and is aspired to elsewhere.[62] Climate change also highlights that the security of future generations is already in our hands. As Henry Shue puts it, by 'keeping as much as possible of the remaining fossil fuel right where it is now', we can leave them a 'legacy of security instead of a legacy of danger'.[63]

Topics and issues for future research within ESS[64]

Scholars of the past three decades have developed the concept of environmental security in several important ways that have been influential in both academic and practitioner communities. There remain gaps in the field, however, and many of the contributors to this volume have already pointed to areas for future research. One such area clearly is the ongoing issue of whether environmental stress (in particular climate change) leads to conflict or to cooperation. In order to answer this question more homogeneity in the language, concepts and variables used is a must. As part of his contribution to this volume Tom Deligiannis (chapter 2) has shown how the research community itself can get caught up in quarrels over interpretation and polemics that might fill academic journals, but otherwise hampers the advancement of environmental security research. One area for future research would then be work on the standardization of key concepts.

It is also the case that the field as a whole continues to use fairly generic understandings of the science of climate change and other forms of environmental change. In some very important measure, however, discussions of the security implications of environmental variables will be enriched as understanding of those variables improves. This is particularly important in the area of climate change, where social science analysis tends to lag far behind natural science research, and uses averages derived from IPCC reports that can be remarkably inaccurate and distorting in terms of what climate science actually reports.

In addition, there is a need to deepen our understanding of the extent to which indigenous adaptation and conflict resolution and mediation mechanisms are able to manage environmental stress or are being overwhelmed by it. This could have significant implications for how well some concept of environmental security may illuminate social patterns in resource-stressed regions of the world, such as in sub-Saharan Africa, South Asia and the Levant.

There is also a need to explore more rigorously the complex relationship among environmental change, security and disaster. Is the concept of double exposure, which suggests that some environmentally stressed societies have heightened vulnerability to violent conflict and disaster, compelling? Can the fields of disaster – which have focused on preparedness, warning, response and recovery behaviours from largely psychological, engineering and planning perspectives – fruitfully be merged with the social science perspectives of conflict and cooperation? In this context it will also be important to (re-)consider the role of the military. To what extent *should* the military be involved in the provision of environmental security?

Human security scholars need to continue to work on policymaking recommendations. More often than not, it is still unclear how exactly emancipatory notions of environmental or climate security are to be realized. One area that needs more work in this context is how scholars of environmental security themselves can, and actually do, influence policymaking and environmental security practice.

There is also a need to probe more deeply into the linkages between environmental change and human security to better understand the drivers of outcomes such as urban riots, population displacement and institutional failure. In particular, environmental security has tended to use a limited range of demographic data in its analysis, but at a time of rapid urbanization, and the massive economic transformations that are accompanying this, a more rigorous investigation into demography is in order.

Feminist environmental security research is destined to become a growth area in ESS in the same way as feminist security studies have become firmly established in security studies more generally. One of the pressing issues feminists will have to address, and Nicole Detraz in chapter 8 touches on this issue, is whether this approach is best as its own stand-alone research programme or whether it should act as a corrective to the human security approach.

Similarly, energy security research will be of increasing importance to ESS. Echoing Adam Simpson's argument in chapter 14, it will be especially important to develop a critical approach to energy security. The environmental security literature is ideally placed here as the intimate connection between environment and security, both as biosphere and as human security, is often ignored in national security statements on energy security.

Concluding remark

This volume has shown that a broader awareness of the size, variety and importance of the environmental security literature is important. Although environmental security now features as part of many undergraduate and postgraduate courses on security studies more generally, and many students are interested in the subject, it is often only aspects of the environmental conflict thesis (notably scarcity theses) that get taught and discussed. As this volume has shown, however, there is more to environmental security studies than simply the study of the possibility and the causes of environmentally induced conflict. The relationship between the natural environment and human and national security plays out in many different issue areas, for example, conservation, food and water, and it is influenced by many issues including climate change, demography and sustainable development. All of these issues are just as important to understanding environmental security as are the many theoretical approaches. Moreover environmental scarcity and conflict theses are heavily contested on ontological and methodological grounds.

I do not wish to claim that one approach to environmental security or any one issue area is most important. Instead, together the various approaches and the many issues make up the subfield of environmental security studies. With the environment in the form of climate change featuring ever more in public and policymaking debates the insights from more than twenty years of environmental security studies need to be heard. This book is one step in this direction; the rest is up to our readers.

Notes

1   I am grateful to Ken Booth, Stuart Croft, Simon Dalby, Tom Deligiannis, Jonathan Floyd, Colin Kilpatrick, Richard Matthew, Chris Paul Methman, Oliver Rathmill and Adam Simpson for valuable comments on earlier drafts of this chapter or for their assistance on some points made.

2   See, for example, S. Dalby, *Security and Environmental Change*, Cambridge: Polity Press, pp. 159–72; P. Dauvergne, *The Shadows of Consumption: Consequences for the Global Environment*, Cambridge MA: MIT Press, 2008.

3   For details see the UK's Department of Energy and Climate Change funding opportunities. Online: www.decc.gov.uk/en/content/cms/funding/en/content/cms/funding/funding_ops/funding_ops.aspx (accessed August 2012).

4   For a different view see W. Sinnott-Armstrong, 'It's not *my* fault: Global Warming and Individual Moral Obligations', in S. M. Gardiner, S. Caney, D. Jamison and H. Shue (eds), *Climate Ethics: Essential Readings*, Oxford: Oxford University Press, 2010, p. 344.

5   P. Singer, *Practical Ethics*, second edition, Cambridge: Cambridge University Press, 1993, p. 285.

6   E. H. Carr, *Conditions of Peace*, London: Macmillan, 1942, p. 118.

7   See also, R. Malnes, 'Climate Science and the way we ought to think about Danger', *Environmental Politics* 17 (4), 2008, 660–72; J. S. Risbey, 'The new climate discourse: Alarmist or alarming?', *Global Environmental Change* 18, 2008, 26–37; M. Hulme, 'The Conquering of Climate: Discourses of Fear and their Dissolution', *Geographical Journal* 174 (1), 2008, 5–16; J. de Wilde, 'Environmental Security Deconstructed', in H. G. Brauch, J. Grin, C. Mesjasz, P. Dunay, N. Chadha Behera, B. Chourou, U. Oswald Spring, P. H. Liotta and P. Kameri-Mbote (eds), *Globalisation and Environmental Challenges: Reconceptualising Security in the 21st Century*, Berlin: Springer-Verlag, 2008, p. 596; E. Swyngedouw, 'Apocalypse Forever?: Post-political Populism and the Spectre of Climate Change', *Theory, Culture & Society* 27 (2–3), 2010, 213–32.

8   A. Giddens, *The Politics of Climate Change*, Cambridge: Polity Press, 2009, p. 12.

9   Ibid., p.33.

10   L. Hansen, 'Theorizing the image for security studies', *European Journal of International Relations*, 17 (1), 2011, 51–74, at 53.

11   B. Buzan, O. Wæver and J. de Wilde, *Security: A new Framework for analysis*, Boulder, CO: Lynne Rienner, 1998, p. 26.

12   O. Wæver, 'Klimatruslen – en sikkerhedspolitisk analyse', *Tidsskriftet Politik* 1 (1), 2009, 5–26; M. J. Trombetta, 'Rethinking the securitization of the environment: old beliefs new insights', in Thierry Balzacq (ed.), *Securitization Theory: How security problems emerge and dissolve*, London: Routledge, 2011, pp. 135–49; M. J. Trombetta, 'Environmental security and climate change: Analysing the discourse', *Cambridge Review of International Affairs* 21 (4), 2008, 585–602.

13   O. Wæver, 'Securitisation: Taking Stock of a Research Programme in Security Studies', unpublished paper, 2003, p.15.

14   Cf R. Floyd, *Security and the Environment: Securitisation Theory and US Environmental Security Policy*, Cambridge: Cambridge University Press, 2010.

15   R. Eckersley, 'Environmental Security Dilemmas', *Environmental Politics* 5 (1), Spring 1996, 140–46, at 142.

16   Floyd, *Security and the Environment*, p. 117.

17   B. Hartman, 'Rethinking Climate Refugees and Climate Conflict: Rhetoric, Reality, and the Politics of Policy Discourse', *Journal of International Development* 22, 2010, 233–46.

18   M. Duffield and N. Waddell, 'Securing Humans in a Dangerous World', *International Politics* 43 (1), 2006, 1–23.

19   Floyd, *Security and the Environment*, p. 155.

20   Easterly cited in Hartman, 'Rethinking Climate Refugees and Climate Conflict', p. 240; see also A. Oels, 'From "Securitization" of Climate change to "Climatization" of the Security Field', in J. Scheffran, M. Brzoska, H. G. Brauch, P. M. Link and J. Schilling

(eds), *Climate Change, Human Security and Violent Conflict: Challenges for Societal Stability*, Hexagon Series on Human and Environmental Security and Peace, Volume 8, Berlin: Springer Verlag, 2012, pp. 186–205; M. Brzoska, 'Climate Change as a Driver of Security Culture', in J. Scheffran *et al.* (eds), *Climate Change, Human Security and Violent Conflict*.

21   I am grateful to Chris Methmann for bringing this point to my attention and for providing me with this example.

22   E. O. Wilson cited in J. Randerson, 'Biodiversity loss is Earth's "immense and hidden" tragedy, Darwin's "natural heir" warns', the *Guardian*, 20 November 2009. Online: www.guardian.co.uk/environment/2009/nov/20/biodiversity-loss-darwin-edward-wilson (accessed August 2012).

23   J. Rockström, W. Steffen, K. Noone, Å. Persson, F. S. Chapin, III, E. Lambin, T. M. Lenton, M. Scheffer, C. Folke, H. Schellnhuber, B. Nykvist, C. A. De Wit, T. Hughes, S. van der Leeuw, H. Rodhe, S. Sörlin, P. K. Snyder, R. Costanza, U. Svedin, M. Falkenmark, L. Karlberg, R. W. Corell, V. J. Fabry, J. Hansen, B. Walker, D. Liverman, K. Richardson, P. Crutzen, and J. Foley, 'Planetary boundaries: exploring the safe operating space for humanity', *Ecology and Society* 14 (2), 2009, 32. Online: www.ecologyandsociety.org/vol14/iss2/art32/ (accessed August 2012).

24   Giddens, *The Politics of Climate Change*, p. 107.

25   D. L. Gauthier *et al.*, 'Assessment of Undiscovered Oil and Gas in the Arctic', *Science* 324 (5931), 2009, 1175–79.

26   See, for instance, J. Barnett, *The Meaning of Environmental Security: Ecological Politics and Policy in the New Security Era*, London: Zed Books, 2001, p. 17; A. T. Price-Smith, *The Health of Nations: Infectious Disease, Environmental Change, and their effects on National Security and Development*, Cambridge, MA: MIT Press, 2002; p. 137ff; D. C. Pirages and T. M. DeGeest, *Ecological Security: An Evolutionary Perspective on Globalization*, Lanham, MD: Rowman and Littlefield, 2004; and B. McDonald, 'Global Health and Human Security: Addressing Impacts from Globalization and Environmental Change', in R. A. Matthew, J. Barnett, B. McDonald and K. L. O'Brien (eds), *Global Environmental Change and Human Security*, Cambridge, MA MIT Press, 2010, pp. 64–67.

27   Buzan et al., *Security: A new Framework for analysis*.

28   Of course, in the environmental sector too different groups of people can clash over resources, or they can intend to harm one another by manipulating the environment. In other words, in this sector too the security dynamic can play out the same as it does in the other sectors of security.

29   O. Wæver, 'Securitization and Desecuritization', in Ronnie D. Lipschutz, *On Security*, New York: Columbia University Press, 1995, p. 63; Buzan, cited in ibid., p. 15; D. Deudney, 'The case against linking environmental degradation and national security', *Millennium* 19, 1990, 461–76 at 464–65.

30   G. Prins (ed.), *Threats without Enemies: Facing environmental insecurity*, London: Earthscan, 1993.

31   C. Soanes (ed.), *The Oxford Compact English Dictionary*, Oxford: Oxford University Press, 2000, p. 1199, emphasis added.

32   Objective existential threats 'are threats to the existence of actors and orders [and other entities] regardless of whether anyone has realized this' (R. Floyd, 'Can securitization theory be used in normative analysis? Towards a just securitization theory', *Security Dialogue*, 42 (4–5), 2011, 427–39, at 430).

33   This is of course an anthropocentric view of the value of the natural environment, whereby the environment is of value only insofar as it supports human life. In eco-centric

formulations of security the biosphere itself is considered objectively existentially threatened.

34 See also T. Villumsen Berling, 'Science and securitization: Objectivation, the authority of the speaker and mobilization of scientific facts', *Security Dialogue*, 2011, 42 (4-5), 385-97; for a different critique of the role of science see Karen O'Brien 'Are we missing the point? Global environmental change as an issue of human security' *Global Environmental Change* 16 (1), 2006, 1–3.

35 J. W. Busby, 'Who Cares about the Weather? Climate Change and U.S. National Security', *Security Studies* 17, 2008, 468–504, at 477.

36 Floyd, 'Can securitization theory be used in normative analysis?', p. 428.

37 The Aberystwyth school of security studies, for example, is more interested in specifying the condition of security which is tantamount to emancipation; while the Copenhagen school is interested in the practice of security, which they refer to as securitization.

38 J. Waldron, Torture, *Terror and Trade-Offs: Philosophy for the White House*, Oxford: Oxford University Press, 2010, chapter 5.

39 This thought is shared by many and the connection between human well-being and security and a functioning natural environment is becoming ever more important. See, for example: Millennium Ecosystem Assessment, *Ecosystems and Human Well-being: A Framework for Assessment*, Washington DC: Island Press, 2003; K. Booth, *Theory of World Security*, Cambridge: Cambridge University Press, 2007; M. McDonald, *Security, the Environment and Emancipation: Contestation over Environmental Change* (PRIO New Security Studies), Abingdon: Routledge, 2001, chapter 2.

40 S. Dalby, *Security and Environmental Change*, Cambridge: Polity Press, pp. 159–72.

41 Ibid., p. 165.

42 Ibid., p. 169.

43 For a generic definition of securitization see M. McDonald, 'Securitization and the construction of security', *European Journal of International Relations* 14 (4), 2008, 563–87.

44 Dalby, *Security and Environmental Change*, p. 170.

45 M. J. Trombetta, 'Rethinking the securitization of the environment: old beliefs, new insights', in T. Balzacq, *Securitization Theory*, p. 143.

46 O. Corry, 'Securitisation and "Riskification": Second-order Security and the Politics of Climate Change', *Millennium: Journal of International Studies* 40 (2), 2012, 235–58; A. Oels, 'From "Securitization" of Climate change to "Climatization" of the Security Field', in J. Scheffran *et al.*, (eds), *Climate Change, Human Security and Violent Conflict*, pp. 186–205; H. G. Brauch and U. Oswald Spring, 'Introduction: Coping with Global Environmental Change in the Anthropocene', in H. G. Brauch, U. Oswald Spring, C. Mesjasz, J. Grin, P. Kameri-Mbote, B. Chourou and J. Birkmann (eds), *Coping with Global Environmental Change: Threats, Challenges, Vulnerabilities and Risks*, Hexagon Series on Human and Environmental Security and Peace, Volume 5, Berlin: Springer Verlag, 2012, pp. 31–60.

47 M. J. Trombetta, 'Environmental security and climate change: analysing the discourse', *Cambridge Review of International Affairs* 21 (4), 2008, 585–602, at 599; see also, Corry, 'Securitisation and "Riskification"'; C. Methmann and D. Rothe, 'Apocalypse Now (and Then): From exceptional rhetoric to risk management in global climate politics', in C. Methmann, D. Rothe and B. Stephan (eds), *(De-)constructing the Greenhouse: Interpretive Approaches to Global Climate Governance*, London: Routledge, 2012; C. Methmann and D. Rothe, 'Politics for the day after tomorrow: The logic of apocalypse in global climate politics', *Security Dialogue* 43(4), 2012.

48 The so-called Paris school of security studies focuses on the work of Didier Bigo and Jef Huysmans. See, for example, D. Bigo and A. Tsoukala (eds), *Terror, Insecurity and*

Liberty: Illiberal Practices of Liberal Regimes after 9/11, Abingdon: Routledge Studies in Liberty and Security, 2008; D. Bigo, 'When Two Become One: Internal and External Securitisations in Europe', in Morton Kelstrup and Michael C. Williams (eds), *International Relations Theory and the Politics of European Integration*, London: Routledge, 2000, pp. 171–204; J. Huysmans, *The Politics of Insecurity: Fear, migration and asylum in the EU*, Abingdon: Routledge, 2006. For an overview and use of the label 'insecuritisation theory' see R. Floyd and S. Croft, 'European non-traditional security theory: From theory to practice', *Geopolitics, History, and International Relations* 3 (2), 2011, 152–79.

49   D. Bigo, 'The Möbius Ribbon of Internal and External Security(ies)', in M. Albert, D. Jacobsen and Y. Lapid (eds), *Identities, Borders, Orders: Rethinking International Relations Theory*, London: University of Minnesota Press, 2001, p. 111.

50   A. Oels, 'From "Securitization" of Climate change to "Climatization" of the Security Field'.

51   A. Oles and A. Carvalho, 'Wer hat Angst vor "Klimaflüchtlingen"? Wie die mediale und politische Konstruktion des Klimawandels den politischen Handlungsspielraum strukturiert', in I. Neverla and M. Schäfer (eds), *Klimawandel in den Medien*, Wiesbaden: VS Verlag für Sozialwissenschaften, 2011, pp. 253–76.

52   J. Huysmans, 'Desecuritisation and the Aesthetics of Horror in Political Realism', *Millennium: Journal of International Studies* 27 (3), 1998, 569–89.

53   C. Aradau, 'Security and the democratic scene: desecuritization and emancipation', *Journal of International Relations and Development* 7, 2004, 388–413; L. Hansen, 'Reconstructing desecuritisation: the normative-political in the Copenhagen school and directions for how to apply it', *Review of International Studies* 38 (3), 2012, 525–46; T. Balzacq (ed.), *Contesting Security: Strategies and Logics*, Abingdon: Routledge, forthcoming in 2013.

54   K. Booth and N. J. Wheeler, *The Security Dilemma: Fear, Cooperation and Trust in World Politics*, Basingstoke: Palgrave, 2008, pp. 225–27 and chapter 9.

55   The latter will of course depend on whether trust will become a key concept in the Aberystwyth school's theorizing. At the present time Booth's co-authored work on trust and that on security remain largely separate. Booth, for one, is wary of overstating the leverage trust can have in improving international relations (which would be the view of English school solidarists like Wheeler). And the Aberystwyth school is more likely to develop the idea that trust will increase and have more causal power in a radically changed international order along the lines suggested in Booth's *Theory of World Security* (2007). To my mind this does not preclude bridge-building between desecuritization and trust, however, because desecuritization is not equivalent to an absence of security as emancipation or human well-being. Desecuritization refers to the unmaking of security practice, not to the absence of the condition of being secure. I am grateful to Ken Booth for discussing with me the relationship between trust and the Aberystwyth school while I was writing up this chapter. I should stress that the idea of bridge-building between the two schools on this point is mine and not shared by Booth.

56   O. Wæver and B. Buzan, 'After the Return to Theory: The Past, Present and Future of Security Studies', in A. Collins (ed.), *Contemporary Security Studies*, Oxford: Oxford University Press, 2007, p. 394.

57   Booth, *Theory of World Security*, p. 112.

58   Hence the recent interest in ethics within critical security studies. See, for example, G. Hoogensen Gjørv, 'Security by any other name: negative security, positive security, and multi-actor security approach', *Review of International Studies* July 2012, 1–25; P. Roe, 'The "Value" of positive security', *Review of International Studies* 34, 2008, 777–94; P. Roe, 'Is securitization a "negative" concept? Revisiting the normative debate

over normal versus extraordinary politics', *Security Dialogue*, 43(3), 249–266, 2012; Journal of International Relations and Development Forum on 'desecuritization and emancipation', 9 (1), March 2006; various contributors in *Security Dialogue, Special Issue on The politics of Securitization*, 42 (4–5), August/October 2011.

59   C. Browning and M. McDonald, 'The future of critical security studies: Ethics and the politics of security', *European Journal of International Relations*, first view (published online October 2011), p. 16.

60   For more on this, see Floyd, 'Can securitization theory be used in normative analysis?'

61   McDonald, *Security, the Environment and Emancipation: Contestation over Environmental Change*.

62   Cf Paul Harris, *World Ethics and Climate Change: From International to Global Justice*, Edinburgh: Edinburgh University Press, 2010.

63   H. Shue, 'Deadly Delays, Saving Opportunities: Creating a more dangerous world?', in S. M. Gardiner, S. Caney, D. Jamison and H. Shue (eds), *Climate Ethics: Essential Readings*, Oxford: Oxford University Press, p. 158.

64   This section was prepared and written with Richard A. Matthew.

INDEX